Energy of the Russian Arctic

Valery I. Salygin
Editor

Energy of the Russian Arctic

Ideals and Realities

palgrave
macmillan

Editor
Valery I. Salygin
Moscow State Institute
of International Relations (MGIMO
University)
Moscow, Russia

ISBN 978-981-19-2816-1 ISBN 978-981-19-2817-8 (eBook)
https://doi.org/10.1007/978-981-19-2817-8

Cover credit: Alex Linch/shutterstock.com

This Palgrave Macmillan imprint is published by the registered company Springer Nature Singapore Pte Ltd.
The registered company address is: 152 Beach Road, #21-01/04 Gateway East, Singapore 189721, Singapore

Contents

Notes on Contributors

Alexey I. Andreev is Associate Professor, Deputy Dean of the Faculty of Global Studies, Lomonosov Moscow State University, Russia. He received his Ph.D. in Biology in 2000 Lomonosov Moscow State University, and further on completed advanced training in economics, sociology, and political science. Currently, his research interests include complex global socio-natural problems, problems of the development of certain regions of the world and Russia, including the Arctic.

Zuzana Čičková is Associate Professor at Faculty of Economic Informatics, University of Economics in Bratislava, Slovakia. Her major research interests include operations research (Game theory mainly).

Simona Chuguryan is Assistant Professor at Faculty of International Relations, University of Economics in Bratislava, Slovakia. Research areas include political sciences, Scandinavian studies and identity, influence of cultures in international relations.

Zarina A. Dadabayeva is Doctor of Political Sciences, Leading Research Fellow at the Institute of Economy of the Russian Academy of Sciences, Russia. Her research interests embrace post-Soviet studies, the Eurasian integration, and modeling of migration flows.

Tatiana Damaskova is Assistant Professor, Faculty of International Relations, University of Economics in Bratislava, Slovakia. She holds Ph.D. in International Economic Relations. His research interests include research into the integration of the European Union, the economy of the Scandinavian countries, the welfare state.

Olga A. Derendyaeva is Junior Researcher at All-Russian Research Institute for Civil Defense and Emergencies, Russia. She holds her master's degree in linguistics from the Shanghai International Studies University. Sustainable development has been the neat focus of her research expertise since 2019. Her main research interests include community sustainability; security and resilience; development of the Sendai framework program, UNDRR projects, and SDGs in Russia.

Vadim O. Evseev is Doctor of Economic Sciences, Professor Plekhanov Russian University of Economics, Professor of RANEPA under the President of the Russian Federation, Russia. His research interests lie in the field of modeling socio-economic, public and political processes. He is author of eight scientific monographs, nine textbooks, and manuals, member of the editorial board of three scientific electronic journals, member of the Scientific Council of the Russian Academy of Sciences on socio-economic problems of Eurasian integration.

Mariia I. Ermilova, Ph.D. in Economics, is Associate Professor, Department of Financial Management, Plekhanov Russian University of Economics, Russia. She is the first Vice President of the Association of Education and Science Managers, as well as a member of several expert communities and scientific journals. She has published over 120 articles in professional journals and twelve books as author, co-author, editor, and co-editor. Her research expertise includes issues related to the housing market and housing construction, investment, and digital technologies in industries.

Matthias P. Finger, Ph.D. in Education and in Political Science, is Professor Emeritus at Ecole Polytechnique Fédérale Lausanne (EPFL), Switzerland and Research Professor at the Northern Arctic Federal University (NArFU) in Arkhangelsk, Russia. His main research interests include management of public enterprises, innovations, urban infrastructures, and smart cities. He is Editor-in-Chief of the Journal Competition and Regulation in Network Industries (Sage) and Director of the international GlobalArctic network project.

Anna V. Firsova is Senior Lecturer of the Department of Shipbuilding Technology at Saint-Petersburg State Marine Technical University (SMTU University), Saint-Petersburg, Russia. She is both an economist and a shipbuilder. Her research interests include shipbuilding, planning, organization, automation of production digitalization, intelligent technologies, and shipbuilding technology.

Sébastien Gadal, Ph.D., Prof., Aix-Marseille University, CNRS UMR 7300 ESPACE (France), B.K.C., is Professor at the North-Eastern Federal University (Russia). His area of research focuses on the processes of human, environmental, and urban adaptations of global warming, globalization, and territorial integrations in the Arctic, Subarctic, and Baltic regions. The analysis combines methods based on GIS, remote sensing, and geo-simulations, with the geographic modeling of the socio-environmental and geopolitical consequences.

Andrei M. Golubchik, Ph.D. in Economics, DBA, FIATA diploma, is Associate Professor at the Department of Oil and Gas Trading and Logistics, National University of Oil and Gas "Gubkin University", Russia. He has been working in the transport industry since 1985, co-author of the project and organizer of the construction of the first private container terminal in post-Soviet Russia. He has been working as a teacher and researcher since 2011. He is author and co-author of more than 170 scientific and practical works, including 14 textbooks on international cargo transportation, transport law, forwarding and transport support of foreign economic activity of enterprises.

Valentina Ignatyeva is Ph.D. Candidate working on renewable energy development in the Russian Arctic, Aix-Marseille University, Aix-en-Provence, France. After working for over a decade in the private sector in the energy, construction, and shipping, she is using her real-world experience and expertise to understand the dynamics of Arctic development and the potential role that renewable energies and information technologies can play in the sustainable socio-economic growth of the region.

Karol Janas is Associate Professor at the Department of Political Science, Alexander Dubcek University of Trencin, Slovakia. He is the Head of this Department. His research areas include history and public administration. He has published numerous vocational articles about history as well as public administration in domestic and foreign periodicals and collections.

Barbora Jánošková is External Postgraduate Student, Department of Political Science, Alexander Dubcek University of Trencin, Slovakia. Her academic expertise focuses on the comparison of the regional level in the countries of the Visegrad Group—the Slovak Republic, Czech Republic, Poland, and Hungary. She compares the competence framework of the regional level in these countries and tries to find recommendations and examples of good practice.

Jūratė Kamičaitytė, Ph.D. is Prof., Kaunas University of Technology, Lithuania. Her main research interests include analysis and evaluation of landscape visual quality, methods of visual impact assessment, evaluation of potential and dynamics of city structures, cultural and landscape identity studies, expression of sustainable development concept in architecture, landscape architecture, urban and territorial planning. Since 2018 she is Member of the Research Group "Cultural and Spatial Environment" at the Faculty of Civil Engineering and Architecture (KTU) and Associate Member of the CNRS research laboratory UMR 7300 ESPACE.

Alexey G. Kazanin, Ph.D. in Technical Sciences, is General Director of OJSC "Marine Arctic Geological Exploration Expedition", Russia. He graduated from Lomonosov Moscow State University and is a well-known Russian geophysicist and manager. He is an author of over 50 scientific publications on geology, mineral exploration technologies, the economics of the oil and gas industry, and the development of the Russian and foreign Arctic.

Marina A. Kazanina is Deputy-General Director of OJSC "Marine Arctic Geological Exploration Expedition", Russia. She graduated from St. Petersburg State University of Engineering and Economics and holds a Ph.D. in Law. In 2019, Dr. Kazanina graduated from the Institute of Business and Business Administration of the Russian Academy of National Economy and Public Administration under the President of the Russian Federation in 2019 under the M.B.A. program, specializing in Entrepreneurship and Company Management. Currently, she is engaged in research of legal and personnel issues of the development of the Arctic region.

Dmitriy S. Khmara is Deputy Head of the Department of Organization of scientific work at Saint-Petersburg State Marine Technical University (SMTU University), Saint-Petersburg, Russia. He holds his Ph.D.

in 2003 from Saint-Petersburg State Marine Technical University. His research interests include the economics of maritime transport and the Northern Sea Route, innovations, and shipbuilding technology.

Angelina A. Kolomeytseva, Ph.D. in Economics, is Associate Professor at the Department of International Economic Relations and Foreign Economic Affairs, MGIMO University, Russia. Her main research interests include energy industry in Russia and the CIS region, industrial development of the Eurasian Economic Union, economic sanctions, digital economy, and international economic organizations.

Natalia Yu. Konina is Doctor in Economics, Professor, and Head of Management, Marketing, and Foreign Economic Activities Department, MGIMO University, Russia. With more than 30 years of experience in research and business consulting, she has published over 100 articles in professional journals and sixteen books as author, co-author, and editor of Russian and foreign publications. Her current research interests are focused on the dynamics of business environment in digital era, globalization and digitalization, competitiveness, international business, geo-economics, and green economy. She is Member of Dissertation Council of the Institute of Europe of the Russian Academy of Science.

Valery N. Konyshev is Professor at the Theory and History of International Relations Department, St. Petersburg State University, Russia. His research interests include American foreign policy, security studies in Arctic region, Arctic policies of BRICS members, military strategy, and International Relations theory.

Andrey K. Krivorotov, Ph.D. in Economics, is Associate Professor at the Department of Innovation Management, MGIMO University, Odintsovo branch, Russia. His main research interests include socio-economic development of the Arctic (with a strong focus on Russia and Norway), oil & gas industry in Russia and the Caspian region, Arctic continental shelf. He is Member of the International Arctic Social Science Association (IASSA), an Expert of the Russian International Affairs Council (RIAC) and of the Arctic Development Project Office (PORA).

Maria L. Lagutina is Professor at the World Politics Department, St. Petersburg State University, Russia. She holds her Doctoral degree in Political Science from St. Petersburg State University. Her current research interests are oriented toward global governance, regional integration, comparative regionalism, Russia's Arctic policy, and international cooperation in the Arctic.

Yana V. Leksyutina is Full Professor at the School of International Relations, Saint-Petersburg State University, Russia. She holds a Ph.D. in Political Science (2008), Doctor of Political Sciences Degree (Post-Doctoral degree in Russian academic classification; 2013), an Accreditation to Supervise Research (Docent, 2012), and an honorary title and academic rank of Professor of the Russian Academy of Sciences (2018). She is Invited Professor at the OSCE Academy in Bishkek (Bishkek, Kyrgyzstan) and Non-Resident Fellow at the Center for the Study of Global Economic Future (CSGEF, Dubai, UAE). She is an author of 2 monographs, co-author of 12 collective monographs, and author of more than 140 academic articles. Her primary research interests are: China's foreign policy, Russia-China relations, regional security and international relations in the Asia-Pacific, US foreign policy in the Asia-Pacific, global governance.

Alexei A. Medkov is Candidate of Economic Sciences, Leading Research Fellow at Market Economy Institute of the Russian Academy of Sciences, Russia. His special research expertise lies in the areas of historical analysis to the formation of a model transport and transit economy.

Natalia A. Nevskaya is Leading Scientific Associate of Department of economic studies of the Institute of Europe of the Russian Academy of Sciences (IE RAS), Russia. She holds a Ph.D. in Economic Sciences from the Pacific National University. She has published over 80 articles in professional journals and six books as author, as well as co-author. She is the author of a textbook for national universities of Russia on Macroeconomics planning and forecasting. Her research interests include a new digital economy and industrial policy. She is Expert of Gromyko Association for Foreign Policy Studies.

Dmitry V. Nikushchenko graduated from St. Petersburg State Maritime Technical University, Russia. In 1999, he obtained the degree of Candidate of Technical Sciences in the area of ship hydrodynamics and propulsion. In 2011, he defended his dissertation for the degree of Doctor of Technical Sciences in the specialty "Mechanics of fluid, gas, and plasma". He is an author of more than 140 research articles in SCI(E) journals in applied mathematics, hydrodynamics, ships, and underwater vehicles' steering ability. Since 1999, he has worked in State Marine Technical University as Assistant Professor, Professor, Vice-rector for research, Director of Institute of Hydrodynamics, and Control Processes of State Marine Technical University.

Vladimir S. Osipov is Doctor in Economics, Professor of Asset Management Department at MGIMO University (Russia), and also Guest Professor of School of Economics and Management at Beijing Jiaotong University (China). He is Member of American Law and Economics Association, Editorial Boards Member of several scientific journals. He has published over 200 articles in professional journals and sixteen books as author, co-author, editor, and co-editor. His primary research expertise embraces institutional policy and public administration. He is Vice-Head of Dissertation Council of Lomonosov Moscow State University (Russia), and also Member of Dissertation Councils at Plekhanov Russian University of Economics (Russia) and MISIS University (Russia).

Alexander N. Pilyasov is Professor of the Department of Socio-Economic Geography of Foreign Countries, Faculty of Geography, Lomonosov Moscow State University, Russia. He is also General Director of the ANO "Institute of Regional Consulting". His research interests include regional economy, innovative development of cities and regions of the Arctic and the North, international economic cooperation in the Arctic, theory of the development of the Arctic regions of the world.

Elena S. Putilova is Geographer, Graduate of the Department of Socio-Economic Geography of Foreign Countries, Faculty of Geography, Lomonosov Moscow State University, Russia. She is also Expert of the ANO "Institute of Regional Consulting". Contributor's research interests embrace regional and urban economy, sustainable development, innovative development of cities and regions, development of Arctic and the North.

Valery I. Salygin is Doctor of Technical Sciences, Professor, Corresponding Member of the Russian Academy of Sciences, twice winner of State Science and Technology Award of Russia. He is Director of International Institute of Energy Policy and Diplomacy at MGIMO University (Russia), and President of the International Fuel and Energy Development Academy. His overall academic and research experience lasts from 1964. His overall publishing portfolio includes over 500 research and academic papers. He has supervised over 50 Doctors and Candidates of

Sciences. He is Member of a Working Group within the State Commission on Arctic Development and an Expert Board under State Duma Energy Committee of the Russian Federation, a Member of an Environmental Safety Working Group under Presidential Commission on Fuel and Energy Industry Development Strategy and Environmental Safety of the Russian Federation, a deputy-chairperson of Rosseti JSPC Science and Engineering Board, a Member of Sustainable Development Council under Gazprom JSPC Science and Engineering Board Chairman, and a co-chairperson of a Steering Board with the Eurasian Information and Analysis Consortium. His primary research interests include energy economics, energy policy and geopolitics, energy diplomacy, system analysis, and international energy regulation.

Vladimir N. Samotuga works at the St. Petersburg State Marine Technical University (SMTU), Saint-Petersburg, Russia. He has a Degree of Candidate of Historical Sciences, with the title of Associate Professor. His research interests include problems of the world economy and the German economy.

Tautginas Sankauskas, Dipl. Eng., is Vice President of Lithuanian National Association of Forwarders and Logistics. He has been working in the transport industry since 1985, an experienced specialist in international transportation and forwarding. He is also CEO and Founder of one of the oldest freight forwarding companies in Lithuania, UAB TEFVilteda, Chairman of the Certification Commission of Vilnius Gediminas Technical University for awarding a bachelor's degree in transport and logistics.

Igor A. Shulyatyev is Leading Researcher at International Law Department, Institute of Legislation and Comparative Law under the Government of the Russian Federation, Russia. He holds his Ph.D. in International Law from Peoples Friendship University of Russia. The thesis was devoted to the functioning of the World Trade Organization. His research interests include international public law, international economic law, and the law of international organizations. In the Institute of Legislation and Comparative Law under the Government of the Russian Federation, he is responsible for the issues concerning with the interaction between international law and the legal system of Russia. He has a wide experience in teaching the international law disciplines in the leading Moscow universities.

Aleksey Y. Smirnov is Professor at Saint-Petersburg State Marine Technical University (SMTU University), Saint-Petersburg, Russia. He holds his Ph.D. in 2013 from the St. Petersburg State University of Cinema and Television. His research interests include the economics of maritime transport and research on the economic problems of the development of the Northern Sea Route.

James C. Towers is an O&G Engineer, USA. Together with Sergey V. Volchkov, he has worked on multiple Arctic and Sub-Arctic projects in Russia. He has also conveyed mega-projects in the Arctic and Sub-Arctic territories of Alaska and Canada. He started working for Humble (later Exxon) Pipeline Company in 1968, as a laborer. After graduation from Texas A&I in 1970, he worked for Exxon Pipeline Company in design, operation, and maintenance functions and was loaned to Alyeska Pipeline Company during the construction, start up and operation phases of the Trans-Alaska Pipeline where he gained extensive Arctic experience. In 1995, he left Exxon to become the Technical Coordinator for the CPC pipeline project. In 2002, he was loaned to Nippon by Technip as an expert consultant to the pipeline and oil terminal design, pipeline route selection, and permitting phases of this project from the Sakhalin Island to De-Kastri (Russia).

Sergey V. Volchkov, Ph.D. is Mechanical Engineer, Independent Consultant, Russia. He holds his Ph.D. from Gubkin State University. During student years, he spent a lot of time working as a laborer at various projects in the former USSR. Some years of his career have been with research institutions and universities. Together with James C. Towers, he has been involved in multiple Arctic and Sub-Arctic projects in Russia. He used to work on Yamal Peninsula, Komi Republic, Sakhalin Island, in Kazakhstan, and other locations. He used to work for Caspian Pipeline Consortium, Sakhalin Energy Investment Company, Yamal-LNG, WorleyParsons, Lukoil Overseas, and others. He is listed in the Russian Federation National Register of construction specialists, ID number 0126489. His research interests include projects management in the Arctic, mega-projects management, systematic permits, and approvals.

Dmitry A. Yakovenko is Honorary President of the self-regulating organization National Institute of Professional Accountants, Financial Managers, and Economists, Russia. The contributor was directly involved in the Arctic problematique as the Head of the Control Service and

Internal Auditor of the Ministry for the Development of the Russian Far East and Arctic from March 2017 to April 2021. He has written several works devoted to both the peculiarities of the development of individual regions and the influence of the state and budget resources on the development of the region.

Moisei Zakharov is Ph.D. Candidate of Geography, Aix-Marseille University, CNRS UMR 7300 ESPACE (France), Senior Lecturer at the North-Eastern Federal University (Russia). His scientific interests relate to geographical modeling and spatial analysis of processes in permafrost landscapes, under conditions of climate change, permafrost degradation on infrastructures, settlements, and human impacts in the zone of continuous permafrost distribution of Eastern Siberia (Yakutia in particular). Research methods are based on the knowledge-based interpretations of remote sensing data and spatial modeling.

Kobilzhon Kh. Zoidov is Associate Professor, Candidate of Physical and Mathematical Studies, and Head of the Laboratory, Market Economy Institute of the Russian Academy of Sciences, Russia. He specializes in the fields of economic and mathematical modeling, evolutionary model of economic system, economic security, and transit economy.

Tetyana Zubro is Assistant Professor at Faculty of International Relations, University of Economics in Bratislava, Slovakia. Her research expertise embraces political sciences, transformation of the post-Soviet countries, and political culture in European post-communist countries.

List of Figures

List of Tables

CHAPTER 1

Introduction: Energy of the Russian Arctic. Quo Vadis?

Valery I. Salygin

This volume is an energy-tailored sequel to the research on the Arctic carried out at Moscow State Institute of International Relations (MGIMO University). Specifically, the book is grounded in the profound academic and practical expertise of the specialized body of MGIMO University—*International Institute of Energy Policy and Diplomacy*, founded in 2000. On the professional side, the Institute has lasting and fruitful cooperation with Russian energy giants such as Rosneft PJSC, Transneft PJSC, and Gazprom PJSC. From the academic perspective in the area of energy policy, for instance, it runs five joint research bodies with its European counterparts broken down to Russian-German, Russian-Italian, Russian-French, Russian-Norwegian, and Russian-Icelandic formats.

V. I. Salygin (✉)
Moscow State Institute of International Relations (MGIMO University), Moscow, Russia
e-mail: miep@mgimo.ru

V. I. Salygin (ed.), *Energy of the Russian Arctic*,
https://doi.org/10.1007/978-981-19-2817-8_1

Major research findings of scholars representing the Institute have been depicted by Salygin, Guliev, and Khubaeva (2015), Salygin, Guliev, and Ryabova (2015), Salygin, Guliev, Akieva, et al. (2019), Salygin, Guliev, Chernysheva, et al. (2019), Salygin, Petrova, and Mustafinov (2019), Salygin et al. (2021), Salygin and Lobov (2021), Salygin and Mustafinov (2018, 2019a, 2019b), Vylegzhanin et al. (2018), Vylegzhanin, Bunik, et al. (2020), Vylegzhanin, Salygin, et al. (2020), Bourmistrov (2015), Kravchenko and Salygin (2015), Guliev et al. (2020, 2021), and Zhiznin et al. (2019), Zhiznin, Timokhov, and Dineva (2020), Zhiznin, Timokhov, and Gusev (2020).

Thus, the research exclusively focuses on energy-related aspects of exploration of the Arctic Zone of the Russian Federation (AZRF). This particular region with its ample oil and gas resources has been comparatively and critically studied by a team of authors representing Russia, USA, France, Switzerland, Slovakia, and Lithuania. Such a multinational group of scholars, engineers, and practitioners has brought greater unbiasedness, inclusiveness, and industry peculiarities to the existing discourse given the deteriorating relations between Moscow and the West and the ongoing economic megatrends heavily influencing the energy sector, especially digitalization and decarbonization.

Rooted in multidisciplinarity, comparative analysis, and critical thinking, the chapters to the volume are intertwined in a way to unfold the following research aims, to:

- put forth a Russian perspective on the region tailored to energy issues covering the obvious gap in the existing research;
- neatly investigate modern trends, challenges, and opportunities of the energy dimension of the Russian Arctic's development, including related policies, given global economic and political turbulence, as well as the greening of the global economy;
- assess the organizational and technical sides of oil and gas extraction, and, consequently, transportation via the Northern Sea Route (NSR);
- enrich the existing research with relevant technical expertise, including transportation and construction, provided by contributors from such specialized bodies as well as industry professionals.

In terms of layout, the volume consists of four parts, broken down to institutional, economic, infrastructural, and greener dimensions of AZRF's exploration.

Editorial introduction opens up the discussion and sheds the light on each contribution to the volume.

The chapter entitled *International Legal Framework of Arctic Exploration* by Igor A. Shulyatyev (Russia) pictures the diversity of the existing framework of the Arctic's governance comprising formal organizations, interstate cooperation, and less rigid but more expert-oriented formats. Clearly, the Arctic Council (AC) and the Barents Euro-Arctic Council appear to be the most decisive and influential areas backed up by international treaties. When broken down to the non-governmental level. The energy interactions, for instance, could be held at Oil Companies International Marine Forum (OCIMF) and The International Association of Oil and Gas Producers (IOGP) platforms. However, the author concludes that there is no universal global, regional, or sub-regional source of international law in the field of mineral resource extraction in the Arctic. In fact, much has been achieved in a bilateral format with Russia-Norwegian case at the front. Thus, the exploration, production, and transportation of energy resources generally represent a combined international legal regime based on international treaties on environmental protection and bilateral interstate agreements.

The chapter called *Russia's Chairmanship in the Arctic Council: In Search of Balance?* by Dmitry A. Yakovenko (Russia) scrutinizes the organizational structure of the Council at large, with a more specific focus on Russia's heading the organization in 2021–2023. The research findings on the Arctic and its Russian part enrich the discourse by an insight legacy, as the author holds solid expertise from the Ministry for the Development of the Russian Far East and Arctic. An interesting perspective has been captured from revisiting the contacts of Russia as a member of the AC with non-Arctic states (i.e., China, South Korea, and Japan), bundled with a trilateral declaration on cooperation in the Arctic. The expert concludes that, despite geopolitical tensions between the Western members of the AC and Russia, the latter during its chairmanship might coin a number of inclusive initiatives.

In their joint chapter *Energy Complex of the Russian Arctic*, Alexey G. Kazanin (Russia), Marina A. Kazanina (Russia), Alexey I. Andreev (Russia), and Vladimir S. Osipov (Russia) position the AZRF in the Arctic's energy complex and present a general overview of oil and gas deposits as well as extraction broken down to the related territories of Russia. Given the dependence of the national economy on energy resources, specific attention has been paid to the related policies, including technological one, centered around the 'Strategy for the Development of the Arctic Zone of the Russian Federation and Ensuring National Security for the Period up to 2035.' It has been stressed that Western sanctions on Russia have severely deprived Russian energy companies operating relevant technologies and equipment on on-shore and (more crucially) off-shore in the Arctic. In essence, Russia would better follow the Norwegian path, fostering an R&D and cluster approach in creating an up-to-date oil and gas industry.

In the chapter headlined *New Trends in the Development of Energy Fields in the Russian Arctic*, Alexander N. Pilyasov (Russia) and Elena S. Putilova (Russia) revisit the existing project-management practices in the Russian part of the Arctic, distinguishing it from the Soviet ones. As of today, the national corporate pattern in the energy field implies the greater presence of non-state actors under the network umbrella, more precise focus on global energy conjuncture, and more intense dependence on agile techniques resulted in the 'temporary operation' regime. More fundamentally, recent energy projects in the AZRF are heavily embarking on green endeavors. The authors precisely trace these trends by using the examples of oil fields such as the Novoportovskoye, Prirazlomnoye, Vankorskoye, and Messoyakhskoye, as well as gas ones—Yamal LNG and Bovanenkovskoye field.

The chapter by Maria L. Lagutina (Russia) and Valery N. Konyshev (Russia), *Rethinking Russia's Energy Policy in the Arctic*, critically tracks the evolution of the related national policy. The scholars capture the changing nature of Russia's energy policy to the region and identify key internal and external factors. The former might be divided into climate, lack of domestic technologies, and poor infrastructure, whereas the latter would be divided into the volatility of energy prices, energy transition, and Western sanctions on Russia. Overall, the major priorities of Russia's energy policy in the Arctic have shifted from a 'resource-based approach' to reviewing Arctic territories as quite promising ones subject to innovations on the principles of sustainable development.

The chapter under the title *China's Participation in Energy Cooperation with Russia in the Arctic* conducted by Yana V. Leksyutina (Russia) investigates the modalities of Beijing's quantitative and qualitative involvement into the exploration of hydrocarbon reserves in the

AZRF. It is here where the dualism of Western sanctions on Russia is mostly vivid. On the one hand, some of the packages imposed starting from 2014 directly hit energy projects in the Russian Arctic. As a result, there was a demise of Western oil and gas giants from the region. On the other hand, these measures laid the grounds for a much closer convergence between Moscow and Beijing. As of today, China is the largest foreign investors into Russian LNG projects in the Arctic (i.e., Yamal LNG and Arctic LNG-2) under the umbrella of the Belt and Road Initiative (BRI). Thus, being a non-arctic state through this springboard China plans to establish its strong presence in the region.

The chapter *Activities of Major Oil and Gas Companies in the Arctic Zone of the Russian Federation* by Maria L. Lagutina (Russia) and Valery N. Konyshev (Russia) carefully goes into corporate detail of energy exploration in the Russian part of the Arctic. Energy companies have been split into two groups: Russian—including state-owned (e.g., Rosneft and Gazprom) and private (Novatek and Lukoil)—and foreign (e.g., Total, CNPC, and CNOOC). Among Russian corporate actors, Rosneft and Gazprom remain the only ones who can execute off-shore projects; however, their business strategies drastically vary. The former undertook large fields in poorly explored areas, counting on closer cooperation with Western companies (for instance, BP remains the company's largest foreign stakeholder), whereas the latter focused on smaller areas nearby the existing deposits. Overall, Russian energy actors appear to be more involved in social corporate and social environmental responsibility. In conclusion, the authors stress the rising ascendancy of the LNG segment, with Novatek being at the front with its Yamal LNG launched in 2018, and Arctic LNG-2 scheduled for 2023. In this regard, Chinese involvement in the region's LNG sector is on the rise.

The chapter by Natalia Y. Konina (Russia) entitled *The Exploration of Arctic Energy Resources and the Competitiveness of Big Russian Oil and Gas Companies* studies the competitiveness in this field from the perspective of the following factors: market factor, resource-production factor, financial factor, and management factor. It has been revealed that the combination of these factors is evolving with digitalization and energy transition in place. Thus, R&D turns out to be edging ahead regarding its competitive advantage as of today, which could be most effectively sustained in vertically integrated energy companies, yet still be subject to the dynamics of global energy prices. The ideals and realities of such a motion in the Russian energy sectors have been tested through the examples of Gazprom, Gazprom-Neft, Rosneft, and Novatek. The scholar concludes by claiming this carbon–neutral path for Russian oil and gas companies is feasible but challenging.

In their joint research Vadim O. Evseev (Russia) and Vladimir S. Osipov (Russia), in the chapter *Expert Systems for Risk Management in the Energy Complex of the Arctic: A Lesson for Russian Oil and Gas Companies*, pioneer a knowledge-based management system for the construction part of energy projects to be executed in the AZRF. In a nutshell, the construction process is broken down into several knowledge-related stages with corresponding risk factors. In the outcome, based on the simulation modeling, a particular construction case might be thoroughly scrutinized from the perspective of the following ten factors: (1) development of feasibility study; (2) design; (3) construction; (4) start-up and adjustment works; (5) commissioning of the facility; (6) installation supervision of the construction process; (7) construction project/process management system; (8) specific Arctic risks; (9) risks of fraud in construction; and (10) political risks.

The chapter entitled *Overview of Practical Experience of Projects' Implementation in the Arctic* by James C. Towers (USA) and Sergey V. Volchkov (Russia) represents a unique professional snapshot from the field of pipeline construction in the Arctic, i.e., in Alaska and the Yamal peninsula. A proposed construction execution plan (CEP) of a pipeline contains consecutive steps of its implementation in direct correlation with soil, labor, and weather conditions. Engineers have enriched the discourse by providing industry-tailored guidelines for field personnel and an exact outline for notes on field observations for pipeline construction estimates.

In their joint chapter *The Northern Sea Route as a New Global Route for Energy Resource Transportation: Problems and Prospects*, Alexey G. Kazanin (Russia), Marina A. Kazanina (Russia), Alexey I. Andreev (Russia), and Vladimir S. Osipov (Russia) critically assess the status of the NSR as the key channel for energy shipments from the Russian Arctic. As of today, LNG and crude oil form the overwhelming share in the route's cargo base. Europe remains the major destination of LNG shipments, while China's direction is on the rise. Major share of shipments is done by vessels of Arc7 ice-class dominated by Panamax and Aframax types. Russia's icebreaker fleet is second to none. However, if Russia is aiming for a more qualitative development of the NSR, it should focus on training highly professional personnel, given the harsh climatic and labor conditions.

In the chapter entitled *The Advantages of the Northern Sea Route as an Energy Transport Artery and its Environment*, Simona Chuguryan (Slovakia), Zuzana Čičková (Slovakia), and Tetyana Zubro (Slovakia) apply the bimatrix game method—*Hawk-Dove* and *Chicken*—to the analysis of Sino-Russian relations on the commercial development of the NSR. Currently, the Arctic at large and its Russian part in particular are seeing the increasing presence of China. The modeling undertaken shows that, given rising geopolitical tensions with the West, both Russia and China might benefit from exploiting the route, granting shorter and secured delivery dates compared to sailing via the Suez Canal. In case of conflict over the NSR and a possible struggle for the region, it should be stressed that an inadequate pursuit of the players' own interests can lead to a global disbalance, which will ultimately affect Moscow and Beijing in the worst possible way. Thus, the overall economic profit of the Sino-Russian joint route's deployment contrasts with rising environmental risks.

The chapter *Global Climate Changes and Their Importance for the Building of Northern Sea Route and the Future of Energy Sector* by Karol Janas (Slovakia) and Barbora Jánošková (Slovakia) links the prospects of commercial shipping of oil and gas to the ongoing global warming. On the one hand, ice-melting in the Russian Arctic might facilitate energy flows via the NSR and reduce the overall logistics costs, as there will be less expense on icebreaker assistance (or even potentially an absence of expenses, under the worst scenario of ice-free navigation). On the other hand, increased shipping might result in more frequent oil spills, greater emissions from shipping fuel combustion, thawing permafrost, and general pressure on the region's biodiversity. Thus, the authors conclude that the ecological factor has become the most decisive one in the NSR's commercial exploitation.

The chapter *Northern Sea Route and the New Energy Agenda* by Natalia A. Nevskaya (Russia) presents a historical and intertwined overview of transport issues of the NSR and energy developments of both the Soviet period and Russia today. Given the increasing oil and gas extraction in the AZRF, it is worth creating an up-to-date stationary transport infrastructure (i.e., ports, railways, reservoirs, and other related facilities), as well as rolling stock (i.e., ice-class vessels and icebreakers). However, such construction must comply with stricter ecological norms and take into account the sanctions of the West. Summing it up, these initiatives are reasonably capital-intensive, but all the necessary institutional framework is not a hurdle anymore.

The chapter by Kobilzhon K. Zoidov (Russia), Alexei A. Medkov (Russia), and Zarina A. Dadabayeva (Russia) titled *Energy Development of the Russian Arctic within the Framework of the Belt and Road Initiative and Greater Eurasia: A View from China and India* draws a holistic connectivity perspective on bridging national transport complex and corresponding systems of China and India when implementing energy projects in the Russian Arctic. The scholars name 'trunk-feeder transport and transit system' and its land communication as the most promising tool for a spatial and inclusive idea of turning the AZRF into a logistics hub. Chinese interests stem from getting access to the Arctic's energy through its Polar Silk Road under the BRI, for instance by the construction of the Northern Latitudinal Railway. In its turn, India is more focused on the formation of the Arctic Ocean–Indian Ocean transport corridor incorporating it into the already functioning 'North–South' landline international transport corridor. The authors conclude that the conjuncture of transport systems of the three should be executed with the help of innovative transport and logistics technologies.

The chapter *Rethinking Year-Round Navigation via the Northern Sea Route* by Vladimir N. Samotuga (Russia), Aleksey Y. Smirnov (Russia), and Dmitriy S. Khmara (Russia) questions the technical side of this idea, broken down to the route's Western and Eastern parts. Year-round navigation is feasible as of today in the former sector of the NSR, which is commercially viable enough. Navigation in the latter sector lasts from January to mid-May, with this segment severely lagging behind in commercial attractiveness. The situation might change once the *Leader* icebreaker is put into operation (planned in 2027). Overall, given the ice situation and more crucially its unpredictability, a rocket of cargo transit via the NSR seems doubtful enough. Besides, a lack of centralized navigation, ice monitoring, and maritime safety systems throughout the route also hampers the prospects of a year-round navigation.

The chapter by Dmitriy S. Khmara (Russia) and Anna V. Firsova (Russia) called *Revisiting the Prospects of a Multihull Icebreaker for Year-Round Navigation via the Northern Sea Route* provides a broad technical and operational overview of the existing and intended types of icebreakers operated by Russia. In this essence, *innovative icebreaker* (IIB) and the *Leader* type are set to sustain the year-round navigation via the NSR. A comparative analysis between the existing types of icebreakers has empirically revealed that IIB and Leader are second to none from the perspectives of fairway done, propulsion and power, and ice-breaking capacity. However, when compared directly, the two models tend to solve different tasks. The IIB is designed to operate in less severe ice conditions (less than 2.5 meters thick) handling wider-hull vessels, whereas the Leader is cut for the most extreme ice situation and is able to sail to any destination within the NSR. Thus, when combined together, these two types can make a year-round navigation via the route possible.

In their joint chapter *Managing the Ice Situation when Ensuring the Safety of Oil Platforms in the Arctic Zone of Russia*, Dmitriy S. Khmara (Russia) and Dmitriy V. Nikushchenko (Russia) propose a promising model to predict and assess the ice situation at off-shore energy platforms in the Russian part of the Arctic. A real-time simulation undertaken has made it possible to build up several scenarios containing models of icebergs and tow-vessels to place them at the digital map of the NSR. The scholars empirically argue that the introduction of the proposed ice-management system might increase the geological exploration phase minimum by two months, and bring additional benefit of around $2 million dollars per platform by eliminating the threat of collision.

Input Specifics of LNG Shipments via the Northern Sea Route by Andrei M. Golubchik (Russia) and Tautginas Sankauskas (Lithuania) depicts another fruitful practical insight from energy professionals. The authors scrutinize the transportation specifics of LNG shipments from both technical and contractual sides. The technical peculiarities of LNG transportation stand for the fact that an LNG consignment daily loses 0.15 percent of the transported mass with a phenomenon of so-called boiled-off gas. As of today, modern LNG carriers can either capture up to 95 percent of this gas and then re-liquefy it or use it as a fuel for ship engines. As a result, a contractual peculiarity is about the inability of any of the parties to know the exact cargo volume before reaching the port of destination. Thus, the most popular delivery terms under Incoterms remain DES and DEQ (Incoterms, 2000), depending on whose pumps are responsible for unloading in the port of destination.

The chapter *Renewable Energy in the Russian Arctic: Energy Transition and Opportunities in the Context of Post-Pandemic Realities* by Valentina Ignatyeva (France) and Sébastien Gadal (France) investigates the ideals and the realities of renewable energy sources' implications in the AZRF given the global energy transition. It is worth mentioning that this notion was not disrupted by the COVID-19 pandemic. However, renewables (excl. large hydro and nuclear) up to date represent less than 2 percent in Russia's energy mix, whereas temperature in these territories on average is rising 2.5 times faster than elsewhere. Besides, Russian Arctic territories are heavily dependent on the *Northern Delivery* responsible for diesel, coal, and food supply carried out only in the summer period. As a result, electricity costs in these regions are on average 4–4.5 times higher than elsewhere. The scholars strongly believe that the formation of renewable energy systems in the AZRF will contribute to secure energy supply and, hence, lower the costs of energy and heat for locals.

The chapter by Olga A. Derendyaeva (Russia) under the title *Climatic Risks of Hydrocarbon Production in the Arctic Zone of the Russian Federation* focuses on the climatic ascendancies when developing oil and gas fields in the Russian part of the Arctic. Climatic risks have been classified in such a way as to present a framework on reducing the average cost of hydrocarbon production in the region executed by Russia's major energy giants, i.e., Gazprom-Neft and Novatek. It is here where the best available technologies (BAT) and best environmental practices (BEP) could be applied to turn their operations sustainable. For instance, applying BAT and BEP fusion might reduce emissions of short-lived climate pollutants (SLCPs) in the AZRF by 2025 almost threefold.

In their joint chapter *Territorialisation, Urbanisation, and Economic Development in the Russian Arctic: Energy Issues*, Sébastien Gadal (France), Moisei Zakharov (France), and Jūratė Kamičaitytė (Lithuania) delve into the essence of Russia's energy policy in its circumpolar regions from the perspectives of its spatial and social dimensions. The research findings are rooted in the profound in-field expertise of the authors gained in Yakutia (Republic of Sakha). As a result, it is argued that the region is experiencing a shift from a Soviet mono-energy model to a mixed one, comprising renewables, but the energy supply is still interrupted. This motion is subject to territorial, urban, and demographic characteristics given scattered settlements and other facilities. Even more fundamentally, access to a reliable energy supply is a sine qua non for revitalization of the related settlements and facilities. This could be achieved by a more active usage of renewables—solar (in the case of Batagaï) and wind (for Tiksi). Greener energy framed into autonomous energy grids can supply the locals with electricity and heat up to 24 hours a day and cut energy costs by almost 30 percent.

The chapter *Nuclear and Hydrogen Prospects for the Russian Arctic* by Angelina A. Kolomeytseva (Russia), Matthias P. Finger (Switzerland), and Andrey K. Krivorotov (Russia) boils the energy discussion down to nuclear and hydrogen perspectives. As of today, Russia is one of the world's leaders in nuclear technologies in the Arctic, used both for domestic and international purposes. It operates the *Kola* nuclear plant in the Murmansk region (440 megawatts) and the world's first floating one—*Akademik Lomonosov* (70 megawatts). Small and very small modular reactors (up to 300 and 15 megawatts, respectively) are counted to be the main energy pillars to sustain energy supply to remote areas of the state. Russia also owns six nuclear icebreakers, making the passage through NSR possible, with plans to have thirteen units by 2035. In terms of hydrogen, it is worth noting that this energy direction is fairly new for Russia, yet there is a related institutional basis. For instance, by 2024 Russia will have seen a green hydrogen production started at the Kola nuclear power plant. Under the base scenario, the state will export up to 2 million tons in 2035 and 15 million tons in 2050.

The chapter *Norwegian Oil Industry Model. Potential for Norwegian-Russian Cooperation in the Arctic* by Tatiana Damaskova (Slovakia) and Mariia I. Ermilova (Russia) outlines the progressive legacy of Norway in managing its energy industry and measures the prospects of energy cooperation between the two in the Arctic. Norway's best practices in linking energy revenues and pension obligations should be carefully studied by Russia. In fact, Russian-Norwegian energy cooperation saw a true breakthrough in 2011 with the delimitation of sea spaces in the Barents Sea and the Arctic Ocean. However, after 2014, the situations are deteriorating, with Russia now being deprived of much-needed Norwegian drilling off-shore technologies and related equipment. Nonetheless, the scholars assume that a green track might revitalize this cooperation. Therefore, the agreement between Equinor and Rosneft on renewable energy, carbon capture, utilization and storage, and low-carbon hydrogen solutions dated 2021 might be a good start.

Overall, the book might be useful for both academics and practitioners. Predominantly, it targets businessmen and analysts in energy, economics, politics, management, transport, and engineering who are engaged in studying the Arctic Zone of the Russian Federation. The scope and contents of the monograph make it distinctive for this large audience and to satisfy their practical needs in any comparative analysis of various energy-related aspects of the region's exploration, given recent tendencies and its up-to-date agenda.

References

Bourmistrov, A., Gudmestad, O., Salygin, V., & Zolotukhin, A. (2015). Norwegian-Russian cooperation on oil and gas education. In A. Bourmisytrov, F. Mellemvik, A. Bambulyak, O. Gudmestad, I. Overland, & A. Zolotukhin (Eds.), *International Arctic petroleum cooperation*. Routledge. https://doi.org/10.4324/9781315768762

Guliev, I., Krivosheeva, E., Akieva, L., & Kruzahilin, P. (2021). The prospects of sustainable energy development in Latin America: Prospects and barriers. *Energy Policy Journal, 24*(3), 161–182. https://doi.org/10.33223/epj/142285

Guliev, I. A., Geladze, S. A., Sokolova, E. S., & Egorova, L. I. (2020). The prospects of conventional and alternative energy production in the countries of the Balkan Peninsula (In the context of developing economic indicators). *International Journal of Energy Economic and Policy, 10*(1), 481–487. https://doi.org/10.32479/ijeep.8290

Kravchenko, S. A., & Salygin, V. I. (2015). A new synthesis of scientific knowledge: The making of interdisciplinary science. *Sotsiologicheskie Issledovaniya, 10*(378), 22–30 (in Russian).

Salygin, V., Guliev, I., Akieva, L., & Krivosheeva, E. (2019). Digital technologies in the field of transportation of oil and petroleum products. *Economics: Yesterday, Today and Tomorrow, 9*(4–1), 438–447 (in Russian).

Salygin, V., Guliev, I., Chernysheva, N., Sokolova, E., Toropova, N., & Egorova, L. (2019). Global shale revolution: Successes challenges and prospects. *Sustainability, 11*(6), 1627. https://doi.org/10.3390/su11061627

Salygin, V., Guliev, I., & Khubaeva, A. (2015). The impact of oil and gas resources in the arctic on the development of the transport system in the region. *MGIMO Review of International Relations, 4*(43), 27–35. https://doi.org/10.24833/2071-8160-2015-4-43-27-35 (in Russian).

Salygin, V., Guliev, I., & Ryabova, M. (2015). Renewable energy in China. *MGIMO Review of International Relations, 4*(43), 36–45. https://doi.org/10.24833/2071-8160-2015-4-43-36-45 (in Russian).

Salygin, V., & Mustafinov, R. (2018). Inclusive model of common electricity markets: The 'Greater Eurasia' concept. *International Trends, 16*(4), 63–73. https://doi.org/10.17994/IT.2018.16.4.55.4 (in Russian).

Salygin, V., & Mustafinov, R. (2019a). The Russian power sector in the context of energy transition. *Russian Journal of Entrepreneurship, 20*(2), 527–541 (in Russian).

Salygin, V., & Mustafinov, R. (2019b). Geopolitical aspects of electric power development: Organizational and strategic approaches. *Journal of International Economic Affairs, 9*(1), 177–194 (in Russian).

Salygin, V., Petrova, R., & Mustafinov, R. (2019). Central Asia: Sustainable development and energy cooperation issues. *Journal of International Economic Affairs, 9*(3), 1563–1574 (in Russian).

Salygin, V., Rybin, M., & Lobov, D. (2021). Operational management systems of leading oil and gas corporations. *Innovations in Management, 1*(27), 48–57 (in Russian).

Salygin, V. I., & Lobov, D. S. (2021). Defining major oil and gas companies' development strategies in the era of energy transition. *MGIMO Review of International Relations, 14*(5), 149–166. https://doi.org/10.24833/2071-8160-2021-5-80-149-166

Vylegzhanin, A., Bunik, I., Torkunova, E., & Kienko, E. (2020). Navigation in the northern sea route: Interaction of Russian and international applicable law. *The Polar Journal, 10*(2), 285–302. https://doi.org/10.1080/2154896X.2020.1844404

Vylegzhanin, A., Salygin, V., Dudykina, I., & Kienko, E. (2018). Positions of the non-Arctic states on the legal regime of the Arctic Ocean. *State and Law, 10*, 124–135 (in Russian).

Vylegzhanin, A., Salygin, V., & Krymskaya, K. (2020). Transboundary hydrocarbon resources political and legal mechanisms for preventing international conflicts. *International Trends, 18*(3), 23–41. https://doi.org/10.17994/IT.2020.18.3.62.2 (in Russian).

Zhiznin, S. Z., Timokhov, V. M., & Dineva, V. (2020). Energy security: Theoretical interpretations and quantitative evaluation. *International Journal of Energy Economics and Policy, 10*(2), 390–400. https://doi.org/10.32479/ijeep.8950

Zhiznin, S. Z., Timokhov, V. M., & Gusev, A. L. (2020). Economic aspects of nuclear and hydrogen energy in the world and Russia. *International Journal of Hydrogen Energy, 45*(46), 31353–31366. https://doi.org/10.1016/j.ijhydene.2020.08.260

Zhiznin, S. Z., Vassiliev, S., & Gusev, A. L. (2019). Economics of secondary renewable energy sources with hydrogen generation. *International Journal of Hydrogen Energy, 44*(23), 11385–11393. https://doi.org/10.1016/j.ijhydene.2019.03.072

PART I

Governing Energy Projects in the Russian Arctic

CHAPTER 2

International Legal Framework of Arctic Exploration

Igor A. Shulyatyev

2.1 Introduction

The interaction between states in the Arctic region is becoming more intensive. Various initiatives of interstate cooperation ensure the formation of favorable conditions for the sustainable development of the Arctic. One of the tools for creating a predictable international environment is the mechanisms of institutional and regulatory regulation of public relations in the Arctic (Torkunov, 2019).

The modern legal regulation of the Arctic regime is carried out mainly at the international and national legal levels (Guterres, 2021). International legal regulation of the Arctic regime is reflected in international multilateral and bilateral agreements. Multilateral international treaties are aimed at legal relations regulating the regime of the Arctic sea areas and regulating environmental protection issues in this region. It

I. A. Shulyatyev (✉)
The Institute of Legislation and Comparative Law under the Government of the Russian Federation, Moscow, Russian Federation
e-mail: iash7@mail.ru

V. I. Salygin (ed.), *Energy of the Russian Arctic*,
https://doi.org/10.1007/978-981-19-2817-8_2

should be noted that the creation and strengthening of international legal mechanisms for the development of oil and gas fields depends on the legal formalization of the delimitation of maritime spaces. Bilateral international agreements are aimed at resolving cross-border issues of neighboring states. This includes settling disputed issues of the delimitation of Arctic sea areas, and cooperating in the fields of fishing and mineral resource extraction. Bilateral international treaties are binding only for their participants and do not create obligations for third countries (Avkhadeev, 2016). This is a normative aspect of international cooperation in the Arctic.

Strategic initiatives formulated in various declarative acts and documents of the working groups are of the most general, situational, and variable nature, so they cannot serve as the only reference point for the development of international law and national Arctic legislation. It can be argued that at the current historical stage of the formation of Russian Arctic legislation, the main direction for improving the legal regulation of the development and development of the Arctic Zone of the Russian Federation in relation to the strategic initiatives of the Arctic states is the implementation of international agreements already concluded with the Arctic states in the Russian legal system. The Prospective development of the international legal framework of cooperation in the Arctic may be aimed at preparing an independent legal regime of the Arctic with its own special principles.

According to Nikolai Korchunov, Russia's Senior Arctic Official:

> Today we do not see the need to conclude any separate universal international treaty on the Arctic – the current international legal framework is quite sufficient to regulate international relations in the region. Rather, agreements of a sectoral nature may be in demand, which would formalize cooperation, for example, on environmental protection, transport development, telecommunications, scientific activities, financing of infrastructure and socio-economic development, responding to potential emergencies, and ensuring contacts between people. (Korchunov, 2021: 5)

This position implies the prospect of diversifying and specializing in the subject of international legal agreements in the Arctic region. This demonstrates the political and economic interest of the international community in the northern region of the Globe. International treaties continue to play a key role in such regulation.

At the same time, the international legal development of the Arctic is under strong political pressure, since, according to Pavel Gudev of the Primakov Institute of World Economy and International Relations,

> interstate competition has not disappeared anywhere: countries participating in maritime activities in the region often have completely different priorities and strictly defend their national interests. The Russian Federation, which is under sanctions pressure, is forced to look for new partners and allies in the Artic. This choice is not easy, since often the overlap of positions on certain issues is minimal. In order to defend its interests, Russia needs to achieve a format of interaction with Arctic and non-Arctic countries that would be extremely mutually beneficial and work for the benefit of our country. (Gudev, 2019: 57)

The international legal vision of the Arctic as a region of peace, harmony, and prosperity creates prerequisites for a careful and balanced approach to the exploitation of the mineral resource base. Without addressing general issues related to the environment, security, the rights of indigenous people, and the development of national legislation, mechanisms for responsible investment, then the formation of international legal foundations for the development, extraction, and transportation of energy resources is difficult.

The legal framework of Arctic cooperation is in transition. A "transition period" in world politics implies a radical change in the principles on which it was built, having lost their effectiveness and the functionality of the previous system of international relations; with transition periods, which may have different duration, characterized by certain regularities (Zagladin, 2016: 27), and a modern regulatory framework, the Arctic also receives a new impetus to its development and, accordingly, new international legal laws.

2.2 Literature Review

Soviet and Russian scholars have traditionally paid much attention to the issues of international and national law in relation to the Arctic region.

In 1928, V. L. Lakhtin's work "Rights to the Northern Polar Spaces" was published. In the Soviet period, such well-known scientists as Pashukanis (1935), Durdnevsky (1950), Lazarev (1974), and others addressed this topic. In his turn, Guslitser (1954) in "The International

Legal Regime of the Arctic and Antarctic" showed that the sectoral principle is a customary rule of international law, and therefore, the legal regime of the Arctic, based on the sectoral principle of the delimitation of spheres of action of polar states, is the interests of peace and security activities in the Arctic and Arctic seas; these seas (Laptev sea, Kara sea, East Siberian sea, Chukchi sea) are of a special flooded type, such as historical bays, the development of which is the merit of the Soviet state.

R.V. Dekanozov's (1975) and M.N. Kopylov's (2013) works also studied the legal regimes of the Arctic sectors, the Arctic Partnership, the 1920 Treaty on the Peaceful Economic and Scientific Use of Svalbard Island, and also the legal status and activities of the Arctic Council. The problems of activities of the Nordic Council were studied by another prominent researcher V.S. Kotlyar (1973)—the author of the first study in the Soviet legal literature "The Nordic Council (structure, operations)", dedicated to the activities of the regional consultative organization of the Nordic countries (Denmark, Iceland, Norway, Sweden, and Finland). The article reveals the history of the Northern Council's creation, and analyzes its international legal status, organizational forms of work, and main practical results of its activities.

The modern science of international law also pays considerable attention to the study of the international legal status of the Arctic. Articles have examined the problems of international cooperation in the Arctic: general issues of international law in relation to the Arctic Zone (Vasiliev, 2012), the legal status of Spitsbergen (Fife, 2004; Sennikov, 2014), issues of international maritime law. International legal aspects of subsurface use and its impact on the protection of the rights of indigenous peoples of the Arctic are studied by MGIMO Professor Vylegzhanin (Khabrieva, 2014). Also, Vylegzhanin, together with, said that there is no basis in international law for an extended interpretation of the powers of a purely technical body—the Commission on the Limits of the Continental Shelf. The 1982 Convention does not provide for any "burden of proof" in the Commission by a coastal state of certain shelf rights beyond 200 miles of baselines (or the "burden of demonstration"). According to general international law, the Commission has no authority to interfere in the efforts of the Arctic states to delineate the Arctic shelf between them (Vylegzhanin & Dudykina, 2017: 302).

A.Y. Kapustin—Scientific Director of the Institute of Legislation and Comparative Law under the Government of the Russian Federation, Professor, and President of the Russian Association of International

Law—pays great attention to general issues of the development of international law in the Arctic. For example, he states:

> Undoubtedly, there are a number of other unresolved international legal problems that hinder the formation of a unified model of the regional mechanism of Arctic international legal regulation. The international legal status of the Arctic remains uncertain, conflicts between the legal positions of the Arctic states have not been overcome, there are different legal regimes of territories and spaces, etc. (Kapustin, 2014: 26)

The institutional leaders of the Russian science of international law in the field of the Arctic research are Moscow State Institute of International Relations, and the Ministries of Foreign Affairs of the Russian Federation and Institute of Legislation and Comparative Law under the Government of the Russian Federation. In University at the Department of International Law, Professor Vylegzhanin formed a whole school for studying international law in the Arctic. The Institute regularly hosts the Arctic Legal Forum and publishes the almanac "Russian Arctic – Territory of Law".

Numerous scientific publications also make other proposals regarding the international legal regime of the Arctic—a common continental shelf, an international consortium, and a comprehensive environmental management regime for the entire region, including the high seas. However, despite the multiplicity of opinions, one thing is indisputable: the Arctic, by virtue of the established and generally recognized international legal custom, is the object of preferential rights of all Arctic states and, unlike Antarctica, cannot be considered as a "common heritage of humanity". The concept of internationalization actually denies the geographical, historical, and ecological features inherent in the Arctic region, which give rise to the preferential rights of the Arctic states to the region in question (Gitsu & Gitsu, 2018: 58). Russian legal thought considers the Arctic region as a complex biosocial system, which serves as a material basis for the claims of the Arctic states to independently determine the vectors and mechanisms of legal development of the Arctic.

In 2013, under the auspices of the Russian International Affairs Council under the general editorship of I.S. Ivanov (Minister of Foreign Affairs of the Russian Federation Russian Federation in 1998–2004), the third volume of the anthology "The Arctic Region: Problems of International Cooperation" was published. This volume presents documents that form the regulatory framework of international cooperation and activities of Russia and other Arctic states.

A general approach to international legal characterization of the Arctic is proposed by I.S. Zhudro. In his opinion, the definition of the "international legal regime of the Arctic" is a historically balanced set of customary and treaty international legal norms and principles that define the behavior and interaction of states (whether land or sea spaces) of the Arctic: global, regional, bilateral, and national legislative levels of legal regulation which, taken together, support the region's legal stability, enabling the formation of steel, including applicable international customs and predefined geographical and climatic characteristics of the Arctic Ocean; this has a marked influence of the trends of globalization, manifested in ocean governance (Zhudro, 2021: 29).

Considering the sustainable development of the Arctic as a holistic and broad concept, Russian science also touches on special issues of Arctic cooperation. For example, establishing the legal basis for the use of public initiatives by businesses, national and cultural associations, indigenous peoples of the North, and other public structures in the development of the Arctic Zone requires systematic law-making monitoring and effective law enforcement activities. When strengthening the sovereignty and state presence of the Russian Federation in the Arctic Zone, active use should be made of existing forms of democracy and public–private partnership (Teplyashin, 2020).

Cooperation in Artic region is also in the focus of foreign researches. For example, after the member states of the Arctic region failed to agree on a long-term strategic plan in May 2019, an idea was expressed in the literature about the need to reform the Arctic Council. In particular, the American researcher David Balton complains about the very structure of the Council, its inability to make effective decisions, and its lack of funding (Balton, 2019). At the same time, he does not take into account that the Arctic Council has a strictly defined mandate, which it fulfills successfully, and for further progress, a consensus of the Arctic states is needed, ensuring a coordinated interstate approach to the development of the Arctic. Also, the political and legal method of strengthening the existing institutional and legal mechanisms of cooperation in the Arctic ensures stability and gradual progress in the development of Arctic spaces and its sustainable development, which is the main goal of intergovernmental Arctic forums. The predictability of interstate mechanisms makes it possible to develop a holistic strategic vision for the future of the Arctic on the basis of existing international agreements at the universal, regional, and sub-regional levels, even if there is currently no unified strategic plan of the Arctic Council.

Kristina Schönfeldt (2017: XXXV–XCVIII) in her book "The Arctic in International Law and Policy" provides "an analytical introduction, a chronology of legally relevant events, and a selection of essential materials covering a wide range of issues—e.g., delineation and delimitation of maritime boundaries, environmental protection, indigenous peoples' rights, shipping, and fisheries".

Michael Byers (2013) also offers a complex approach to the international law dimension of Arctic issues in his "International Law and the Arctic", including such issues such territory, maritime boundaries, Beaufort Sea boundary, extended continental shelves, arctic straits, environmental protection, indigenous peoples, and security.

Thus, a brief review of the scientific literature demonstrates the diversity of research in the field of legal measurement of the Arctic both at the present stage and in the historical aspect, with a tendency to show interdependence of all arctic aspects.

2.3 Strategic Context

The Arctic region is home to the political and legal efforts of organizations such as the Arctic Council (1996), the Council of the Baltic Sea States (1992), the Barents Euro-Arctic Council (1993), the Nordic Council of Ministers (1971), and the Conference of Arctic Parliamentarians and the European Union (Northern Dimension) (Mordvinova et al., 2017). States also cooperate in the Arctic region by coordinating their policies without creating any rigid and formal interstate structures. One example is the coordination of cooperation within the framework of the Northern Dimension, which functions through four partnerships: the Northern Dimension Partnership for Public Health and Social Welfare (NDPHS), Northern Dimension Environmental Partnership (NDEP), Northern Dimension Transport and Logistics Partnership (NDPTL), and Northern Dimension Cultural Partnership (NDPC). The structure, nature, and objectives of partnerships range from project-oriented financing to expert-oriented collaboration. This is the institutional aspect of interstate cooperation in the Arctic region.

The Arctic Council and the Barents Euro-Arctic Council have the most developed program documents and acts that are important for studying the main directions for improving the legal framework and development of the Arctic Zone of the Russian Federation/Euro-Arctic region.

Members of the Arctic Council (AC) are Canada, Denmark, Finland, Iceland, Norway, the Russian Federation, Sweden, and the United States of America (the so-called Arctic states). The Declaration on the Establishment of the Arctic Council of 1996 sets out the general objectives of cooperation between the Arctic states: issues of common interest; sustainable development and environmental protection; emergency prevention, preparedness, and response; dissemination of information; promotion of education; and promotion of interest in issues related to the Arctic.

Separately, it is emphasized that the Arctic Council should not deal with issues related to military security. Decisions of the AC are made by its members on the basis of consensus.

At the first meeting of the Arctic Council at the Ministerial level in Iqaluit (Canada, September 17–18, 1998), it adopted the Rules of Procedure of the Arctic Council (hereinafter referred to as the Rules of Procedure), which were further on altered at the eighth ministerial meeting of the Arctic Council held in Kiruna (Sweden, May 15, 2013).

According to paragraph 26 of the Rules of Procedure:

> An Arctic State or a Permanent Participant may propose joint initiatives. All proposed programs and projects that do not have a valid mandate issued by the ministers are submitted to the Arctic Council for adoption by the Council of the relevant decision. Program and project proposals should include information in accordance with the points set out in Annex 1. The submission of a proposal for consideration by the Council must be made by putting it on the agenda in accordance with these Rules.

In accordance with the guidelines for making proposals for projects and programs of the Arctic Council (Annex 1 to the Rules of Procedure of the Arctic Council):

> When preparing proposals for projects and programs it is encouraged to include the following items, if applicable: the main problems or questions; reasons why the Arctic states should review and approve a proposal; other important recommendations related to the offer, including recommendations about the body or bodies, which should implementation, coordination, or promoting the implementation of relevant activities; information on expenditures and sources of financing relevant events; a plan of work, including the start and end dates of the event; relationship to

other programs or activities of the Arctic Council and other significant events regional or international forums; assessment of the impact on the environment; other important information related to the proposal.

The Arctic Council may establish working groups, task forces, or other subsidiary bodies to prepare and implement programs and projects in accordance with the instructions and guidance of senior officials. The composition and mandates of such bodies are agreed upon by the Arctic states at Ministerial Meetings. The activities of these bodies are carried out in accordance with the Rules of Procedure of the Arctic Council.

Official sources containing strategic initiatives, common goals, and plans of the states of the Arctic Council in relation to the development of Arctic spaces include the Declaration on the Establishment of the Arctic Council (1996), declarations of ministers of the Arctic states, international legal agreements concluded jointly by the Arctic states.

In accordance with the Iqaluit Declaration (1998), the Arctic states established a program in the field of sustainable development on such issues as children and youth in the Arctic, health care, telemedicine, environmental management (including fishing and cultural and ecological tourism), and scientific and technical exchange to improve the health control system in the Arctic.

In the Declaration on the results of the Second Meeting of the Arctic Council (2020), the Ministers of the Arctic states noted that environmental protection in the Arctic is one of the most important national policies of the Arctic states, and that the Council intends to take an active part both in the assessment of environmental risks and in the implementation of measures to prevent environmental degradation.

The fifth ministerial session of the Arctic Council in Salekhard (2006) adopted the Salekhard Declaration, which noted that the Arctic climate is an extremely important component of the global climate system, with all its consequences for the whole world. The Arctic Ministers stressed the need to improve living conditions, expand economic opportunities in the Arctic, and strengthen Arctic communities by taking measures to train personnel, promote education, and promote research.

Declaration Tromsø, on the occasion of the Sixth Ministerial Session of the Arctic Council (2009), recorded such areas of cooperation between the Arctic states as climate change in the Arctic, the Arctic marine environment, human health and human development, energy, pollutants, biodiversity, and the management and organization of the Arctic Council.

The Nuuk Declaration on the occasion of the Seventh Ministerial meeting of the Arctic Council (May 12, 2011) reaffirmed the commitment of the Arctic Council to promote environmental protection and sustainable development of the Arctic. The Arctic states noted the need for a comprehensive overview of human development in the Arctic and called to assess the current state in the field of human development in the Arctic and its relationship with climate change and other factors influencing community of the Arctic.

The Kiruna Declaration (2013) recognizes the central role of business in the development of the Arctic and sets out the decision to strengthen cooperation and interaction with the business community in order to promote sustainable development in the Arctic.

The Iqaluit Declaration in 2015 focuses on general issues of maintaining the viability of Arctic communities, protecting the unique Arctic environment, and strengthening the Arctic Council.

In 2016, the Declaration on the occasion of the 20th anniversary of the Arctic Council notes that the Council is involved in important joint practical actions, such as the Framework Plan for Cooperation in the Field of Preventing Oil Pollution in the Arctic Marine Areas as a result of Oil and Gas Activities and Shipping, the Framework Document on the Creation of The Pan-Arctic Network of Marine Protected Areas, the Framework Document on Actions to Reduce Black Soot and Methane Emissions, and the Circumpolar Biodiversity Monitoring Program; the Council also protects the mental health of the Arctic population. The launch of the Project Support Tool in July 2014 helped to strengthen the protection of the Arctic environment by providing financial support for projects aimed at eliminating pollution. The Arctic Council has contributed to the creation of additional structures for regional cooperation and collaboration, such as the University of the Arctic, the Forum of the Sustainable Arctic Observation Network, the Arctic Coast Guard Forum, the Arctic Economic Council, and the Forum of Regulators of Oil and Gas Activities on the Arctic Shelf.

The Fairbanks Declaration (2017) highlights the issues of safety and security in the Arctic Ocean and the rational management of its resources, improving economic and living conditions, addressing the consequences of climate change, and strengthening the Arctic Council. The Ministers separately noted that they have instructed senior officials of the Council to develop a strategic plan based on the fundamental documents of the

Council and on the strategies and guidance documents of the subsidiary bodies, and submit it to the Ministers for approval in 2019.

However, in 2019, the Arctic Council's Strategic Plan was not developed and approved. Moreover, 2019 was marked by the fact that the Arctic states failed to agree on a joint declaration, which indicated serious interstate contradictions among them. In a joint statement, the ministers said that senior officials were instructed to continue working on strategic planning.

In general, the documents adopted by the member states of the Arctic Council are very numerous and not systematized, which makes initiatives to deepen interstate cooperation in the Arctic difficult to implement in practice. Declarations of the Arctic states are not legally binding, so the choice of sources of regulatory regulation of issues reflected in the declarations remains at the discretion of the states themselves.

The development and strengthening of the existing formats of Arctic cooperation and the Arctic Council is a complex, long-term intergovernmental process involving interested states. Following the meeting of the Foreign Ministers of Russia and Iceland, Russian Foreign Minister Sergey Lavrov stated:

> There are no plans to increase the number of members of the Arctic Council, no one has put forward such ideas. Everyone can get observer status... We do not see any contraindications against obtaining observer status for countries that are really ready to participate in the work in a non-politicized, pragmatic way, while fully respecting the basic decisions that the Arctic Eight makes. (TASS, 2019)

This view was supported by the Icelandic Foreign Minister, Gudlaugur Thór Thórdarson:

> I believe that the balance that currently exists among the Arctic Council members and observer countries is already quite stable and good... I agree that there is no reason to expand the membership of the Arctic Council. The current system is very successful. (TASS, 2019)

An important milestone in the strategic vision of the Arctic region was adopted in 2021 with the Strategic Plan for the Arctic Council for the Years 2021–2030, which consists of the following blocks: protection of the natural environment, sustainable social and economic development, and the strengthening of the Arctic Council. Its aims include: protecting the climate of the Arctic, healthy and viable Arctic ecosystems, healthy Arctic marine environment, sustainable social development, sustainable economic development, knowledge and information exchange, and a stronger Arctic Council. It is important to note that it is necessary to "promote action on issues critical to maintaining the health of Arctic ecosystems, as well as the Arctic population, and encourage cooperation among Arctic states in the field of an ecosystem approach to Arctic management in order to promote nature conservation and sustainable use of natural resources based on the best available scientific evidence" (paragraph 2.5, Strategic Plan). Special attention was paid to the need to comply with international law while striving for ambitious policy recommendations and expanding cooperation with institutions and initiatives that reflect the relationship between the Arctic and the rest of the world.

Thus, the activities of the Arctic Council today act as the main institutional and legal forum for the development of the Arctic region. However, the Arctic Council is not the only institutional and legal form of interstate cooperation in the Arctic region.

In accordance with the "Declaration on Cooperation in Barents/Euro-Arctic Region", the Barents Euro-Arctic Council was established in 1993. Denmark, Finland, Iceland, Norway, Russia, Sweden, and the European Union are members to this Council.

The purpose of the Council is to serve as a forum for cooperation among its members and to promote the sustainable development of the region, bearing in mind the principles and recommendations contained in the Rio Declaration and Agenda 21 of the Conference of Public Associations Department of Environment and Development. In this regard, the Council will serve as a forum for discussing bilateral and multilateral cooperation in the fields of economy, trade, science and technology, tourism, environment, infrastructure, education, and cultural exchange, as well as projects specifically aimed at improving the situation of the indigenous population of the North. Decisions in the Council and its bodies are made by consensus. According to the Declaration on the results of the meeting of the Heads of Government of the Barents Euro-Arctic Council countries (Kirkenes, June 3–4, 2013), the priorities of cooperation in this format are the economy, financial issues, people-to-people contacts, education, environment, culture, and health.

At the level of international agreements between Barents Member States and The Euro-Arctic Council is an Agreement on Cooperation within the Field of Emergency Prevention, Preparedness and Response in the Barents Region (2008), the subject of which is cooperation in the field of emergency prevention, preparedness, and response in the Barents Sea and in the Euro-Arctic region, including cross-border cooperation and joint training and exercises to improve the efficiency and responsiveness of emergency response. The contracting parties shall develop methods and implement measures that enhance the effectiveness of international cooperation in the field of emergency prevention, preparedness, and response.

There is no comprehensive document reflecting strategic initiatives and legal mechanisms for implementation in the Barents Region/Euro-Arctic Council. However, among the declarative acts of the Council, documents on the development of transport infrastructure stand out in particular: the Declaration of the Transport Ministerial Meeting of the Barents Euro-Arctic Council (2019), and the Joint Barents Transport Plan (2019).

Thus, currently in the Arctic Council and the Barents Sea/Euro-Arctic Council, there are no full-package conceptual and strategic documents containing long-term goals and mechanisms for the development of the Arctic region agreed by states. Cooperation among states in the Arctic region is of a political and declarative nature, without binding specific legal mechanisms and formats of interaction.

The strategic documents contain the most general description of the approach to Arctic cooperation (so to speak, starting points for convergence of positions). No one disputes that the Arctic is a territory of peace and prosperity. However, there is an intertwining of interests of the international community, states, and the private sector, which makes public relations in the Arctic region very difficult to coordinate due to different interests. This implies a variety of acts of different legal force: from soft law to international legal acts. In this regard, it is very important to determine the competence of various legal regimes in Arctic matters.

2.4 Legal Regimes and Regulation

It should be noted that the concepts of "Arctic" or the "Arctic Zone" do not have a single and precise definition. The Arctic is considered as a geographical feature. Therefore, for the legal regulation of public relations in the Arctic region, the specific scope of various legal rules of different nature is important.

The Arctic territory is defined as a part of the globe, including its land and water spaces, subsurface, and air space. Within these areas, international law distinguishes between the territories of states, international public territories, and mixed-regime territories. Depending on the type of territory, each of them has a certain international legal status and regime. A distinctive feature of the legal regime of the state territory of the Arctic is that it is under the sovereignty of the specific state to which it belongs and within which the state exercises its territorial supremacy.

International public territories usually include the open sea, the airspace above it, the seabed outside the continental shelf of states, and outer space. The legal regime of such territories is determined by international law on the basis of the principles of non-appropriation in any form and general use. State sovereignty does not extend to such territories.

In modern international law, Arctic territories with a mixed regime include the following areas of the Ocean: exclusive economic zones, the continental shelf, and adjacent zones. They are not part of the state territory of the Arctic countries, but coastal states exercise their sovereign rights and jurisdiction for the purposes of exploration, development, and conservation of natural living and mineral resources, the use of energy, water, and wind, and protection of the natural environment (exclusive economic zone and continental shelf) or the control and jurisdiction in specially designated areas (surrounding areas).

The legal regime of the Arctic is determined by the applicable customary and treaty norms of international law and the national law of the Arctic states. Starting with the Russian–English (1825) and Russian–American (1867) conventions, which for the first time provided for the sectoral (along the meridian to the North Pole) principle of dividing rights to the northern polar possessions, the Arctic states themselves determined the status of the Arctic at the bilateral and regional levels, which traditionally received the tacit consent of most non-regional states.

There is a system of international legal acts of various levels (universal, regional, bilateral) and legal natures (international treaties, international customs, declarations, recommendatory acts, and other acts of "soft law") that apply to the regulation of relations between states in the Arctic (Kapustin, 2015: 39–40).

Thus, in the legal sense, the Arctic space is a set of legal regimes that provide a regulatory framework of the development of the Arctic.

With the revision of the legal regimes governing activities in the Arctic, today the rules, norms, and procedures of various international organizations and conferences occupy an important place. The activities of such actors create the prerequisites for the international legal formalization of Arctic cooperation. The Arctic Council has examples of successful consolidation of forms and procedures of multilateral interstate cooperation in the provisions of international treaties that are legally binding on states, which are implemented in national legal systems in accordance with the legislation of the states-parties to the agreement:

(1) "Agreement on Cooperation on Aeronautical and Maritime Search and Rescue in the Arctic" (2011)
The Agreement aims to strengthen cooperation and coordination in the field of aviation and maritime search and rescue in the Arctic.
(2) "Agreement on Cooperation on Marine Oil Pollution Preparedness and Response in the Arctic" (2013)
The purpose of the Agreement is to strengthen cooperation, coordination, and mutual assistance between the parties in the field of preparedness and response to oil pollution in the Arctic in order to protect the marine environment from oil pollution.
The Agreement applies to oil pollution incidents that occur in any maritime area or may pose a threat to any maritime area over which a state whose government is a party to this Agreement exercises sovereignty, sovereign rights, or jurisdiction in accordance with international law, including its internal waters, territorial sea, exclusive economic zone, and continental shelf.
(3) "Agreement on Enhancing International Arctic Scientific Cooperation" (2017)
The purpose of the Agreement is to strengthen cooperation in the field of scientific activities to increase the effectiveness and efficiency of the development of scientific knowledge about the Arctic.

Thus, certain strategic initiatives of the AC have taken the form of an international treaty, which is mandatory for Arctic states to comply with at the interstate and domestic levels. At the same time, the Arctic states determine the legal forms of domestic implementation of the provisions of such international treaties independently.

Implementation of the main tasks in the field of international cooperation development is ensured through international legal registration of the external border of the continental shelf and maintaining cooperation with the Arctic states in order to protect national interests and implement the rights of the coastal state in the Arctic provided for in international acts, including those related to the exploration and development of the continental shelf resources and the establishment of its external borders.

The Russian approach assumes that the Arctic is the northern polar region of the Earth, including the northern margins of Eurasia and North America (except for the central and southern parts of the Labrador Peninsula), the island of Greenland (except for the southern part), the Arctic Ocean seas with islands (except for the eastern and southern parts of the Norwegian Sea), and adjacent parts of the Atlantic and Pacific Oceans.

It seems that a regulatory legal act regulating the development of the Russian Arctic Zone should have a comprehensive character: instead of duplicating the provisions of numerous legal acts at the federal and regional levels, it should rather reflect the specifics of the region as much as possible. Special attention should be paid to issues related to public administration; requirements for the implementation of economic and other activities; development of infrastructure and transport network; features of nature management; environmental protection; extraction and processing of natural resources and minerals; and the preservation of the original culture of indigenous small-numbered peoples of the North. Given the important geopolitical significance of the region, it is necessary to work out in detail all aspects of the implementation of international cooperation (Savenkov, 2018: 37).

International law creates an interstate framework of cooperation in the Arctic, the goal of which is the sustainable development of the Arctic in the era of energy transition and climate change, increasing the role of the social factor, and ensuring peace and security.

Despite the global importance of the Arctic for world politics, in the legal sense it remains a zone of cooperation between practical states and their institutional and legal regulators. For the energy sector, this approach is a cornerstone for investment in the exploration, production, and transportation of mineral resources, since there is no special legal regime for such activities and it is necessary to use the norms of general international law or special regional agreements that are not directly related to the fuel and energy complex.

The role of the private sector cannot be ignored. Among the most "authoritative" international non-governmental organizations and professional associations in relation to the issues under consideration, the following can be distinguished: Oil Companies International Marine Forum (OCIMF), the International Chamber of Shipping (ICS), The International Association of Oil & Gas Producers (IOGP), International Petroleum Industry Environmental Conservation Association (IPIECA), The International Tanker Owners Pollution Federation Limited (ITOPF), The International Association of Classification Societies (IACS), and Lloyd's Register Group Limited (LR). Such structures can develop their own regulations aimed at standardizing certain aspects of Arctic activities, while not replacing legally binding rules.

2.5 Overview of the International Legal Framework

It should be noted that the term "Arctic" has no single universally accepted definition in the scientific literature; however, in international legal doctrine entrenched behind the definition, "the Arctic is the part of the terrestrial spheroid, the center of which is designated the geographic North pole, and the marginal border of which is the Arctic circle" (Vylegzhanin, 2015: 109). The Arctic region includes the Arctic Ocean, the shelf of which is rich in minerals and natural resources. In addition, the Northern Sea Route (NSR) passes through the Arctic Ocean—the shortest route connecting the Atlantic Ocean through the Arctic with the Pacific Ocean (Vylegzhanin, 2017: 12). It is estimated that the use of the NSR reduces the distance of navigation from China to Western Europe by up to 40 percent compared to the distance covered by the Suez Canal (Li & Huang, 2016). Moreover, in 2016, the volume of cargo transportation along the NSR, including transit cargo, reached a record of 7265 million tons (Starodubtsev, 2017: 14), while in 2017, at the beginning of December, it had reached 9694 million tons (TASS, 2018).

The sources of international law in the field of Arctic cooperation and establishing the legal regime of the Arctic Ocean are diverse. In the field of general international law, the framework is built out of the following documents: International Convention on Civil Liability for Oil Pollution Damage (1969), Convention on Wetlands of International Importance especially as Waterfowl Habitat (1971), Convention on the Prevention of Marine Pollution by Dumping of Wastes and Other Matter (1972),

Declaration of the United Nations Conference on the Human Environment (1972), Convention on Long-Range Transboundary Air Pollution, often abbreviated as Air Convention or CLRTAP (1979), Convention on Biological Diversity (1992), United Nations Framework Convention on Climate Change (1992), the Rio Declaration on Environment and Development (1992), the Kyoto Protocol (1997), United Nations Framework Convention on Climate Change (1992), Stockholm Convention on Persistent Organic Pollutants (2001), and the Paris Agreement under the United Nations Framework Convention on Climate Change (2015).

A special place is occupied by Article 234 "Ice-covered areas" of the Convention of the UN Law of the Sea (1982), which states that:

> Coastal states have the right to adopt and enforce non-discriminatory laws and regulations for the prevention, reduction and control of marine pollution from vessels in ice-covered areas within the limits of the exclusive economic zone, where particularly severe climatic conditions and the presence of ice covering such areas for most of the year create obstructions or exceptional hazards to navigation, and pollution of the marine environment could cause major harm to or irreversible disturbance of the ecological balance. Such laws and regulations shall have due regard to navigation and the protection and preservation of the marine environment based on the best available scientific evidence.

The text of this norm is formulated in the most general form, without reference to the Arctic spaces and water areas of the Arctic Ocean. Therefore, any reference to the text of this article in interstate relations when condemning the Arctic issues should have a strict argumentation. This is especially important when considering that the United States of America has not ratified this convention. The interpretation of this norm is currently changing due to the registration of melting processes of polar ice. However, reliable scientific data on the dynamics and content of climate change in the polar altitudes are not yet available.

To regulate navigation issues in the Arctic Ocean, the International Code for Ships Operating in Polar Waters (the Polar Code) is used, which is a document that is not uniform in its legal force since it contains both legally binding requirements and recommendations. As parts of the International Convention for the Safety of Life at Sea (SOLAS) and the International Convention for the Prevention of Pollution from Ships (MARPOL) clarify, the mandatory parts of the Code are a source of international law; on the other hand, Parts I-B and II-B, which include only recommendations on the application of the provisions of this document, can be attributed to auxiliary sources of international law, since the Polar Code was adopted by resolutions of the International Maritime Organization (IMO) Committees.

One of the main principles of creating the Code is to develop it on the basis of previously adopted IMO legal regulations. During the preparation of the Polar Code, there was a gradual transition from creating advisory provisions for navigation in Arctic waters to expanding the scope of such provisions to include Antarctic waters and making such requirements legally binding.

The purpose of the Code is to ensure the safe operation of ships and the protection of the polar environment by addressing risks specific to polar waters that are not adequately addressed in other documents of the Organization. The International Code for Ships Operating in Polar Waters has been developed as a supplement to the existing International Maritime Organization's (IMO) instruments in order to enhance the safety of ship operations and reduce the negative impact of navigation on humans and the environment in remote and vulnerable polar waters, which are potentially dominated by harsh conditions. The Code recognizes that, when operating in polar waters, vessels, ship systems, and operations may be subject to additional requirements, in addition to those prescribed SOLAS of 1974 and MARPOL of 1973—as firstly amended by the 1978 Protocol and later by the Protocol of 1997 (MARPOL Convention)—as well as other relevant mandatory IMO instruments.

An independent international legal institution for Arctic cooperation should consider peaceful settlement of disputes.

Summarizing the sections of international treaties makes it possible to identify the following mechanisms: direct negotiations, direct consultations, and negotiations among the competent authorities. A dispute resolution mechanism related to hydrocarbon production is provided in the Treaty between the Kingdom of Norway and the Russian Federation concerning Maritime Delimitation and Cooperation in the Barents Sea and the Arctic Ocean. If the parties are unable to conclude the Merger Agreement referred to, such a disagreement should be resolved as soon as possible through negotiations or through any other procedure agreed between the parties. If the dispute is not resolved within six months from the date on which one of the parties' requested negotiations with the other party, either party may refer the dispute to a three-member arbitral tribunal.

Among the mechanisms of interstate cooperation in the Arctic, international interagency cooperation between the competent authorities of various states is also important for the Russian Federation. One example of such an authority is the Arctic Coast Guard Forum, which is an independent, informal, non-treaty organization to promote safe, reliable, and environmentally responsible maritime activities in the Arctic. Participants of the forum are all Arctic countries: Canada, Denmark, Finland, Iceland, Norway, Russia, Sweden, and the United States. The Russian Federation is represented by the Federal Security Service of the Russian Federation. The functions of the forum chair change every two years in accordance with the chairmanship of the Arctic Council. The importance of this forum lies in the formation of organizational mechanisms for cooperation between states on a specific issue of the development of the Arctic region. The forum does not have the authority to adopt legal acts binding on states.

The activities of international non-governmental forums also contribute to the interpretation of the agreement and mechanisms for resource development in the Arctic. For example, the Arctic Economic Council is considering the possibility of using the UN principles on responsible investment and implementing ESG methodologies to assess the effectiveness of economic projects.

The resource and raw material potential of the Arctic Zone of the Russian Federation has a high degree of projected demand, which is the basis for the competitiveness of Russian hydrocarbons from the Arctic Zone of the Russian Federation in the global energy market and an undoubted competitive advantage of Russia in the global energy markets. In addition, the hydrocarbon resources of the Russian Arctic are essential for ensuring reliable supplies to the domestic market and stable supply of the needs of the national economy in the long term (Symbolokov, 2021: 143–144).

There is no global, regional, or sub-regional source of international law in the field of mineral resource extraction. However, such mechanisms appear at the bilateral interstate level. One example is the "Agreement between the Russian Federation and the Kingdom of Norway on Maritime Delimitation and Cooperation in the Barents Sea and the Arctic Ocean" (together with "Issues of Fisheries" and "Cross-border Hydrocarbon Deposits"), signed in Murmansk on 15.09.2010.

According to this Agreement (Para. 2, Part 2 of Article 5),

> if the existence of a hydrocarbon deposit on the continental shelf of one of the Parties is established, and the other Party believes that this deposit extends to its continental shelf, the latter Party may notify the first Party about this and must provide data on which it bases its opinion.
> If such an opinion is provided, then the Parties begin to discuss the issue of the contours of the hydrocarbon field and the possibility of operating such a field as a single whole. During such discussion, the initiating Party should provide a justification for its opinion, supported by geophysical and / or geological data, including any existing drilling data, and both Parties should make every effort to ensure that all relevant information is made available for such discussion. If a hydrocarbon Deposit extends over the continental shelf of each of the Parties, and the field on the continental shelf of one Side may be partially or fully operated from the continental shelf of the other Party, or exploitation of hydrocarbon deposits on the continental shelf of one Side may affect the possibility of exploitation of hydrocarbon deposits on the continental shelf of the other Party, at the request of one of the Parties in accordance with Annex II of the Agreement on the exploitation of this Deposit of hydrocarbons as a whole, including its distribution between the Parties (hereinafter referred to as the Agreement to merge). The exploitation of any hydrocarbon field that extends over the continental shelf of the other Party can only be initiated in accordance with the provisions of the Association Agreement.

According to Fodchenko,

> Another legal inconsistency in Article 1.6 (d) is the use of the term "cross-border hydrocarbon field" in the Russian and Norwegian copies of the 2010 Cooperation Agreement. In the Norwegian text ("engrenseoverskridende petroleumsforekomst"), as in the English text ("trans-boundary hydrocarbon deposit"), the term is translated as "trans-boundary hydrocarbon deposits", while in the Russian text "trans hydrocarbon" ("boundary hydrocarbon field") is used. In Norway, discovered hydrocarbon deposits receive the status of a "field" (or become part of an existing field) when the licensing group has decided to put it into operation and the Development and Operation Plan has been approved by the relevant government authorities. In Russian legislation, obtaining the status of a deposit is less clearly defined. Article 10 of the Law "On Subsurface Resources" replaces the term "subsoil plots/deposit" with the term "field" as the transition from the stage of geological exploration to the stage of mining. (Fodchenko, 2018: 186)

An important scientific issue and scientific task is the formation of an independent legal system of international cooperation in the Arctic: the so-called autonomous regime.

There are two approaches to the concept of a self-contained regime. One is aimed at consolidating the position of a group of international treaties that establish the objective mode, characterized by the fact that they contain obligations *erga omnes* the interests of the entire international community. The other focuses on the agreements forming autonomous modes, generally in the framework of modern international organizations in the form of legal complexes, seeking to separate from General international law (Kapustin, 2018: 34–35). This framework confirms the basic responsibility of practical states.

The creation of an independent autonomous international legal regime in the Arctic, in cooperation with national legislation, is essential for the balanced and fair distribution of rights and freedoms and the distribution of freedoms. This focuses attention on treaties that form autonomous regimes, usually within the framework of modern international organizations in the form of legal complexes that seek to separate from general international law (Kapustin, 2018: 34–35). For example, the Agreement on Strengthening International Arctic Scientific Cooperation improves the legal environment for scientific activities in the Arctic and goes beyond what is provided for in general international law, including the law of the sea in relation to marine scientific research, and relevant regional and bilateral agreements. Freedom of scientific research is an important part of the legal regime in relation to spaces that are not under the sovereignty of states. A different legal regime in the Arctic does not prevent the use of experience in conducting scientific research in areas that do not fall under the sovereignty of a particular state. At the same time, the development of such research requires special international legal regulation (Sokolova, 2020: 126–127).

2.6 Conclusions

The international legal framework of Arctic cooperation has its own history of development. To date, no single international legal definition of the term "Arctic" has been developed, and each source of international law can independently determine the legal boundaries of the Arctic region.

Russian law studies a wide range of issues related to international Arctic cooperation, providing a doctrinal justification for the rights of Russia as an Arctic state.

The main source of modern international law governing legal relations in the Arctic is the international treaty. Institutional structures are not explicitly empowered by states to adopt legally binding acts. The most representative international structure with a broad mandate in the Arctic can be considered to be the Arctic Council. Adopted in 2021, the Arctic Council's strategic plan calls for promoting compliance with international law, including considering the United Nations Convention on the Law of the Sea as the legal framework within which the Arctic marine environment should be managed, while emphasizing the role and responsibility of coastal states.

The legal regimes and regulations of Arctic cooperation are diverse. The balance of international and national legal regulators—taking into account the standards of international intergovernmental and international non-governmental organizations—create an independent legal regime in the Arctic without a single universal international treaty.

Forming the international legal framework of Arctic cooperation is the responsibility of the Arctic states. This approach is justified by the role of national legislators of coastal states in promoting an ecosystem-based approach to Arctic management, nature conservation, and sustainable development in the region.

Cooperation of the Arctic states in the fields of the exploration, production, and transportation of energy resources comes from a combined international legal regime consisting of global agreements in the areas of environmental protection and bilateral international treaties that emphasize the importance of effective and responsible management of each state's own hydrocarbon resources.

References

Avkhadeev, V. R. (2016). International legal regulation of cooperation between the Russian Federation and neighboring countries in the Arctic on the basis of bilateral agreements. *Journal of Russian Law, 9*, 138–146 (in Russian).

Balton, D. (2019). *A strategic plan for the Arctic council: Recommendations for moving forward*. Harvard Kennedy School. Belfer Center. https://www.belfercenter.org/publication/strategic-plan-arctic-council-recommendations-moving-forward

Byers, M. (2013). *International law and the Arctic*. Cambridge University Press. https://doi.org/10.1017/CBO9781107337442

Convention of the UN Law of the Sea. (1982). https://www.un.org/depts/los/convention_agreements/texts/unclos/unclos_e.pdf

Dekanozov, R. V. (1975). Some questions of the legal nature of territories (spaces) withdrawn from the sphere of state sovereignty. *Soviet Yearbook of International Law*, 203–214 (in Russian).

Durdnevsky, V. N. (1950). Problems of the legal regime of the circumpolar regions. *Bulletin of Moscow State University, 7*, 112–120 (in Russian).

Fife, R. E. (2004). Subject and objectives of the treaty on Svalbard from the point of view of International Maritime Law. *Moscow Journal of International Law*, No. 4 (in Russian).

Fodchenko, I. P. (2018). Operator of a cross-border hydrocarbon field-legal analysis of the agreement between the Russian Federation and the kingdom of Norway on Maritime delimitation and cooperation in the Barents Sea and the Arctic Ocean. *Matters of Russian and International Law, 8*(4a), 177–197 (in Russian).

Gitsu, M. A., & Gitsu, V. D. (2018). International legal regime of the Arctic (conceptual confrontation). *Gosudarstvo i Pravo, 6*, 45–58 (in Russian).

Gudev, P. A. (2019). New risks and opportunities for interstate cooperation in the Arctic. *Arctic and North, 36*, 47–68. https://doi.org/10.17238/issn2221-2698.2019.36.57

Guslitser, G. M. (1954). *The international legal regime of the Arctic and Antarctic*. Abstract of a Ph.D. Thesis. Lomonosov Moscow State University.

Guterres, A. (2021). The UN secretary-general on the global digitalization and international cooperation. *Digital Law Journal, 2*(2), 10–13. https://doi.org/10.38044/2686-9136-2021-2-2-10-13

Kapustin, A. Y. (2014). International legal regulation of the Arctic: History and modernity. In T. Khabrieva (Ed.), *Arctic law conception of the development*. Publishing House "Jurisprudence" (in Russian).

Kapustin, A. Y. (2015). International law and the Arctic. In S. Naryshkin, T. Khabrieva, & D. Kobylkin (Ed.), *Protection and sustainable development of the Arctic: Legal aspects*. Publishing House "Jurisprudence" (in Russian).

Kapustin, A. Y. (2018). *International treaty and international legal order: Some conceptual approaches*. In V. Avkhadeev, V. Astashova, L. Andrichenko et al. (Eds.), *Contract as a general legal value*. Statut Publishers (in Russian).

Khabrieva, T. Y. (Ed.). (2014). *Arctic law conception of the development*. "Jurisprudence" (in Russian).

Kopylov, M. N. (2013). On the legal regime of the Arctic. *Legal Initiative*, No. 12 (in Russian).

Korchunov, N. (2021). There are no problems in the Arctic that require a military solution. *Kommersant*. https://www.kommersant.ru/doc/4641929 (in Russian).

Kotlyar, V. S. (1973). *Nordic council: Structure, operations*. International Relations (in Russian).

Lakhtin, V. L. (1928). *Rights to the northern, polar spaces*. Literary Publishing House of the People's Commissariat for Foreign Affairs (in Russian).

Lazarev, M. I. (1974). International legal regime of the Arctic space. *Modern International Maritime Law* (in Russian).

Li, W., & Huang, Y. (2016). International code for ships operating in polar waters: Challenges to polar shipping safety rules in China. *Advances in Polar Science, 27*(3), 146–153.

Mordvinova, T. B., Skaridov, A. S., & Skaridova, M. A. (Eds.). (2017). *Polar law*. Justicija (in Russian).

Pashukanis, E. B. (1935). *Essays on international law*. Sov.zak-vo (in Russian).

Savenkov, A. N. (2018). Arctic: Legal support for sustainable development and cooperation. *Proceedings of the Institute of State and Law of the Russian Academy of Sciences, 13*(1) (in Russian).

Schönfeldt, K. (2017). *The Arctic in international law and policy*. Hart Publishing.

Sennikov, S. A. (2014). International treaties of the Russian Federation as a legal basis for fishing in the sea area of the Svalbard Archipelago. *Eurasian Law Journal, 7*(74) (in Russian).

Sokolova, N. A. (2020). International Arctic scientific cooperation: Institutionalization of the legal regime. *Actual Problems of Russian Law, 15*(9), 120 130. https://doi.org/10.17803/1994-1471.2020.118.9.120-130 (in Russian).

Starodubtsev, V. (2017). Latitudes of high importance. *Kommersant*. https://www.kommersant.ru/doc/3254502 (in Russian).

Symbolokov, O. A. (2021). The main trends and factors of strengthening the competitive advantages of Russia in the world energy market and meeting the growing needs of the national economy, exports in hydrocarbon raw materials through the use of the resource base of the Arctic zone of the Russian Federation. In T. Ya. Khabrieva, S. A. Bogolyubov, & V. P. Emelyantsev (Eds.), *Russian Arctic—the territory of law: Almanac. Scientific support of lawmaking for the development of the Arctic* (Issue V, pp. 143–144) . Publishing House "Jurisprudence" Government of the Yamalo-Nenets Autonomous Okrug.

TASS. (2018). *The volume of traffic along the Northern sea route is expected to reach 10.2 million tons by the end of the year*. http://tass.ru/ekonomika/4786154 (in Russian).

TASS. (2019). *S.V. Lavrov: There are no plans to increase the membership of the Arctic Council*. https://tass.ru/politika/7201641 (in Russian).

Teplyashin, I. V. (2020). Arctic: Institutional and legal public participation. *Actual problems of Russian law, 2*, 36–43. https://doi.org/10.17803/1994-1471.2020.111.2.036-043

Torkunov, A. V. (2019). International studies: Chaos or Pluralism? *Polis: Political Studies, 5*, 7–18 (in Russian). https://doi.org/10.17976/jpps/2019.05.02

Treaty between the Kingdom of Norway and the Russian Federation concerning Maritime Delimitation and Cooperation in the Barents Sea and the Arctic Ocean. (2010). https://www.un.org/depts/los/LEGISLATIONANDTREATIES/PDFFILES/TREATIES/NOR-RUS2010.PDF

Vasiliev, A. V. (2012). International cooperation in the Arctic and Russian approaches. *Arctic Statements*, No. 1 (in Russian).

Vylegzhanin, A. N. (Ed.). (2015). *International law: Textbook for academic bachelor's programme*. Yurayt Publishing House (in Russian).

Vylegzhanin, A. N. (Ed.). (2017). *The political and legal junction of the Arctic and the Pacific Ocean: An interdisciplinary study of the status of the Bering Strait region*. IMI MGIMO MFA of Russia (in Russian).

Vylegzhanin, A. N., & Dudykina, I. P. (2017). UN convention on the law of the sea and the legal regime of the Arctic Shelf. *VSU Bulletin. Series: Pravo, 1*, 284–302 (in Russian).

Zagladin, N. V. (2016). Transition periods in world politics. *World Economy and International Relations, 60*(12), 27–39. https://doi.org/10.20542/0131-2227-2016-60-12-27-39 (in Russian).

Zhudro, I. S. (2021). *International legal justification of the boundaries of the Arctic shelf of the Russian Federation*. Doctorate Degree Thesis (in Law). RAS Institute of State and LAW. http://igpran.ru/bitrix/%D0%94%D0%B8%D1%81%D1%81%D0%B5%D1%80%D1%82%D0%B0%D1%86%D0%B8%D1%8F%20%D0%96%D1%83%D0%B4%D1%80%D0%BE%20%D0%98.%D0%A1.pdf (in Russian).

CHAPTER 3

Russia's Chairmanship in the Arctic Council: In Search of Balance?

Dmitry A. Yakovenko

3.1 Introduction

The Arctic Council (AC) was created to develop a common policy and approaches to stimulate the development of the Arctic, taking into account the need to preserve the environment and its other features. The work of the Arctic Council, the strategies for the development of the Arctic in individual countries, and the issues of interaction of various countries (including non-Arctic countries) in the Arctic development were covered by Karaganov et al. (2021), Vylegzhanin (2021), Vylegzhanin and Kienko (2021), Voronenko and Greizik (2019), Zhuravel (2019), Zagorskii (2012), Konyshev and Sergunin (2011), Sakharov (2015), Todorov (2019), Gudev (2019), Graczyk and Koivurova (2015), Konyshev et al. (2017), Sergunin (2021), Bhagwat (2020), Koivurova and VanderZwaag (2007).

D. A. Yakovenko (✉)
Self-regulating organization "National Institute of Professional Accountants, Financial Managers, and Economists", Samara, Russia
e-mail: cafmo@yandex.ru

V. I. Salygin (ed.), *Energy of the Russian Arctic*,
https://doi.org/10.1007/978-981-19-2817-8_3

The author worked on the problems of Arctic development in the Ministry for the Development of the Russian Far East and Arctic from March 2017 to April 2021 as the head of the Control Service and the internal auditor of the Ministry. He has written several works devoted to both the peculiarities of the development of individual regions and the influence of the state and budget resources on the development of the region, including in co-authorship with S.V. Tskhovrebov (Yakovenko & Tskhovrebov, 2020, 2021).

The research is centered around the following methods: observation, fact collection, analysis, and synthesis. This provided for a systematic and integrated approach to recommendations and conclusions.

3.2 Organizational Features of the AC

Many organizations around the world set themselves the goal of developing specific macro-regions. Such organizations are created to achieve the goals of regional development, environmental protection, distribution of zones of influence, production quotas, etc. These include the Arctic Council, which can be considered a vivid example of successful cooperation.

The President of the USSR Mikhail Gorbachev was one of the first to promote the cooperation between the Arctic states. The organization was created in 1989 when the Finnish government invited the representatives of eight Arctic countries (*Denmark, Iceland, Canada, Norway, Russia, USA, Finland, and Sweden*) to Rovaniemi (Finland) to discuss environmental protection issues in the Arctic. The meeting was called the "Rovaniemi Process". The "Rovaniemi Process" led to the creation of a Strategy for the Protection of the Arctic Environment in 1991. During the implementation of the strategy, the need to organize close cooperation between the Arctic states became obvious, which was implemented on September 19, 1996, in Ottawa (Canada), by creating an intergovernmental forum—*The Arctic Council* (hereinafter referred to as the AC). Its founders and *permanent members* were the eight Arctic States listed above.

Six organizations formed by the indigenous peoples of the Arctic take part in the work of the AC as *permanent participants*:

- the *Association of Indigenous Peoples of the North, Siberia, and the Far East of the Russian Federation* (hereinafter the AIPNSFE) was established in 1990. It unites 41 nations, the population of which is about 250,000 people.
- the *Sami Union* was founded in 1959 as a cooperation body of the Sami organizations of Norway, Sweden, Russia, and Finland.
- *Inuit Circumpolar Council* was established in 1977 in Barrow, Alaska. It unites Inuits (Eskimos) of four countries: Canada, the USA (Alaska), Denmark (Greenland), and Russia (Chukotka).
- *The International Association of Aleuts* was established in 1998 in Alaska by the Association of the Aleutian Islands/Pribilof Islands.
- *The Arctic Council of Athabascans* unites about 45 thousand Athabascans living in the circumpolar regions of Alaska (USA) and Canada.
- *The International Council of Gwich'ins* was founded in 1999. It unites about 9 thousand Gwich'ins living in Alaska (USA) and Canada.

In addition, *AC permanent observers* include 26 international and non-governmental organizations: *the Conference of Parliamentarians of the Arctic Region, the International Federation of Red Cross and Red Crescent Societies, the World Union for Nature Conservation, the Council of Ministers of the Nordic Countries, the Northern Environmental Finance Corporation*, and *the Northern Forum*.

The AC Founding Declaration implies assigning *observer status* to non-Arctic countries. The Arctic Council has granted permanent observer status to the following 13 countries: *Great Britain, the Netherlands, Poland, Germany, France, Spain, India, Italy, Switzerland, China, the Republic of Korea, Singapore*, and *Japan*. Several countries are currently expressing their interest in obtaining observer status (Estonia, Turkey, Mongolia, etc.).

All in all, the AC includes different participants:

- Founding States, Participants (permanent members);
- Observer States;
- Permanent Participants (national non-governmental associations);

- International non-governmental organizations who are permanent observers.

The supreme body coordinating the AC's activities is the sessions of foreign ministers representing the participating countries, held twice a year. The presiding country is constantly changing as it is elected by voting. It is *the Committee of Senior Officials that prepares sessions and deals with the current issues related to the AC's activities*. This working body meets at least twice a year. There are *6 thematic groups in the AC*. Each of them carries out its activities according to a special mandate. These groups are managed by the Chairman, the Management Board (which may be a Steering Committee), and the Secretariat. These divisions of the Council develop the mandatory documents (reports, manuals, etc.) and implement specific projects. The groups carry out various projects and develop non-binding documents (manuals, reports, assessments, etc.) on the issues related to their mandate.

The Working Group of the Arctic Monitoring and Assessment Program (AMAP) monitors and assesses pollution in the Arctic region (e.g., persistent organic pollutants, heavy metals, radionuclides, nutrients provoking ocean acidification, petroleum hydrocarbons) and climate change, documenting their levels and trends, paths, processes, and the impact on ecosystems and people, and also proposes the actions to mitigate the related threats to the consideration of the government.

AMAP participates in the initiative "Network of Arctic reference observations" together with the International Arctic Scientific Committee.

The *Arctic Council Action Plan to Eliminate Pollution of the Arctic (ACAP)* contributes to the efforts to reduce environmental risks and prevent environmental pollution in the Arctic. ACAP is developing projects that contribute to reducing the emissions of pollutants in cooperation with the authorities of the Arctic states.

The *Working Group on the Program for the Conservation of Arctic Flora and Fauna (CAFF)* deals with the problem of conservation of Arctic biodiversity. Within the framework of the CAFF, there are expert groups for the conservation of flora (CAFF Flora Group), Circumpolar Seabird species (Circumpolar Seabird Group), and the Circumpolar Protected Areas Network; the Circumpolar Biodiversity Monitoring Program is being implemented.

Sustainable Development Working Group (SDWG) develops measures that Arctic states can take to promote sustainable development in the region. This includes the realization of opportunities to protect and promote the development of the environment and economy, culture, and health of indigenous peoples and Arctic communities. The Group has carried out projects in the field of health protection of the people living in the Arctic, the development of reindeer husbandry, tourism, transport infrastructure, and cooperation in the field of oil and gas development.

The *Working Group on Emergency Prevention, Preparedness, and Response (EPPR)* was designed to promote prevention, preparedness, and response to environmental emergencies and other incidents, as well as search and rescue in the Arctic; its members carry out the projects to address gaps, prepare strategies, exchange information, collect data, and collaborate with relevant partners on opportunities and research needs. The projects and activities of the group include the development of methodological principles and methods of risk assessment, coordination of response exercises, training, and the exchange of information on best practices regarding prevention, preparedness, and response to incidents and threats from unintentional emissions of pollutants and radionuclides, as well as the consequences of natural disasters.

The working Group on the Protection of Arctic Marine Environment Program (PAME) develops policy and other measures related to the conservation and sustainable use of the marine and coastal environment of the Arctic. Within the framework of the group, projects are being fulfilled to implement the Strategic Plan for the Protection of the Arctic Marine Environment, assess Arctic shipping and port facilities for processing ship and cargo waste, study large marine ecosystems, and update the regional action program for the protection of the Arctic marine environment from anthropogenic activities. Several temporary task forces are currently operating in PAME:

- The Task Force on Maritime Cooperation was established in 2015 during the American presidency of the AC. Initially, the United States actively promoted the initiative to establish a regional supranational mechanism for managing the Arctic seas within the framework of the task force on maritime cooperation, but this idea was not supported by the other AC Member States. The Russian side claimed that one of the main principles of such a body would be the respect for the sovereignty and sovereign rights of the Arctic Coastal States under the 1982 UN Convention on the Law of the Sea, in particular concerning the continental shelf.

- The Task Force on Black Carbon (Soot) and Methane prepares reports on the activities of Arctic States to reduce emissions of these pollutants. In 2015, Russia was one of the first to submit a corresponding national report on this topic to this group.
- In 2013, a permanent *AC Secretariat* (Tromso, Norway), which provides administrative support for the development of international cooperation under the auspices of the AC, was established. In 2016, the Secretariat of Indigenous Peoples joined the AC Secretariat.

The Arctic Council has a joint fund for financing environmental projects (mainly in Russia)—the *Project Support Instrument* (PSI). The Fund is about $16 million. It is managed by the Northern Environmental Finance Corporation. Russia is its largest donor.

The AC working groups have prepared several reports on such areas as the prospects of navigation in the Arctic, trends in biodiversity, and the spills of oil and other dangerous and toxic substances in polar waters. Several non-binding guidelines have been developed, including the Guidelines for the Development of Offshore Oil and Gas Fields; the Guidelines for Pumping Refined Oil and Petroleum Products in Arctic Waters; the Guidelines for the Safe Transportation of Oil in Arctic Waters; and the Practical Guidelines for Eliminating the Consequences of Oil Spills in the Arctic Waters.

The signing of the Agreement on Cooperation in Aviation and Maritime Search and Rescue in the Arctic (Pan-Arctic Agreement) in 2011 by the Arctic States can be considered one of the most important achievements of the Arctic Council. In 2013, the Arctic states signed an Agreement on Cooperation in the field of Preparedness and Response to Oil Pollution of the Sea in the Arctic. In April 2015, a framework plan for cooperation on the prevention of oil pollution from hydrocarbon production and the intensification of shipping in the Arctic and a framework document on reducing soot and methane emissions were adopted at a ministerial meeting in Iqaluit (Canada).

The Arctic Coast Guard Forum was established in 2015.

In 2017, the AC members signed an agreement to strengthen international scientific cooperation in the Arctic. In 2014, the "*Arctic Council Project Support Instrument*" (PSI) began its work. PSI finances developments to reduce environmental pollution approved by the AC. The funds are provided by AC members, observers, and other interested parties. It is the Northern Environmental Finance Corporation (NEFCO) which manages the funds. NEFCO is an international financial organization founded in 1990 by five Nordic countries: Denmark, Iceland, Norway, Finland, and Sweden. NEFCO provides loans and investment capital to improve the environment of the Northern Region. NEFCO finances about 400 small and medium-sized environmental projects in the Northern and Baltic regions.

As of 2019, the PSI had €15.9 million provided by Iceland, Norway, Russia (€10 million; the largest contribution among all participants), the USA, Finland, Sweden, and NEFCO, as well as the Sami parliaments (the elected bodies of Sami in their countries of residence). Other states and organizations can also contribute.

Until 2019, 40 percent of the PSI funds were used for projects dealing with hazardous waste and persistent organic pollutants as well as environmental pollution; about 25 percent of funds went toward measures to eliminate mercury pollution; and about 20 percent were for clean production and energy efficiency projects.

In 2014–2016, about 30 projects were supported within the framework of the PSI. One of the latter is aimed at countering the decrease in populations of migratory birds nesting in the Far North (approved in October 2016). Projects such as Valdai and Tundra (in Karelia and the Murmansk Region), aimed at reducing soot emissions entering the atmosphere due to the operation of power plants, as well as the production of clean energy for Sami reindeer herders in Lovozero (Murmansk region), is being implemented on the territory of Russia with the help of PSI.

In 2014, it was established that the *Arctic Economic Council* (AEC) is an independent business membership organization representing companies that work with and within the Arctic. It consists of three representatives from the Member States and permanent participants of the AC. An important feature of this organization is that the AEC is independent of the AC. The Arctic Economic Council is entrusted with the mission of bringing a business perspective to the activities of the Arctic Council and business development in the Arctic. Deputy managers represent the Russian Union of Industrialists and Entrepreneurs, PJSC Rosneft, and PJSC Sovcomflot in the AEC.

The Russian Federation plays a key role in the activities of the Arctic Council. This was influenced by factors such as the significant length of the coastline, the scale of minerals, the volume of their development (more than 70 percent of all oil and gas resources of the Russian Federation are extracted in the Arctic), the territory located beyond the Arctic circle, and its large icebreaking fleet. Possessing such rich resources, the Russian Federation does not only take an active part in the implementation of the projects developed by the participants of the Arctic Council, but also offers its relevant initiatives.

Russia is chairing the Arctic Council in 2021–2023. Nikolay Korchunov, Ambassador-at-Large of the Russian Ministry of Foreign Affairs, Senior Official of the Arctic Council from the Russian Federation, represents the Russian Federation in the Arctic Council. He noted that the issues of sustainable development should be highlighted in 2021–2022, since sustainable development is the only policy that combines three factors—social, economic, and environmental—and does not lead to the accumulation of environmental damage.

3.3 Russia's Cooperation as a Member of the AC with Non-Arctic Countries

Non-Arctic countries take an active part in the work of the Arctic Council. Of the observer countries, the Asian countries—several of which are strategic partners of the Russian Federation—are taking a particularly active role in the work of the AC. However, some countries are also creating intergovernmental bodies to coordinate their Arctic initiatives. For example, in 2018, China, Japan, and the Republic of Korea adopted a joint declaration on trilateral cooperation in the Arctic.

Japan received permanent observer status in the Arctic Council in 2013 and immediately established a Coordination Conference of National Ministries and Agencies on Arctic issues (Iida, 2020).

China is one of the most interested parties. China's interest in the Arctic has been evident since the beginning of the twentieth century: in 1925, China joined the Svalbard Treaty, and in the late 1990s it opened the Yellow Research Station on the territory of Svalbard. It should be noted that in 2002, it was South Korea that opened its polar research station in Svalbard. India joined them in 2007. Several other countries, including Poland, have their research stations in Svalbard.

China published a "White Paper"—China's Arctic policy—in 2019. According to this document, China has the following interests in the Arctic: scientific research, environmental protection and climate change, security and legal regulation of the use of natural resources, international cooperation and management of the Arctic, and the peace and security of the region.

In one of his speeches, Chinese President Xi Jinping said that China is determined to develop sea corridors in the Arctic together with Russia (Pak, 2021). The idea of creating a "Polar Silk Road" was put forward by China in 2017. The inclusion of the new Arctic Economic Corridor in the architecture of the "One Belt, One Road" initiative should accelerate the development of the Northern Sea Route (Pak, 2020). China is ready to jointly finance icebreaking projects with Russia and is set to create its own icebreaking fleet (Qin & Qi, 2021). It is not only representatives of China but also the new Chinese icebreaker "Xuelong-2" ("Snow Dragon") that are taking part in the work of the MOSAiC international polar expedition among the 20 participating countries. China is currently implementing a plan to build its first nuclear icebreaker.

The key interest of China and other Asian non-Arctic countries is the Northern Sea Route (NSR). The Northern Sea Route reduces the distance for the delivery of goods from China to Europe by almost 10 thousand kilometers (compared to the currently used route through the Indian Ocean, the Suez Canal, and the Mediterranean Sea). In addition, the distances between other ports are also significantly reduced: the one between Yokohama and Rotterdam along the Northern Sea Route is shorter than the path through the Suez Canal by 7 thousand kilometers, the path from Busan to Rotterdam via the Northern Sea Route is shortened by 7.4 thousand kilometers, and the path from Rotterdam to Vancouver along the Northern Sea Route is shorter than the one through the Panama Canal by 3.6 thousand kilometers. To promote transportation via the NSR, in 2015 the South Korean government developed a program to subsidize Korean companies 50 percent of the costs for using port services, provided that cargo ships are sent along the Northern Sea Route.

The first commercial cruise started on August 8, 2013, when the cargo ship "Yong Sheng" from the port of Dalian, passing through the Northern Sea Route, arrived in Rotterdam on September 10. It took 33 days, which is 15–17 days less than if the route ran through the Indian Ocean and the Suez Canal. At the same time, transportation costs decreased by $820 thousand. According to the forecasts of Chinese experts, in the next 5 years, about 1/6 of all cargo from China should be sent via the Northern Sea Route. The Chinese company COSCO is exploring the possibilities of regular operations on the Northern Sea Route and plans to buy or build its icebreaker transport vessels.

Many companies from Asia have taken part in the construction of offshore facilities in the Russian Arctic. For example, the Japanese company Mitsui O.S.K. Lines and the Chinese company China COSCO Shipping are involved in the operation of the Sabetta seaport, while the Korean company Hyundai Merchant Marine Co. is going to participate in the project for the reconstruction of the Arkhangelsk deep-water seaport and the development of the Murmansk transport system together with the Chinese Poly Group.

In June 2019, Mitsui & Co. Ltd. signed an agreement with PJSC NOVATEK on the acquisition of 10 percent of the Arctic LNG 2 LLC package. As part of this project, natural gas will be produced at one of Russia's onshore fields, and a liquefied gas production plant will also be built. It is planned that Total (France, a subsidiary of the Chinese National Oil and Gas Company) will also take part in this project. The Japanese bank JBIC is also involved in the project (Vylegzhanin & Kienko, 2021: 313).

Connecting the Northern Sea Route with internal Russian waterways is also very promising. An alternative solution is the transshipment of cargo from sea vessels to river vessels. In 2016, such a route was tested by the Korean logistics company SLK Kukbo and the shipping company Pan Ocean, which organized the delivery of large-capacity goods from Ulsan (Republic of Korea) and Shanghai (China) to Pavlodar (Kazakhstan) via the Ob-Irtysh river arteries.

China's interests are not limited to the transport corridor. NOVATEK—together with the Chinese National Petroleum Corporation (CNPC) and the Chinese Silk Road Fund—has been financing the Yamal LNG project, which is engaged in the production of liquefied natural gas, since 2013. The share of Chinese investments in the cost of the project is 29.9 percent. Under the terms of the agreement, 80 percent of the equipment used in the Yamal LNG project should be made in China. In particular, the Chinese shipbuilding group of companies "Wuchang" supplies modules and other equipment for LNG production. The volume of deliveries is 36 modules weighing 180 tons. To implement the project, the Exim Bank of China and the China Development Bank provided loans for 15 years in the amount of €9.3 billion and CNY 9.8 billion, respectively (which together amounts to more than $13 billion), under 3.3–3.5 percent per annum.

China plans to take part in another NOVATEK project—Arctic LNG 2—to build a new plant on the Gydan Peninsula. Investments in this project are estimated at $10 billion. NOVATEK has signed a binding agreement with Chinese companies: the subsidiary of the Chinese National Petroleum Corporation (CNPC), CNODC, and CNOOC. Both agreements provide for 10 percent participation of each of the Chinese companies in the project.

Other projects are being prepared:

- the development of oil and coal deposits in the Taimyr Peninsula at the Krasnoyarsk Territory (expected cargo turnover of up to 20 million tons);
- the development of the Payakh group of oil fields in the Krasnoyarsk Territory (Neftegazholding company);
- coal mining on the Taimyr Peninsula (VostokUgol company).

China shows great interest in the development of transport and logistics facilities. In particular, China intends to invest in the project of a deep-water hub port in the Arkhangelsk region and take part in the construction of the Belkomur railway (White Sea–Komi–Ural) connecting Siberia, the Urals, and the Komi Republic with Arkhangelsk, with a length of about 800 kilometers.

China does not only invest in the territory of Russia, but has also made significant investments in the economy of Iceland and Greenland.

China is actively engaged in research aimed at studying the Arctic. In 2009, a specialized Center for Polar and Oceanological Research began operating in Shanghai.

India is the second non-Arctic state, acting as a partner of the Russian Federation in the international arena, which is actively participating in the development of the Arctic. The international branch of the Indian state oil company ONGC Videsh Limited (OVL), together with Rosneft, are involved in the development of the Vankor oil and gas condensate field. This is why India has every reason to become the first non-Arctic state extracting natural resources in the Arctic. Russia has offered Indian companies a share in the second stage of the Yamal LNG project.

India's Arctic strategy is very similar to China's. Among the priority areas, one can single out the assessment of Arctic resources and the study of their possible exploitation. *Japan*, *Republic of Korea*, and *Singapore* are facing similar goals. The interest of these countries also extends to the issues of climate change, security, etc. In July 2010, the "Japanese Arctic Meeting" was established, which set the task of Japan's entry into the Arctic Council, and the "Arctic Task Force" was created in the Japanese Foreign Ministry.

The Republic of Korea is interested not only in Arctic research projects. For example, the icebreaker "Aaron" was launched at the shipyards of Hanjin Heavy Industry Co. Ltd. in 2009. The strategy of the Republic of Korea's participation in the development of the Arctic is formulated in the "Measures to promote the policy of the Polar region" adopted in 2012. Following this, South Korea developed the documents "The plans to promote a comprehensive Arctic policy" and "The Arctic Policy of the Republic of Korea".

As part of participation in the implementation of the Yamal LNG project, Daewoo Shipbuilding and Marine Engineering is building 15 ice-class tankers for LNG transportation at the shipyards of the Republic of Korea. Daewoo Shipbuilding and Marine Engineering was the first company in the world to build icebreaking tankers. The Republic of Korea is helping Russia to create and operate an Arctic container shipping line from Petropavlovsk-Kamchatsky to Murmansk.

The South Korean company Samsung Heavy Industries has built six Arc7 shuttle tankers that meet the requirements of the Polar Code for the implementation of another project—the "Gates of the Arctic", dealing with the development of the Novoportovskoye oil and gas condensate field on the Yamal Peninsula. It should be borne in mind that 2/3 of gas carriers in the world are built in Korean shipyards (Table 3.1).

All things considered, non-Arctic Asian countries (China, India, the Republic of Korea, Japan, etc.) have significant interests in scientific research, exploration and development of minerals, and laying trade routes in the Arctic.

European non-Arctic countries are also showing interest in participating in the work of the Arctic Council. These include *Germany, Great Britain, France, Italy, Spain, Poland*, and *the Netherlands*. Several research projects in the Arctic are funded from the budget of the *European Union*, which has been actively striving to increase its role and importance in the Arctic for the last decade. The European Commission approved a Comprehensive Arctic Policy in 2016, and then adopted a new Arctic Strategy of the European Union in 2021. The new Strategy defines the priorities, goals, and objectives of the European Union in the

Table 3.1 Russia and the countries of Northeast Asia: basic approaches to Arctic policy

Country	*Key documents related to Arctic policy*	*Interests in the Arctic and priorities of Arctic policy*
China	China's Arctic Policy (2018)	• Expanding exploration and deepening understanding of the Arctic • Protecting the Arctic environment and combating climate change • Using the Arctic resources legally and rationally • Actively participating in international governance and cooperation in the Arctic • Promoting peace and stability in the Arctic
Japan	Ocean Policy Plan (2013); Japan's Arctic Policy (2015)	• Making full use of Japan's advantages in science and technology • Considering the Arctic environment and ecosystem • Ensuring the rule of law and promoting international cooperation in a peaceful and orderly manner • Respect for the rights of indigenous peoples, including the traditional economic and social foundations of their life • Studying the security situation in the Arctic • Making efforts to ensure economic and social compatibility with climate change and the environment • Studying the possibilities of using sea transport routes and exploring natural resources in the Arctic

(continued)

Table 3.1 (continued)

Country	*Key documents related to Arctic policy*	*Interests in the Arctic and priorities of Arctic policy*
Republic of Korea	Arctic Policy Master Plan (2013)	• Creating an Arctic partnership within the international community to solve modern problems in the region • Expanding research activities to better understand the Arctic • Ensuring sustainable development of economic activity in the Arctic
Russia	"Fundamentals of the state policy of the Russian Federation in the Arctic for the period up to 2020 and beyond" (2008); "Strategies for the development of the Arctic Zone of the Russian Federation and ensuring national security for the period up to 2020" (2013)	• Socioeconomic development of circumpolar territories through the development of natural resources and commercial use of the NSR • Preserving the Arctic ecosystem • Maintaining peace and stability in the region • Ensuring state sovereignty on the continental shelf under the jurisdiction of Russia and along the NSR

Source Compiled by the author

Arctic macro-region at the present stage. The Strategy addresses some important issues for the Arctic region: sustainable development, climate and its change, international and regional cooperation, security, search and rescue operations, research, permafrost melting, natural resources, indigenous peoples and their culture, health and languages, women, and youth. The European Union has not yet been granted observer status in the Arctic Council.

Germany has developed the Arctic Strategy, which addresses the issues of the opportunities, risks, and potential the Arctic may have for Germany. The Otto Schmidt Russian-German Laboratory for Polar and Marine Research was established at the Arctic and Antarctic Research Institute (FSBI "AANI") in 1999. The economic cooperation between the German concern BASF (through its subsidiary Wintershall) and Gazprom in the development of the Yuzhno-Russkoye oil and gas field (Yamalo-Nenets Autonomous Okrug) seems very promising. German policy in the Arctic is quite similar to French policy. These countries often work together to implement Arctic policy.

The UK also has its own Arctic strategy. British oil companies are involved in the Arctic projects. For example, British Petroleum owns 19.75 percent of Rosneft's shares and has joint projects with Rosneft in the Arctic. The UK is a world leader in insurance of ship companies, including in the Arctic. The Lloyd's Ship Register inspects about 25 percent of the world icebreaking fleet. The UK is interested in studying the effects of the Arctic on the climate, actively participates in NATO exercises in the Arctic.

3.4 Conclusions

The Arctic Council was established 25 years ago. Since then, the Council has expanded to include representatives of the small peoples of the North, international non-profit non-governmental organizations, and non-Arctic countries interested in the development of the Arctic. Non-Arctic countries are playing an increasingly significant role in the development of the region: they conduct research in Svalbard and other Arctic regions, build icebreaking fleets, and invest in the development of the Northern Sea Route and the development of mineral deposits both on the mainland and on the shelf. The exploration and development of the Arctic are currently being conducted at a brisk pace.

References

Bhagwat, J. (2020). Russia and India in the Arctic: A case for greater synergy. *Arktika i Sever, 38*, 73–90 (in Russian).

Graczyk, P., & Koivurova, T. (2015). The Arctic Council. In L. C. Jensen & G. Hønneland (Eds.), *Handbook of the politics of the Arctic* (pp. 298–327). Edward Elgar.

Gudev, P. A. (2019). New risks and opportunities for interstate cooperation in the Arctic. *Arktika i Sever, 36*, 57–83. https://doi.org/10.17238/issn2221-2698.2019.36.57 (in Russian).

Iida, J. (2020). Digital transformation vs. COVID-19: The case of Japan. *Digital Law Journal, 1*(2), 8–16. https://doi.org/10.38044/2686-9136-2020-1-2-8-16

Karaganov, S. A. (2021). *Russia's Arctic policy: International aspects*. Higher School of Economics Publishing (in Russian).

Koivurova, T., & VanderZwaag, D. (2007). The Arctic Council at 10 years: Retrospect and prospects. *University of British Columbia Law Review, 40*(1), 121–194.

Konyshev, V., Sergunin, A., & Subbotin, S. (2017). Russia's Arctic strategies in the context of the Ukrainian crisis. *The Polar Journal, 7*(1), 104–124. https://doi.org/10.1080/2154896X.2017.1335107

Konyshev, V. N., & Sergunin, A. A. (2011). International organization and cooperation in the Arctic. *International Organization Research Journal, 3*(34), 27–36.

Pak, E. (2020). Transit potentials of Russia and Kazakhstan. *World Economy and International Relations, 64*(11), 132–138 (in Russian). https://doi.org/10.20542/0131-2227-2020-64-11-132-138

Pak, E. V. (2021). Transport and logistics in the EAEU. In N. A. Piskulova (Eds.), *The economic dimension of Eurasian integration*. Palgrave Macmillan. https://doi.org/10.1007/978-3-030-59886-0_6

Qin, B., & Qi, S. (2021). Digital transformation of urban governance in China: The emergence and evolution of smart cities. *Digital Law Journal, 2*(1), 29–47. https://doi.org/10.38044/2686-9136-2021-2-1-29-47

Sakharov, A. (2015). Arctic Council as a regional governance institution. *International Organization Research Journal, 10*(4), 40–53. https://doi.org/10.17323/1996-7845-2015-04-72

Sergunin, A. (2021). Thinking about Russian Arctic Council chairmanship: Challenges and opportunities [Special issue]. *Polar Science*. https://doi.org/10.1016/j.polar.2021.100694

Todorov, A. A. (2019). Future work of the international seabed authority in the context of the Arctic Governance. *Arktika i Sever, 34*, 90–109. https://doi.org/10.17238/issn2221-2698.2019.34.90 (in Russian).

Voronenko, A. L., & Greizik, S. V. (2019). Prospects of cooperation between Russia and North-East Asian countries in the Arctic region. *Arktika i Sever, 34*, 90–109. https://doi.org/10.17238/issn2221-2698.2019.35.60

Vylegzhanin, A. N. (Ed.). (2021). Arctic Council: Status and activities. *Russian International Affairs Council*. https://russiancouncil.ru/activity/publications/arkticheskiy-sovet-status-i-deyatelnost/ (in Russian).

Vylegzhanin, A. N., & Kienko, E. V. (2021). Consultative meetings of non-Arctic states on the status of the Arctic. *Vestnik of Saint-Petersburg University Law, 2*, 296–318 (in Russian).

Yakovenko, D. A., & Tskhovrebov, S. V. (2020). Russia chairing the Arctic Council: Standards, plans, prospects. *Herald of Professional Accountant, 10–12*, 9–47 (in Russian).

Yakovenko, D. A., & Tskhovrebov, S. V. (2021). Arctic Zone of the Russian Federation: Preliminary results, prospects, and third countries. *Herald of Professional Accountant, 1–3*, 15–28 (in Russian).

Zagorskii, A. V. (2012). *Internationalization of the Arctic.* http://old.imemo.ru/ru/sc/2012/24102012.pdf (in Russian).

Zhuravel, V. (2019). USA taking over the chairmanship in the Arctic Council from Finland. *Analytical Papers of the IE RAS, 16*(167), 1–5. https://www.instituteofeurope.ru/images/uploads/analitika/an85.pdf (in Russian).

PART II

Energy Economics of the Russian Arctic

CHAPTER 4

Energy Complex of Russian Arctic

Alexey G. Kazanin, Marina A. Kazanina, and Alexey I. Andreev

4.1 Introduction

A new stage in the development of strategic governance in Russia was marked by the Decree of the President of the Russian Federation of November 8, 2021 No. 633 "On Approval of the Foundations of State Policy in the Sphere of Strategic Planning in the Russian Federation", setting tasks primarily to strengthen the scientific foundations of this area of public policy. Currently, Russia is strengthening its role of a global leader in energy sector; the severity of the global energy-ecological

A. G. Kazanin (✉) · A. I. Andreev
Lomonosov Moscow State University, Moscow, Russia
e-mail: a.kazanin@mage.ru

A. I. Andreev
e-mail: andreev@fgp.msu.ru

M. A. Kazanina
St. Petersburg State University of Engineering and Economics, Saint Petersburg, Russia
e-mail: marina.ca@mage.ru

V. I. Salygin (ed.), *Energy of the Russian Arctic*,
https://doi.org/10.1007/978-981-19-2817-8_4

problem is increasing, as are requirements for appreciating environmental issues when developing hydrocarbon production. In these conditions, new challenges arise for the strategic governance of the development of the Arctic. State policies relayed to the development of the Arctic are closely related to the development of hydrocarbon reserves in this region. Commercial development of oil and gas has been carried out in the Arctic region for over 80 years and has particularly intensified since the 1960s. The largest milestones in the history of oil and gas development in the Russian (then, of course, Soviet) Arctic were the discoveries of the Timan-Pechora (1930–1957) and West Siberian (1958–1968) oil and gas provinces. In the Arctic regions of the USSR, hydrocarbon production expanded to include more and more territories. Thus, in 1972 the exploitation of oil fields began in the Yamal-Nenets Okrug; in 1980, oil production began in the Nenets Okrug. Other Arctic states also began developing oil and gas deposits in their respective Arctic territories. Canada turned its attention to the delta of the Mackenzie River, and later to the Bent Horn deposit. By 1977, the United States had completed the construction of the Trans-Alaska oil pipeline to the port of Valdez, which made it possible to start oil production in the Alaskan Prudhoe Bay field and adjacent fields. In the 1980s and 1990s, the oil and gas industry began to develop in other Arctic territories as well. In Alaska, more and more attention is paid to offshore oil deposits; oil and gas activities in Norway reached the Barents Sea, where large deposits were discovered.

In the turbulent 1990s, according to a number of researchers, Russia "left" the Arctic—however, it would be more accurate to say that it limited its Arctic policies to a narrow range of internal problems of the Russian Arctic territories (Laruelle, 2014: 16). Moreover, some decisions made at that time now seem to have been very controversial. The North was presented not as an actively developing subject, but rather as an object in need of total state support. The colossal potential of growth was completely overlooked, and plans for the systematic development of this potential were totally absent. As a result, the entire Russian Arctic was pushed to the side of the economic development, and could not become an organic part of the Russian economic structure. This contradicted the interests of the state and could not form the basis of state policy.

The new stage in the Russian Arctic state policy in general, and in the development of the Arctic oil and gas sector in particular, began in the late 1990s to early 2000s. The perception of the Arctic was changing, and the need for a systematic approach to its development and the solution of the problems facing this region—as well as the need for Arctic policy to be outlined and implemented in the general context of the global strategic development goals of Russia—became obvious (Torkunov, 2018). This approach was further developed and entrenched in the legal framework through a number of legislative initiatives of the late 2000s–2010s. The strategic vision of the new Arctic policy of Russia was reflected in the "Foundations of State Policy of the Russian Federation in the Arctic for the period up to 2020 and beyond", which came into force in 2008. In this document, the formation of the Arctic Zone as a strategic resource base for Russia was recognized as one of the key interests of the national Arctic policy.

4.2 Energy of the Russian Arctic: Main Dimensions

According to estimates made in 2007, the Arctic produced approximately 10 percent of world oil production and 25 percent of world gas production; of these volumes, approximately 80 percent of oil and 99 percent of gas were produced in the Russian Arctic. A new surge of global interest in the circumpolar region came with the results of the US Geological Survey (Gautier & Moore, 2017), which estimated the undiscovered reserves of oil and gas resources at almost 90 billion barrels of oil, 47 trillion cubic meters of natural gas, and 44 billion barrels of gas condensate liquids (Table 4.1).

At the same time, the researchers emphasized that the resource potential of the Arctic remains largely uncertain, since, outside of several well-studied areas, the marine Arctic is still largely unexplored in terms of oil reserves. Geological evidence still suggests that significant amounts of as-yet undiscovered oil remain outside developed areas. Available data and estimates indicate that 44–157 billion barrels of conventional oil and 22–85 trillion cubic meters of conventional natural gas can be found in the northern circumpolar region. In addition, large resources of shale oil, shale gas, heavy oil, gas hydrates, etc., can also be found in the Arctic.

Geologically, the most likely locations for significant volumes of undiscovered Arctic oil are onshore on the North Slope of Alaska, on the shelves in the Chukchi and Beaufort Seas, in the Mackenzie River Delta in northwestern Canada, in northeastern Greenland, in the north of the Kara Sea, and in Western Siberia. Of all the Arctic states, Russia is the most likely to have the largest undiscovered resources of conventional

Table 4.1 Top-20 Arctic provinces with the highest mean estimates of undiscovered petroleum resources of the 2008 Circum-Arctic Resource Appraisal

Province	*Oil, million barrels*	*Total gas, billion cubic feet*	*Natural gas liquids, million barrels*	*Total barrels of oil and oil-equivalent natural gas*
West Siberian Basin	3659.88	651,498.56	20,328.69	132,571.66
Arctic Alaska	29,960.94	221,397.60	5904.97	72,765.52
Eat Barents Basin	7406.49	317,557.97	1422.28	61,755.10
East Greenland Rift Basins	8902.13	86,180.06	8121.57	31,387.04
Yenisey-Khatanga Basin	5583.74	99,964.26	2675.15	24,919.61
Amerasia Basin	9723.58	56,891.21	541.69	19,747.14
West Greenland-East Canada	7274.40	51,818.16	1152.59	17,063.35
Laptev Sea Shelf	3115.57	32,562.84	867.16	9409.87
Norwegian Margin	1437.29	32,281.01	504.73	7322.19
Barents Platform	2055.51	26,218.67	278.71	6704.00
Eurasia Basin	1342.15	19,475.43	520.26	5108.31
North Kara Basins and Platforms	1807.26	14,973.58	390.22	4693.07
Timan-Pechora Basin	1667.21	9062.59	202.80	3380.44
North Greenland Sheared Margin	1349.80	10,207.24	273.09	3324.09
Lomonosov-Makarov	1106.78	7156.25	191.55	2491.04
Sverdrup Basin	851.11	8596.36	191.20	2475.04
Lena-Anabar Basin	1912.89	2106.75	56.41	2320.43
North Chukchi-Wrangel Foreland Basin	85.99	6065.76	106.57	1203.52
Vilkitskii Basin	98.03	5741.87	101.63	1156.63
Northwest Laptev Sea Shelf	172.24	4488.12	119.63	1039.90

Source Compiled by the author based on Gautier, D.L., Moore, T.E. (2017). *Introduction to the 2008 Circum-Arctic Resource Appraisal (CARA) professional paper—U.S. Geological Survey Professional Paper*. No. 1824-A, P.8

natural gas—most of these are located in the country's Arctic territories. Of the total volume of undiscovered gas reserves in the Arctic, about 70 percent are in the Arctic territories, in particular on the continental shelf (up to 90 percent of Russian Arctic gas lies underwater). Certain parts of the basins of the Kara and Barents Seas look the most promising.

It was already clear at that time that the exploration and development of the Arctic continental shelf and its resources would be one of the priority areas for the development of the Russian oil and gas sector, not only in the short term, but also in the medium and long-term perspective. Therefore, tasks were already posed in 2008, within the framework of the development of the Arctic resource base, concerning the comprehensive study of the continental shelf and coastal territories. Such a study was expected to ensure a significant increase in the balance reserves of Arctic offshore hydrocarbon deposits and to allow the formation of a reserve fund of deposits in the Arctic Zone of the Russian Federation. Another task was related to developing the Timan-Pechora oil and gas province and hydrocarbon deposits on the continental shelf of the Barents, Pechora, and Kara Seas, as well as the Yamal and Gydan peninsulas.

The initial total volume of recoverable hydrocarbon resources in the Russian Arctic (located both on land and in the waters of the Arctic seas of Russia) as of 2014 was estimated at 258 billion tons (approximately equal to 60 percent of all oil and gas resources of the country). However, this figure included only the reserves explored at that time—along with these reserves, the Russian Arctic also had a colossal potential of undiscovered hydrocarbon energy resources (more than 90 percent of resources on the shelf and slightly more than half of the resources on land are still undiscovered). Of course, not all of the explored reserves were recoverable—the initial estimate of recoverable oil reserves in the Russian Arctic amounted to about 7.7 billion tons (of which about 500 million lay on the Arctic shelf); for gas, the initial estimate of recoverable reserves amounted to 67 billion cubic meters, of which about 10 billion were offshore.

Among all the water areas belonging to the Russian Arctic continental shelf, the southern regions of the Barents and Kara Seas are the most well-studied; the geological exploration of the Russian Arctic shelf is much more concentrated in its western part. This fact was largely predetermined by geological features, since the waters of these seas were more easily accessible for exploration than the seas of the eastern part of the Russian Arctic (they will be discussed below). Moreover, even according to preliminary estimates, it seems that the oil and gas potential of the Arctic basins are very unevenly distributed over the Russian continental shelf, and the bulk of these resources are to be found in its western part. In total, 84 wells were drilled in the seas of the western part of the Russian

Arctic by 2012 at different times, of which 54 were in the Barents Sea and another 30 were in the Kara Sea (mainly in the Ob and Taz Bays). Exploration drilling in this region turned out to be extremely productive—more than 20 fields were discovered in this part of the shelf.

The largest fields were discovered in the western part of the Russian Arctic shelf relatively long ago, back in the late 1980s and early 1990s. Among them, the unique Shtokman gas condensate field deserves special mention—the probability of the existence of this field was established back in 1981, before the field itself was discovered in 1988–1989. Among the largest deposits in the western part of the continental shelf of the Russian Arctic, the Rusanovskoye and Leningradskoye gas condensate fields should also be noted—both of these giant fields were discovered at the beginning of the 1990s in the southwestern part of the Kara Sea. The total initial reserves of natural gas in these three fields exceeded 10 trillion cubic meters. However, the development of the Rusanovskoye and Leningradskoye fields was hampered by the fact that the technologies required to work with these fields were in their infancy. As for the Shtokman field, other obstacles arose: its depth, difficult ice conditions, extremely difficult logistical accessibility, and an almost complete absence of onshore infrastructure.

Since the beginning of the twenty-first century, further geological and geophysical exploration activities of "Gazprom", concentrated in the waters of the Ob and Taz bays of the Kara Sea, have made it possible to substantiate and prove the gas content of shelf "extensions" of a number of gas condensate fields located in the Arctic territories (for example, Antipayutinsky, Semakovsky, and some other gas fields, as well as gas condensate fields in the Yamalo-Nenets Autonomous Okrug). Exploration of these offshore "extensions" made it possible to increase natural gas reserves in the Ob and Taz Bays by about 2 trillion cubic meters. In total, the existence of 12 deposits was proved in the Kara Sea and the Ob and Taz bays. In this context, it seems quite logical that the development of the Arctic continental shelf in the Kara Sea basin was started from these shelf "extensions" of the Arctic gas fields and gas condensate fields (it should be noted that the Kara Sea has been studied much less than the Ob and Tazovaya bays).

However, a number of exogenous factors intervened in the initial plans for the development of hydrocarbon resources of the Arctic continental shelf of Russia. These factors require a significant adjustment of the state policies in this region.

Firstly, the sharp drop in oil prices on the world market in the early 2010s made the development of most of the Arctic oil fields unprofitable due to high production costs; a number of offshore development projects were halted.

Secondly, the geopolitical factor played its role, too. Given the strategic importance of the energy sector for the current and future development of Russia, it is not surprising that this sector (along with the banking sector) was particularly targeted from the very beginning of the imposition of sanctions by Western countries in 2014. On July 31, 2014, the European Union imposed "a ban on the supply of high-tech equipment to Russia for oil production in the Arctic, on deep-sea shelf, and shale oil" (Guliev, 2015). Exactly a week later, on August 6, 2014, the United States introduced an essentially similar ban on supplying Russia with equipment for the development of hydrocarbon resources of the Arctic shelf, as well as deep production in general. For deep-sea, Arctic, and offshore projects, there is a whole list of equipment that the United States prohibited to import into Russia. Since September 2014, the sanctions have extended not only to the supply of equipment, but also to the provision of services, the exchange of information with Russian partners, as well as the participation of Western companies in the most technologically advanced production projects.

It was calculated that, due to the sanctions, the Russian oil industry might lose about $1 trillion in investments during 2015–2045. This would lead to tangible losses for the budget—lost revenues could amount to $27–65 billion in the period up to 2020. In order to avoid such consequences in the Russian oil and gas sector, it was necessary to accelerate the development of innovative equipment, new offshore technology, equipment, and materials, and to introduce them into practice as soon as possible. The need for Russian oil field service companies to immediately enter the offshore sector also became apparent.

This challenge was extremely quickly recognized by the country's leadership, and the development of an adequate response began immediately. First of all, it should be noted that the current edition (08/31/2017) of the state program "Socio-economic development of the Arctic Zone of the Russian Federation for the period up to 2020" was the first document to put forward the priority of scientific and technological advances in the development of the Russian Arctic. The efficiency of various operations of the Arctic oil and gas sector should increase through enhancing its own scientific and technical activities and introducing the results of its own research and development and R&D into these operations (Decree 31.08.2017). At the same time, subprogram No. 3 of this version of the state program is fully devoted to the creation of competitive equipment and technical and technological systems necessary for the development (exploration, production, and processing) of mineral resources in the Russian Arctic, including those on the Arctic continental shelf (Decree 31.08.2017).

The analysis of the practical response to this challenge in the form as it developed—firstly within the framework of the legal framework of the Russian Federation, and then directly in production—should be divided into two parts. The first part concerns work on the production of the equipment for the oil and gas sector that used to be purchased abroad before the imposition of sanctions. The second part concerns the intensification of research and development work in the oil and gas sector in general and in offshore field development in particular.

Noting that among all Russian industries in the oil and gas sphere, the highest degree of dependence on imports of equipment and technologies was observed in shelf projects, such a situation makes it extremely relevant to study the strategy for the development of the Arctic oil and gas sector (including the development of the Arctic continental shelf of Russia) in recent years, as well as promising approaches to further improving the policies regulating this sphere. Taking into account the above estimates of Arctic hydrocarbon reserves, such a study is undoubtedly significant for assessing the further dynamics of the Russian Arctic oil and gas sector.

Our analysis shows significant discrepancies in the plans on increasing import substitution in the oil and gas sphere developed by the Ministry of Industry and Trade and by the largest Russian oil and gas companies. The plans of the Ministry of Industry and Trade did not sufficiently take into account the need to develop import substitution in a number of the most important areas outlined in the plans of the companies. This leads to the following practical recommendations. In addition to the areas identified in the Plans of the Ministry of Industry and Trade, two more areas should be included:

1. The widespread introduction of digital technologies for analyzing seismic data and monitoring work processes (including subsea production) and the state of systems, including the development of software in the field of processing and interpretation of seismic

data and well surveys; geological, hydrodynamic, and geomechanical modeling, technological modeling of hydrocarbon collection and treatment systems, well design and support; the development of robotics for performing work on drilling rigs, subsea rigs, and remotely controlled technological solutions, including autonomous underwater vehicles and drones used to inspect hard-to-reach structures and timely identify problems; the development of high-speed communication systems and systems of security and operational response;
2. The development of constructive protection of mining complexes and the development of services for the installation and maintenance of equipment in the Arctic.

It is necessary to develop these areas not only in the scientific production aspects, but also in the practical aspects too (Adžić, 2021; Inozemtsev, 2021; Pichkov & Ulanov, 2021). For this, it is necessary to provide clear mechanisms for translating scientific developments in priority areas into prototyping and industrial production.

Moreover, clear mechanisms should be provided for interaction between industry representatives directly involved in the development of Arctic hydrocarbon resources and several parties: with representatives of government agencies to coordinate and jointly develop plans for the development of Arctic hydrocarbon resources; with representatives of the scientific community, who form a fundamental and practical scientific groundwork for the subsequent development of proprietary technical and technological complexes; with representatives of other industries who are ready to undertake the production of proprietary technical and technological complexes in the required volume; and with representatives of newly created scientific and educational centers and centers of advanced vocational training to form an order for training personnel capable of successfully working with newly created technical and technological complexes.

With only 1.5 percent of Russian population, the Arctic Zone accounts for almost one-tenth of Russian GDP (Ria Novosti, 2020). "The Strategy for the development of the Arctic Zone of the Russian Federation and ensuring national security for the period up to 2035" envisages an increase in the share of oil produced in the Arctic to 22 percent of the total oil extraction in Russia by 2030. By 2035, this figure should rise to 26 percent. The share of natural gas produced in the Russian Arctic was originally planned (in the draft Strategy) to increase to 92 percent by 2035 from 82.7 percent in 2018. The volume of production of liquefied natural gas (LNG), according to the draft strategy, should grow almost 7 times by 2030, up to 73.5 million tons, and by 2035 by another 63 percent, up to 120 million tons (Decree 26.10.2020).

The impact of the oil and gas industry on the economy is multifaceted, multiplicative, and strategically important. According to recent estimates by Rosstat, the share of mining operations has increased over the past decade (from 34.1 percent in 2010 to 38.9 percent in 2018), while the share of manufacturing industries has decreased from 53.2 to 50.7 percent. At the same time, the share of the oil and gas industry also increased in the structure of processing industries (the share of petroleum products increased from 17 to 23 percent), while the share of metallurgy and machine building decreased.

In 2018, the share of the Arkhangelsk region (excluding the Nenets Autonomous Okrug) in the total Russian mineral extraction was 0.2 percent, while its share in the country's GDP was 0.6 percent; the respective figures for Chukotka Autonomous Okrug were 0.47 and 0.1 percent, the Murmansk region 0.78 and 0.6 percent, the Nenets Autonomous Okrug 1.92 and 0.4 percent, the Komi Republic 2.26 and 0.8 percent, the Republic of Sakha (Yakutia) 4.31 and 1.3 percent, and the Yamalo-Nenets Autonomous Okrug 13.7 and 2.8 percent. This is clear evidence of the growing contribution of the oil and gas industry in the Arctic to both the country's extractive industry and the economy of the Arctic macroregion as a whole. The dominant role of the Yamalo-Nenets Autonomous Okrug is noteworthy; however, exploration and production is being carried out in other Arctic and subarctic regions as well (Strunitskiy & Chalganova, 2019).

As can be seen from the name of the Strategy for the Development of the Arctic Zone of the Russian Federation and Ensuring National Security for the Period until 2035 (adopted in 2020), the issues of sustainable development are rightly combined with issues of national security, which is a definite step forward in Russian strategizing. Indeed, national security is synonymous with sustainable development in a rapidly changing world.

The Strategy pays significant attention to the role of the oil and gas industry in the development of the Arctic Zone of the Russian Federation. Thus, paragraph 12, which deals with the main tasks in the field of economic development of the Russian Arctic, postulates its provision through the implementation of measures, including (subparagraph a) "the introduction of a special economic regime that contributes to the transition to a circular economy"; the development of private investment in exploration; the development of the economy as a whole, including especially high-tech industries; "development of new oil and gas provinces", "deposits of solid minerals and hard-to-recover reserves of hydrocarbons"; the development of industries with deep oil refining, production of LNG (liquefied natural gas) and gas chemistry products; and so on. The same clause 12 (subparagraph i) implies "the provision of state support measures aimed at the creation and development of technologies for the development of oil and gas fields (including technologies used on the continental shelf), the production of liquefied natural gas, as well as at ensuring the production of corresponding industrial products" (Decree 26.10.2020: 12).

The Arctic region is extremely sensitive to environmental disturbances. The oil and gas industry is a source of very significant environmental risks that have been growing in recent years and in the strategic perspective due to the globally shared policy of "all-sphere" decarbonization. Clause 15 of the Strategy states the need to solve the problems of environmental protection and ensuring environmental safety through (clause 3) "the development of a unified state system for the prevention and elimination of emergencies in order to implement measures to eliminate spills of oil and oil products".

The oil and gas theme is clearly traced in the provisions of the Strategy concerning the goal-setting of the development of individual regions. Thus, in accordance with clause 21 in the Nenets Autonomous Okrug (pp. C), "the development of the Varandeysky, Kolguevsky, Kharyago-Usinsky and Khasyreysky oil mineral resource centers" is strategically significant, as well as the creation (pp. G) of "gas condensate mineral resource centers", including "the development of the Korovinskoye and Kumzhinskoye gas condensate fields, Vaneivisskoye and Layavozhskoye oil and gas condensate fields".

For the Yamalo-Nenets Autonomous Okrug, (pp. e) speaks of the strategic importance of the development of the Novoportovskoye oil and gas condensate and Bovanenkovskoye gas condensate mineral resource centers, as well as the development of the Tambey group of fields; in addition, it provides for preparation for the development of shelf deposits. (Pp. f) provides for the development of oil and gas chemical production in the area of the villages of Sabetta, Yamburg, and the city of

Novy Urengoy, where it is planned to create a diversified industrial and technological complex for gas processing and petrochemistry.

For the Komi Republic (clause 25), also a powerful center of the mining industry, the Strategy foresees (clauses c) "the formation and development of oil and gas mineral resource centers on the basis of the Timan-Pechora oil and gas province".

In the Arctic municipalities of the Republic of Sakha (Yakutia), the Strategy provides for (clause 26) "integrated development ... of the West Anabar oil and mineral resource center" (pp. b).

Clause 27 deals with the Arctic territories of the Krasnoyarsk region, providing for the formation (pp. d) of an oil and mineral resource center on the basis of the deposits of Western Taimyr; its products will be focused on the export of manufactured products through the Northern Sea Route. An oil terminal will be built in the port of Dikson (pp. h).

Clause 28, associated with the strategizing of the development of the Arkhangelsk region (its individual municipalities), foresees an increase in the capacity of LNG and gas chemistry production.

Russian vertically integrated oil companies (VIOCs)—Gazprom, Gazprom Neft, Rosneft, Novatek, and Lukoil—are active in the Russian Arctic. Gazprom holds 30 licenses for fields on the Arctic shelf. The Arctic fields of Gazprom Neft (Novoportovskoye, Prirazlomnoye, and Vostochno-Messoyakhskoye) account for 27 percent of the company's total production. Rosneft has 28 licensed areas on the Arctic shelf, and their cumulative resource is 34.6 billion tons of oil equivalent. Basically, this is the Western Arctic (Barents, Pechora, and Kara Seas), where the company has 19 projects; 9 more in the Laptev Sea, East Siberian, and Chukchi Seas belong to the Eastern Arctic. Novatek pioneered the production of LNG for export directly in the Arctic, supplying LNG abroad for the first time via the Northern Sea Route through the Far East in 2018. Lukoil develops the onshore part of the Vostochno-Taimyr area and is the operator of the Varandey terminal. Vertically integrated oil companies develop shipbuilding, infrastructure, and research projects—thus, Rosneft builds Aframax vessels, ice-class Arc4 tankers, Arc7 ice-class LNG tankers, and a nuclear icebreaker, ensuring the development of the Zvezda shipbuilding complex. Gazprom is supporting the construction of the Obskaya-Bovanenkovo railway (with a length of 572 kilometers).

2020 was marked by the adoption of several strategic documents at once that are directly related to the regulation of various aspects of the development of the Arctic Zone of the Russian Federation. The first is

"The Foundations of State Policy in the Arctic until 2035", in which the goals and objectives of Russia's Arctic policy for the specified period are set in close conjunction with issues of national security in the region. The document is interesting because for the first time in domestic practice, it directly unites the issues of security and development of the Arctic, not in any individual aspects, but in the implementation of these two policy areas as a whole under the auspices of observing Russia's national interests.

Attention is drawn to the very positioning of the new Foundations as a document of strategic planning in the field of ensuring the national security of the Russian Federation. For comparison, in the previous document, adopted in 2008, there was no such clear positioning of the Foundations of State Policy of Russia in the Arctic in the context of the directions for the implementation of Russian internal policy. Our analysis of the corpus of documents of the 2000s and 2010s, which in one way or another touched upon the development of the Russian Arctic, allows us to consider the development of Arctic policy, touched upon (albeit briefly and in a very general way) in the 2000 National Security Concept; the Arctic in general, and the Barents Sea shelf in particular, were mentioned among the places of the most likely concentration of international political interests in the near future due to the presence of significant reserves of energy resources in the National Security Strategy of the Russian Federation until 2020 (approved in 2009). The updated National Security Strategy of the Russian Federation No. 683 of December 31, 2015 also mentioned the Russian Arctic and the priorities of its development in the context of ensuring Russia's national security—in particular, in the sense that leadership in the development of Arctic resources is of particular importance in competition between states. However, in all these documents, the conjugation of the development of the Russian Arctic and issues of national security of Russia had the character of separate "points of contact"; the elaboration of complex, multidimensional links between these two strategically important directions of Russian policy lacked a systemic character. In the current edition of the Foundations of State Policy in the Arctic until 2035, this drawback is corrected.

The new Foundations seem to be a much more balanced document from the point of view of the postulated priorities of the Arctic policy—for example, unlike the 2008 edition of the Foundations, the priority of ensuring a high quality of life in the Arctic appears among the stated Russian national interests, which is then developed as a whole set of goals and objectives aimed at increasing the accessibility and quality of healthcare, education, housing, transport infrastructure, ICT infrastructure, and social support for the population of the Russian Arctic.

With regard to the strategic prospects for the development of the oil and gas industry in the Arctic region in the context of the new Foundations, a thorough analysis of the document reveals a number of issues. Some of them can be resolved at the level of subsequent documents (strategies, state programs, etc.) and should be taken into account when drawing them up, since leaving them unattended is fraught with an insufficient development rate or even the stagnation of important areas of development of the mineral resource complex of the Arctic. The other part relates to the mechanisms for the implementation of the Foundations and the very structure of the Russian governance of the Russian Arctic. Further, an attempt will be made to classify these issues according to the possible administrative level of their solution and proposed solutions that seem optimal in the context of a complex combination of external and internal challenges to Arctic development.

Among the main interests of Russia in the implementation of its national Arctic policy is the use of the Arctic Zone as a strategic resource base; in this paragraph, the document seems to be completely consistent with the 2008 Foundations, where it was present in a similar wording. This is not surprising—the share of oil and gas produced in the Russian Arctic is 17 and 80 percent of the total Russian production, respectively.

Meanwhile, the list of key threats does not note ensuring mineral security in the Arctic—neither in the context of ensuring economic security, nor as an independent point. With regard to mineral security, only one threat is highlighted: the low pace of geological exploration. Of course, some other threats outlined in the Foundations are also relevant for the mineral resource complex (for example, those related to the backlog of transport infrastructure and ICT infrastructure)—however, this does not negate the fact that the list fails to highlight a number of strategic threats specific to mineral resources: foremostly, threats related to technological development and human resources.

The baggage of proprietary technologies in the field of oil and gas exploration and production—accumulated and constantly updated and improved by oil and gas TNCs—will most likely be in demand in the development of Arctic fields, although it will require revision, taking into account the extreme Arctic conditions. In this light, the current "cutoff" of Russia from technological cooperation in the oil and gas sector (especially with regard to offshore production) due to the imposed sanctions poses a threat of the inconsistency of the developed technologies with the challenges of the development of the Arctic oil and gas resources.

Moreover, it is quite obvious that the accumulation of proprietary technical and technological baggage in the field of exploration and production in the Arctic fields must necessarily be based on Russia's own knowledge and expertise—which, of course, must be in constant interaction with the global scientific and technological frontier, but also must essentially consist of Russian R&D, since this is the only way to ensure a confident leadership in the development of Arctic oil and gas. The development of science and technology for the Arctic is highlighted in the new Foundations as a special goal with its own subsystem of tasks; however, none of the tasks relates directly to the creation of a proprietary technical and technological complex for the development of oil and gas resources in the Arctic (although certain tasks, of course, can be related to this, such as the development of a research fleet can be related to seismic exploration of shelf areas). The lack of a special goal to create a fundamental and practical scientific groundwork to formulate a proprietary technical and technological complex for the development of oil and gas resources in the Arctic within the framework of the development of science and technology in the development of the Arctic seems to be an important omission.

The Strategy for the Development of the Arctic Zone of the Russian Federation and Ensuring National Security for the Period until 2035 lists some more important threats—not only the low rates of exploration and study of Arctic resources (including those lying on the shelf), but also insufficient rates of the development of Russian technologies, as well as fundamental and applied research necessary for the development of the Arctic. The technical and technological complex required for the development of the Arctic hydrocarbon resources, and the knowledge base of this complex, are again not highlighted as a separate priority. Furthermore, the tasks set in the Strategy provide for "state support for the creation and development of technologies and industrial products for the development of oil and gas fields, including on the continental shelf of the Russian Federation, for the production of liquefied natural gas" (Decree 26.10.2020). However, the most important questions remain outside the scope of attention: namely, who is responsible for the creation and development of these technologies; which institutions are responsible for the formation of their knowledge base; how is the message of fundamental and practical knowledge built; and which structures are responsible not only for development but also actually for applying these technologies in production. For some unknown reason, only the creation and development of technologies receives attention. At the same time, research and knowledge generation, on the basis of which technologies will be created, remain "outside the scope". The creation and testing of industrial prototypes and the establishment of the production of Russia's own equipment, instruments, materials, and suchlike as necessary and sufficient to fulfill the tasks of developing the hydrocarbon reserves of the Russian Arctic does not receive attention. In this case, we consider it expedient—and even necessary at the Strategy level—to present this entire chain in its indissoluble sequential unity, since breaking it up into separate links can significantly worsen coordination and slow down the strategic development of this extremely important area.

Moreover, the problem of human resources receives insufficient attention as well. The Strategy sets out the task of organizing training and retraining human resources, but only from the angle of those human resources needed for the Northern Sea Route, not the Arctic oil and gas sector. At the same time, the emergence of new technical and technological complexes in this sector will obviously require the retraining of human resources and the search for new qualified human resources. A separate acute problem is the training of highly qualified scientists and researchers to create the actual fundamental and practical knowledge base of these technologies. The Strategy provides for the creation of a system of research and educational centers, the formation of a register of critical technologies, and, most importantly, their coordination; already at the level of the Strategy, a number of priority research areas have been identified. However, none of these activities is specifically tied to the needs of the Arctic oil and gas complex.

In the Fundamentals of State Policy of the Russian Federation in the Arctic for the period up to 2035, as well as in the Strategy for the Development of the Arctic Zone, considerable attention is paid to a set of tasks in the field of science and technology development, the solution of which is necessary for the development of the Arctic. As mentioned above, there seems to be a strategic omission: there is no specific task to create a fundamental and practical scientific groundwork to form a proprietary technical and technological complex, which would be used in developing oil and gas resources of the Arctic in the framework of the development of science and technology in the Arctic. Nevertheless, a number of the tasks set can contribute to the creation of such a reserve—in particular, certain areas of the comprehensive program of fundamental and applied research in the interests of the development of the Arctic. The need to develop these areas not only in scientific, but also in practical, production aspects has already been highlighted: this will provide mechanisms for the translation of scientific knowledge into industrial production. Below, we will focus on the existing human, institutional, knowledge base, which should form the basis for a comprehensive program.

The publication efficiency of research by scientists in the Arctic and the adjacent regions is close to the average Russian level; however, at the global level, Russia is not the leader in the number of publications on the topic of Arctic oil. Moreover, Russia has a noticeably smaller share of publications on technical sciences and ecology than the more successful Arctic countries in terms of economic development. It is obvious that international cooperation plays an important role in the development of Arctic research. However, for Russia, this path in a number of key areas for the Arctic oil and gas sector turns out to be very limited—not only because of sanctions, but because of the dual use of a number of foreign technical means and technologies, which puts Russian Arctic science in front of the need for its own fundamental research and independent search for applied solutions in these areas. It is recommended that these technical areas be included in the list of strategically significant solutions, as well as priority ones in the comprehensive program of fundamental and applied research proposed in the Strategy (at present, priority is given instead to earth sciences).

Highly qualified scientists and researchers in the Russian Arctic have their work reproduced at a low level, which is in fact more than two times lower than that in Russia as a whole; for doctors of sciences this is at a much lower level than the rest of the country too. The Strategy is supposed to create a network of new Research and Education Centers (RECs) in various regions of the Arctic Zone of the Russian Federation. However, in terms of the volume of Arctic research in Russia, the country's main university, Moscow State University, leads the way, followed by the Moscow Institute of Oceanology named after P.P. Shirshov; none of the leading research centers located in the Arctic itself are ahead in this regard (although the role of such, of course, is growing). Accordingly, it is recommended to work out (1) mechanisms for ensuring the reproduction of scientific personnel in Arctic universities and research centers;

and (2) mechanisms of cooperation between research centers located in the Arctic itself with research centers outside the Arctic Zone (firstly, the leading institutes of the Russian Academy of Sciences, including the Ural Branch of the Russian Academy of Sciences, the Siberian Branch of the Russian Academy of Sciences, and the Moscow State University), which will provide a multilateral transfer of knowledge and competencies required, taking into account the comprehensive strategic importance of the Arctic region for Russia.

The oil and gas industry has a long history of digital technology, especially in production, but there remains significant potential for leveraging entire groups of new technologies to further improve operations (Salygin & Lobov, 2021; Savoskin & Rozhkova, 2021). The widespread use of such technologies can reduce production costs by 10–20 percent, including through high-tech processing of seismic data, the use of sensors, and advanced modeling of oil and gas reservoirs. Technically recoverable oil and gas reserves could be increased by about 5 percent globally, with the largest increase expected to come from shale gas (IEA, 2017: 17).

Compared to the past twenty-five years, the world is indeed changing quite markedly in the ways in which humanity is trying to meet its growing energy needs (Osipov et al., 2021). The leading role is taken by the extraction and use of natural gas, as well as two more important trends—the rapid growth of renewable energy sources, and the desire to increase energy efficiency. The latter trend plays a colossal role in slowing down the growth in energy demand and, accordingly, the load on energy sources. Renewables are already satisfying 40 percent of the increase in primary demand, and their explosive growth in the energy sector, according to some experts, marks the end of the coal era. The role of oil and gas will vary greatly across regions, sectors, and over time.

The issues of technological re-equipment and staffing, which are extremely relevant for the development of Arctic oil and gas resources, found their reflection in the current Strategy of Scientific and Technological Development of the Russian Federation until 2035. The issue of human resources is undoubtedly posed very astutely: it is noted that the research and development sector interacts extremely weakly and haphazardly with the real sector of the economy, and the mechanisms of interaction between these two sectors, in fact, are not debugged (Dun et al., 2020; Potapova & Potapov, 2021). As a result of this failure, there is not a closed innovation cycle in the Russian economy, which leads to the drain of the most competitive human resources capable of creating breakthrough products into the economies of other countries; such people move to other countries where they feel that they and the fruits of their labor are more in demand. In other words, Russian investments in increasing human capital essentially "plays into the hands" of the competitiveness of the economies of other countries (Strategy for Scientific... 01.12.2016).

The list of big challenges of the new Strategy includes "the exhaustion of the opportunities for Russia's economic growth based on the extensive exploitation of raw materials" and "strengthening Russia's position in the field of economic, scientific, and military development ... of the World Ocean, the Arctic and Antarctic" (Strategy for Scientific ... 12.01.2016). It seems that the solution to both major challenges may lie at their "intersection". One of the most important areas of strengthening Russia's position is the search, exploration, and assessment of the recoverability of hydrocarbons from Arctic fields, especially on the Arctic shelf (in this case, not the start of large-scale production, primarily due to the unprofitability of this at the current level of hydrocarbon prices).

Nevertheless, it is necessary to provide substantial support for Russia's own R&D in creating technologies and machines for the Arctic oil and gas sector—this will create the necessary proprietary reserve for the future, and will help to overcome dependence on imports of equipment and materials. Moreover, these processes will create attractive jobs for highly qualified human resources (who, in all likelihood, will have to undergo additional training in large Russian companies or as part of international internships in states developing approaches to developing their own Arctic hydrocarbons).

The example of Norway in the 1970s can be considered very inspiring here. Norway was almost 100 percent dependent on imports of oil and gas equipment and technologies. However, after just 20 years, the country not only managed to get rid of import dependence, but firmly took the place of one of the world's leading exporters of high-tech equipment intended for the development and operation of offshore hydrocarbon deposits (the volume of such exports of Norway in 2015 was estimated at 20 billion US dollars). According to researchers, this impressive progress was achieved thanks to large-scale investments of oil and gas companies in R&D, as well as the principle of localizing production (attracting Norwegian technology suppliers to the developed oil and gas projects) (Mirzoyev & Skripnichenko, 2015). Other researchers note that for the development of new fields in the Arctic Zone of the Russian Federation, the experience of the formation and development of oil and gas clusters in Norway, formed around the development of offshore oil and gas fields as a new system-forming direction for the oil and gas industry, may be extremely relevant. Thus, Stavanger became the center of a cluster of downhole technologies, which included administrative centers (both public and private), technology firms, research institutes and centers, and universities; a cluster of subsea equipment has developed around Bergen; Mor has "pulled" around itself a cluster of maritime shipping, Christiansend became the center of a cluster of drilling, and so on (Larichkin et al., 2012).

4.3 Conclusions

Easily accessible hydrocarbon reserves are shrinking; as this process slowly but inevitably becomes global, so the issue of extracting oil and gas from more complex and hard-to-reach fields will increasingly come up on the agenda (Fidler & Noble, 2012; Keil, 2014; Mikkelsen & Langhelle, 2008; Morgunova, 2020; Offerdal, 2010; Pelaudeix, 2017; Schiermeier, 2012). This means that the oil and gas sector will become more and more knowledge-intensive, requiring significant investments in R&D and the creation, development, and implementation of a significant number of innovations. Training (and retention) of staff will also be required to work with these technologies. Our analysis shows that, as these imperatives are only reflected in the latest editions of documents of both the Arctic policy and energy policy and the policy for the development of science and technology, they are now relatively fully realized.

Therefore, in the current edition (of August 31, 2017) of the state program "Socio-economic development of the Arctic Zone of the Russian Federation for the period up to 2020", for the first time, a clearly formulated task has been presented, which puts forward the priority of scientific and technological development in the development of the Arctic, its resources, and the intensive development of resource base, increasing the efficiency of various operations in the Arctic oil and gas sector by stepping up the area's own scientific and technical activities and introducing the results of its own research and development and R&D into these operations.

Hydrocarbon deposits in the Arctic shelf waters of the Russian seas are, without a doubt, the most important strategic reserve of the oil and gas complex of the Russian Federation, and the lack of coherence in the policies of different levels aimed at the development and use of this reserve can carry serious risks of reducing the efficiency of resource use and shortfalls in Russia's benefits from its natural resources.

References

Adžić, S. (2021). Everything is digital: Getting ready for the inevitable changes. *Digital Law Journal, 2*(1), 94–96. https://doi.org/10.38044/2686-9136-2021-2-1-94-96

Decree of the Government of the Russian Federation of August 31, 2017 No. 1064 "On the new version of the state program 'Socio-economic development of the Arctic zone of the Russian Federation'". http://government.ru/docs/29164/ (in Russian).

Decree of the President of the Russian Federation of December 1, 2016 No. 642 "On the strategy of scientific and technological development of the Russian Federation". http://kremlin.ru/acts/bank/41449 (in Russian).

Decree of the President of the Russian Federation of October 26, 2020 No. 645 "On the strategy for the development of the Arctic Zone of the Russian Federation and ensuring national security for the period until 2035". http://www.kremlin.ru/acts/bank/45972 (in Russian).

Dun, L., Yuan, G., & Lunqu, Y. (2020). The age of digitalization: Tendencies of the labor market. *Digital Law Journal, 1*(3), 14–20. https://doi.org/10.38044/2686-9136-2020-1-3-14-20

Fidler, C., & Noble, B. (2012). Advancing strategic environmental assessment in the offshore oil and gas sector: Lessons from Norway, Canada, and the United Kingdom. *Environmental Impact Assessment Review, 34*, 12–21.

Gautier, D. L., & Moore, T. E. (2017). *Introduction to the 2008 Circum-Arctic Resource Appraisal (CARA) professional paper—U.S. geological survey professional paper*. No. 1824-A.

Guliev, I. A. (2015). Energy policy of Russian oil and gas companies in the Newest conditions of imposed sanctions. *Vestnik NGUEU, 2*, 283–290 (in Russian).

Inozemtsev, M. I. (2021). Digital law: The pursuit of certainty. *Digital Law Journal, 2*(1), 8–28. https://doi.org/10.38044/2686-9136-2021-2-1-8-28

International Energy Agency. (2017). *Digitalization and energy*. OECD/IEA.

Keil, K. (2014). The Arctic: A new region of conflict? The case of oil and gas. *Cooperation and Conflict, 49*(2), 162–190.

Larichkin, F. D., Fadeev, A. M., & Cherepovicyn, A. E. (2012). Problems of study and development of mineral resources of the Arctic region. *The Arctic: Ecology and Economics, 1*(5), 8–15 (in Russian).

Laruelle, M. (2014). *Russia's Arctic strategies and the future of the far north*. ME Sharpe.
Mikkelsen, A., & Langhelle, O. (Eds.). (2008). *Arctic oil and gas: Sustainability at risk?* Routledge.
Mirzoyev, M. R., & Skripnichenko, V. A. (2015). The main problems of the development of offshore oil and gas resources in the Arctic and ways to solve them. *Vestnik Severnogo (Arkticheskogo) Federal'nogo Universiteta. Serija: Gumanitarnye i Social'nye Nauki, 4*, 115–123 (in Russian).
Morgunova, M. (2020). Why is exploitation of Arctic offshore oil and natural gas resources ongoing? A multi-level perspective on the cases of Norway and Russia. *The Polar Journal, 10*(1), 64–81.
Offerdal, K. (2010). Arctic energy in EU policy: Arbitrary interest in the Norwegian High North. *Arctic, 63*(1), 30–42.
Osipov, V. S., Bulgarov, M. A., Krivtsov, O. F., & Vagin, S. G. (2021). Financial provision of energy security of regional economy in transition to industry 4.0. In E. B. Zavyalova & E. G. Popkova (Eds.), *Industry 4.0*. Palgrave Macmillan. https://doi.org/10.1007/978-3-030-75405-1_34
Pelaudeix, C. (2017). Governance of offshore hydrocarbon activities in the Arctic and energy policies: A comparative approach between Norway, Canada and Greenland/Denmark. In *Governance of Arctic offshore oil and gas*. Routledge.
Pichkov, O. B., & Ulanov, A. A. (2021). Regulation of robotics: Analysis of the leading countries' experience. *Digital Law Journal, 2*(2), 31–44. https://doi.org/10.38044/2686-9136-2021-2-2-31-44
Potapova, N. D., & Potapov, A. V. (2021). On the issue of digitalization of labor relations: Theoretical and practical aspects. *Digital Law Journal, 2*(2), 45–64. https://doi.org/10.38044/2686-9136-2021-2-2-45-64
Ria Novosti. (2020, May 21). *Rosstat: Residents of the Arctic Zone provide a tenth of Russia's GDP*. https://ria.ru/20200521/1571761572.html (in Russian).
Salygin, V. I., & Lobov, D. S. (2021). Defining major oil and gas companies' development strategies in the era of energy transition. *MGIMO Review of International Relations, 14*(5), 149–166. https://doi.org/10.24833/2071-8160-2021-5-80-149-166
Savoskin, A. V., & Rozhkova, N. A. (2021). Digitalisation of state companies. *Digital Law Journal, 2*(1), 83–93. https://doi.org/10.38044/2686-9136-2021-2-1-83-93
Schiermeier, Q. (2012). The great Arctic oil race begins. *Nature News, 482*(7383), 13.
Strunitskiy, A. V., & Chalganova, A. A. (2019). Study of the dependence of the share of GRP on the share of mining in the Arctic Zone of the

Russian Federation in 2018. *Integratsiya Nauki i Proizvodstva, 6*, 133–136 (in Russian).

Torkunov, A. V. (2018). Great Eurasian power in the stream of changes. *International Trends, 1*(52), 6–13 (in Russian).

CHAPTER 5

New Trends in the Development of Energy Fields in the Russian Arctic

Alexander N. Pilyasov and Elena S. Putilova

5.1 Introduction: The Novelty of the Situation

In the last 20 years, Russia has embarked on a new development of the Arctic resources. This corporate development is radically different from the Soviet period, being based on other actors and drivers, and forming a different territorial structure (bases and development routes) (Pilyasov & Putilova, 2020a, 2020b; Pilyasov & Zamyatina, 2019; Zamyatina & Pilyasov, 2018, 2019).

If before, the state was the only one organizer of economic development, now the network of actors has expanded significantly, but resource private and state corporations have become the most important among

A. N. Pilyasov (✉)
Lomonosov Moscow State University, Moscow, Russia
e-mail: pelyasov@mail.ru

E. S. Putilova
Autonomous Non-Commercial Organisation "Institute of Regional Consulting", Moscow, Russia
e-mail: es_putilova@mail.ru

V. I. Salygin (ed.), *Energy of the Russian Arctic*,
https://doi.org/10.1007/978-981-19-2817-8_5

them. The situation on the global markets has become a new driver for the development of many resource projects in the Arctic, whereas earlier, in the Soviet industrial time, it was mainly the internal national economic needs for natural resources. Previously, economic development was based on the creation of a permanent network of settlements (single-industry towns and villages) and a permanent network of roads and railways; now it is based on the active use of seasonal transport schemes, winter roads, sea and river development routes.

Integrally, one can say that the new Arctic projects approve absolutely new spatio-temporal schemes for the development and distribution of productive forces. For example, in the past, the creation of any processing industry in the Arctic and the North was practically prohibited, because it was believed that high production and labor costs here make only mining possible. However, the new realities of the construction of LNG plants in the Arctic, including the localized connection of mining and processing industries, show that this is becoming possible now.

The temporal organization of the development of resource deposits is also changing. Under the Soviet regime, a new project was created "for centuries", but at the same time the phase of its construction dragged on for years and the commissioning time was constantly postponed: the postponement did not affect the position of the organization in the market in any way, since the demand for its products was provided by the state order. The construction progress depended on the supply of the necessary materials, the timing of which was unpredictable, so the project was carried out as it was delivered. On the other hand, now the new project and all its key elements are initially of a temporary nature—the term of development of the field, the shift camp, the main production modules, etc.—and no one else is ashamed of this temporality. This is evidenced by the "temporary operation" regime, meaning the exploitation of the field according to a temporary scheme, which is favored by some companies (for example, Varandey terminal of Lukoil in Nenets Autonomous Okrug or exploitation of Talakan oil and gas deposit in Yakutia by Surgutneftegaz during the first years etc.); however, at the same time, after making a decision to start development, these companies do their best to meet the construction deadlines and put the facility into operation as soon as possible.

Many projects are implemented not by one development company, but on the basis of cooperation and partnership relations between Russian and foreign companies. Among such investors in projects for the extraction and processing of hydrocarbons are often companies from countries that consume the final product.

Comparing these projects with the Soviet era highlights the exceptional, global scale, as well as the breadth of the network of suppliers and contractors. The company that holds the license for the development of the field, in fact, becomes a network leader, acting as a component manager, attracting tens and hundreds of organizations for the production, supply, and operation of equipment, construction of transport and engineering facilities, operators of the production process, recruitment agencies for the search and hiring personnel, etc. The markets for the final products of Arctic projects are also significantly more diversified compared to Soviet times: as a rule, now these are the countries of the Asia–Pacific region, the countries of the European Union, and domestic consumers in Russia itself.

The main idea of the authors of this chapter is to give the reader a gist of the new nature of energy projects for the development of the Russian Arctic: new maritime logistics on which they rely, new spatial effects that resource corporations use, and new general features of these resource projects of the development of the Russian Arctic 2.0. To this end, the chapter gives a brief description and summarizes the features of about a dozen of the most iconic energy projects of the modern development of the Russian Arctic: the Novoportovskoye project, the Prirazlomnaya platform, the Vostok Oil project, Yamal LNG, Arctic LNG-2, and the Obskiy gas chemical complex, the development of the Messoyakhsky group of fields, and the Bovanenkovsky and Syradasay deposits.

The chapter is based on the results of work carried out by the team of authors in the last years within the framework of a project funded by the Russian Foundation for Basic Research № 18-05-00600 “A New theory of the development of the Arctic and the North: multi-scale interdisciplinary synthesis”.

5.2 The Novoportovskoye Field as a Vivid Example of How the New Maritime Logistics of Arctic Projects Brings Soviet-Era Projects Out of Oblivion

The main feature of the current period of development for Russian Arctic projects is the removal of previous prohibitions on the maritime logistics. In Soviet times, in the development of the North and the Arctic, given the choice between alternative land and sea routes for the export of final products, priority was unambiguously given to the first, which was considered safer and more guaranteed. Therefore, permanent roads were built and pipelines were laid, which transported the extracted resource products to the south.

In the 1990s, the corporatization and globalization of the development of Arctic resources led to new technological opportunities. Together with the mitigation of the severity of the climate in the Arctic and, as a result, a sharp decrease in the ice cover of the Northern Sea Route, they became powerful factors that led to a revolutionary rediscovery of the possibilities of maritime logistics. Therefore, coastal resource projects—long-known, but postponed for decades—got a chance to be developed.

Among them, perhaps, the most striking example is the Novoportovskoye oil and gas condensate field, located 30 km from the coast of the Ob Bay in the Yamal region of the Yamal-Nenets Autonomous Okrug, which was postponed for half a century. The field was the first to be discovered on the Yamal Peninsula, with significant oil reserves proven back in 1964; however, due to difficult geological conditions, its development began only after 2010. Another important reason for the half-century delay in development is the lack of a logistics scheme adequate to the complex transport and geographic conditions of export.

The construction of a pipeline to the field with a length of more than 700 km was not economically feasible. The option of seasonal export along the winter road to the Payuta railway station was considered, but the existing capacity of the railway infrastructure is not enough for the export of the required volume of raw materials.

As a result, Gazprom Neft's proposed offshore scheme for the export of extracted raw materials became a real innovative breakthrough and reflected the fact that, under new climatic conditions and new technological opportunities, the marine Arctic is no longer a barrier or obstacle to the development of natural resources to such an extent as it was

30–50 years ago. Instead of the usual "southern" route of pipeline transport, the "northern route" of export by sea through the "Gates of the Arctic" oil terminal near Cape Kamenny was proposed as a solution for transporting oil from the Novoportovskoye field.

Oil produced at the field is transported through a 105-km pipeline to the terminal, where it is loaded onto Arc7 ice-class shuttle tankers, which operate trips from the Novoportovskoye field and the Prirazlomnaya platform to the Kola Bay for transshipment through the Gazprom Neft offshore terminal for further transportation to Rotterdam. Year-round transportation by tankers is supported by two diesel-electric icebreakers with an innovative maneuvering and energy saving system, designed by the Vyborg Shipyard for Gazprom Neft.

The "Gates of the Arctic" oil loading terminal is located on a remote platform over 89 m high, 3.5 km from the coast. It took three years to build a pressure pipeline and an oil terminal. The construction of the terminal's onshore infrastructure was impossible due to shallow depths limiting tankers' approach to the shore and alluvial currents in the area of the cape.

The terminal is a technical structure with a height of more than 80 m and is designed for year-round operation in the extreme conditions of the Arctic. It is connected to the onshore infrastructure by a 7.9-km subsea pipeline. The terminal is controlled remotely using a fiber-optic cable. It operates year-round, which allows shipping about 6 million tons of oil annually.

The settlement of Novy Port, which has existed since 1921 on the bank of the Gulf of Ob, is used as a local base for the development of the field. The Novy Port is the base for the material and technical supply of the field; it is also planned to build a warehouse infrastructure in the village for the needs of the field. In 2016, the village and the field were connected by a new year-round road 22-km long.

The rear intellectual base for drilling the complex reserves of the Novoportovskoye field is the Gazprom Neft Science and Technology Center in St. Petersburg. With the help of a remote telemetry system, the center's specialists are able to collect information and simulate the process and further progress of drilling in real time.

"Taking many Arctic projects out to sea" requires new infrastructural elements that were not needed under the previous Soviet model of predominantly onshore development and transportation of Arctic resources. First of all, these are the numerous and technologically diverse docking production and transport hubs that resource companies create in the Arctic to save costs and speed up the transshipment/transportation of extracted natural resources. Usually, these are docking points of two types: at the place of production, and at the place of reloading from vessels of reinforced ice class to ordinary ones. Docking points at the place of production are created due to the shallow waters of the Arctic seas and the inability for large tankers to approach directly to the shore.

The first among large companies to establish a year-round permanent sea ice-resistant shipping terminal Varandey in the Pechora Sea in 2008, 20 km from the coast, was Lukoil. Lukoil's experience was taken up by Gazprom Neft, which in 2016 launched its year-round "Arctic Gate" loading terminal in the Kamenny Cape area for transshipment of oil from the Novy Port field. Now, Rosneft plans to use this experience in the Vostok Oil megaproject: oil from the Vankor cluster is to be transported to world markets through the new port-terminal "Bukhta Sever" in Taimyr near Dikson.

Russian resource corporations operating in the Arctic are implementing their own expensive programs for the construction of Arctic tankers and gas carriers. Often, the implementation of these programs turns out to be cheaper than paying daily for icebreaker escort services.

5.3 Spatial Effects in New Energy Projects of Corporate Development of the Russian Arctic

New energy projects implemented by Russian corporations in the process of new development in the Russian Arctic rely on specific spatial effects that provide them with additional economic rent and increase the overall profitability of the project. Despite the fact that very often companies actualize all the possible effects of the rational distribution of the productive forces of a new project in the Arctic space, it is not uncommon for each energy project to be characterized by just one specific effect—localization, regional, or intercorporate. This paper will consider new energy projects in the Russian Arctic from the standpoint of that specific, special spatial effect on which it primarily relies.

5.3.1 *The Effect of Spatial Localization—Platforms and Development Islands: Prirazlomnaya, Yamal LNG, Arctic LNG-2*

In recent years, in the Russian Arctic, a new scheme of the spatial organization of productive forces in pioneer development projects has emerged—platforms and economic "islands": the Prirazlomnaya platform in the Nenets Autonomous Okrug, a port and a liquefied natural gas plant in Sabetta, and the Center for the construction of large-scale offshore structures on artificial islands in the Kola Bay (a plant of factories that manufactures offshore complexes for liquefying natural gas, storing and shipping LNG).

Developers of platform solutions for the development of Arctic resource projects focus on the use of robotic mining complexes, unmanned trucks, remote unified control from production to shipment through the use of intelligent telecommunication systems, and the use of ready-made modular structures (assembled at the production facility itself as soon as possible). Reliance on maritime logistics saves time and costs (for example, offshore LNG complexes reduce the loading time of gas carriers), and simplifies the construction process due to the absence of a large number of thermal stabilizers and piles, which are necessary for working in permafrost conditions.

Extreme localization and compact packaging of all elements of a new project on a separate island platform provides savings in transport and production costs, minimizes the negative environmental impact of the project, and leads to positive externalities from localized integration of technologically conjugate elements. Cost savings and space savings are directly linked in this case.

The Prirazlomnaya project is interesting in that the platform effect of the compact localization of the production site is very visible here, just as there is little else in the Arctic. After all, the main production facility here is the Prirazlomnaya offshore ice-resistant oil production platform, which was specially designed for the development of the field.

The Prirazlomnaya oil field on the shelf of the Pechora Sea, 60 km from the Varandey settlement in the Nenets Autonomous Okrug, was discovered in 1977. The development of the field was planned from the beginning of the 1990s. However, due to financial and organizational difficulties, oil production only began 30 years later, in 2013. The Prirazlomnaya oil field is the first and, at the moment, the only one being developed on the Russian Arctic shelf.

The platform complex includes the functions of drilling, oil production, and storage; completely autonomous existence of production, including power supply and shipment of produced oil; the principle of "zero discharge" (all waste is disposed of by removal or pumped into the reservoir); an automatic control system that monitors the state of operational activities; and remote monitoring—based on the digital model of the platform equipment, specialists from different offices remotely monitor and analyze technological processes, and predict the result.

The platform relies on a system (or network) of supply bases: the village of Varandey (60 km to the platform) is the onshore subdivision of the platform (a heliport is also located here); Usinsk has a residential complex for platform workers; Murmansk is a base for servicing offshore operations: at this location Gazprom Neft has created an onshore base to ensure uninterrupted operation of the Prirazlomnaya platform; Arkhangelsk is the base from where shift personnel are delivered by charter An-24 aircraft to Varandey, from where Mi-8ATM helicopters deliver personnel to the platform.

Oil production and transportation are carried out all year round. All oil wells are developed using the modern horizontal method, as well as the "fishbone" technology. During operation, a significant number of innovative automation systems are used. The produced oil is supplied for storage in tanks with a total volume of 124 thousand meters square located in the lower part of the platform. For its reloading into tankers, a complex of devices for direct oil offloading is used. Due to new wells being drilled, oil production has been expanding in recent years.

Two ice tankers, Mikhail Ulyanov and Kirill Lavrov, specially created for the project, are used to export oil. The tankers transport oil to the storage tanker and transshipment complex "Umba" with a deadweight of 300 thousand tons, installed on the roadstead in the Kola Bay near Murmansk. From there, oil is delivered to Europe by standard tankers.

The ARKO Prirazlomnaya grade is distinguished by its high sulfur content and high density. This narrows down the list of oil refineries that are technologically ready for its processing. Therefore, not all European ports are suitable as final destinations for tankers from Prirazlomnaya, but only those that are located near refineries capable of processing a high-density grade of oil produced at the field, characterized by an increased sulfur content and a reduced paraffin content.

Novatek's new LNG projects—Yamal LNG, Arctic LNG-2, and Obskiy GKH—have similar platform features of extreme localization, compactness, seaward capabilities, and modularity of construction solutions.

Yamal LNG is a project for the production, liquefaction, and supply of liquefied natural gas based on the South Tambeyskoye gas condensate field in the Yamal-Nenets Autonomous Okrug. The field development project includes the creation of all the necessary production, transport, and living infrastructure on an extremely localized territory: the seaport of Sabetta, an LNG plant with a capacity of 16.5 million tons per year, including a navigable approach channel in the Gulf of Ob, with an oil terminal, an airport, a gas power station, and a shift camp.

The utmost compactness and localization of production operations is combined with broad international links of the project. Yamal LNG is actually an international project, distinguished by a large number of foreign partners, both in the capital structure and among contractors, suppliers, and consumers. The project is multi-country in terms of ownership structure: the owner of the controlling stake (50.1 percent) in Yamal LNG is PJSC Novatek, approximately 20 percent belongs to the French oil and gas company Total, and the remaining 30 percent of the project belongs to investors from China (Kommersant, 2020). The airport serving the Yamal LNG project has an international status and is connected by regular flights to Novy Urengoy, Moscow, and Samara.

For the transportation of LNG, tankers of the reinforced ice class "Arc7" were specially designed. Gas carriers are capable of passing through ice up to 1.5 m thick, which makes it possible to do without icebreaker assistance all year round when navigating to the west. To the east, which is the key for the project (more than 85 percent of LNG is supplied to the Asian Pacific Region countries), tankers do without icebreaker assistance during the Arctic summer. Delivery of goods to the east along the NSR (for example, to China) is carried out twice as fast as along the traditional route through the Suez Canal with 19 days versus 35 days (TASS, 2016). The liquefied gas export route includes two transshipment points from ice-class gas carriers to conventional ones: in the Kola Bay during transportation to the West, and in Kamchatka in the Bechevinskaya Bay during transportation to the Asia–Pacific region and Southeast Asia.

Platform compactness is characteristic even more for the second LNG project of Novatek—that is, Arctic LNG-2: if Yamal LNG is all located onshore, in the Sabetta shift camp, then Arctic LNG-2 will use the land-sea contact zone. This is a project for the production, liquefaction, and transportation of natural gas based on the Salmanovskoye (Utrennee) field in the Yamal-Nenets Autonomous Okrug. The project includes the construction of three LNG production lines. In 2022, the maximum number of employees on the project will be 11 thousand people. Start-up is scheduled for 2023, with full capacity by 2025.

The main delivery of materials for the development of the field takes place in the summer, by the river. Each block of construction has its own contractor and there is a general contractor for all construction. The facility provides for the creation of not one but several shift camps for each category of workers: builders, engineers, and technical workers. For year-round navigation in the shallow waters of the Gulf of Ob, dredging works are planned. Within the field, three production zones are clearly distinguished, which are called the Kupol: North, Central, and South.

The main destination for the export of products, as well as for other gas liquefaction projects, are the Asia–Pacific countries, among which the bulk of consumption falls on Japan. For transportation along the NSR, it is planned to build five Arc7 class LNG carriers for the project.

The technological base that provides a platform effect on the maximum localization of the location of LNG plants is the Center for the construction of large-scale offshore facilities in the Murmansk region. On the basis of this Center, PJSC Novatek has organized the assembly of gravity platforms that accommodate the production and storage of LNG. These platforms are "mobile" factories that are installed close to the mining site.

The Obskiy gas chemical complex of Novatek in the north of the Yamal-Nenets Autonomous Okrug, which relies on the resource base of the Verkhnetiuteyskoye and Zapadno-Seyakhinskoye fields, is the third in a series of "platform" projects for the maximum localization of gas processing. Instead of the previously planned production of liquefied natural gas, the production of ammonia, hydrogen, and methanol will be organized. The first stage of the ammonia plant with a capacity of 2.2 million tons is planned to be launched in 2026.

A common feature of all the projects considered: they do not have the usual and obligatory icebreakers before. Both the delivery of production platforms and the export of manufactured products are carried out either year-round by ships of reinforced ice class or seasonally, but almost always without icebreaker escort. Another common feature is that transport problems are solved without the need to build permanent roads.

Many management issues are resolved remotely and automatically, and due to this, there is the minimum number of personnel on the platform itself.

5.3.2 *District Effect: From the Vankor Development Project to the Vostok Oil Megaproject*

The district effect was the main one in the spatial development of new territories and resources under the Soviet era, and was realized in the form of industrial territorial production complexes (TPKs) and economic regions. In the context of the modern corporate development of the Russian Arctic, its nature is changing. Now it is received not by the state, but by corporations, due to the close location of several license areas (production facilities), connected to each other not by year-round physical infrastructure facilities, but by seasonal, temporary telecommunication and transport infrastructure facilities. The district effect from the new intra-corporate conjugation takes on either a center-peripheral form—the best practices of the first project are replicated in subsequent ones—or a network form of simultaneously implemented neighboring projects, without the "seniority" of one of them.

The desire to obtain a district effect in new development in the Arctic is clearly demonstrated by many companies: for example, Novatek, which created a corporate "empire" in the Gulf of Ob from the autonomous (but relatively closely located) mining and license areas of the Yamal and Gydan peninsulas. Gazprom plans to create Megaproject Yamal, a new gas production center that will unite 32 fields in total (Kharasaveyskoye, Tambeyskoye, etc.) by transport infrastructure. The Yamal gas production center based on the Bovanenkovskoye field will become the largest for Gazprom and will replace the Nadym-Pur-Tazovsky region with depleting fields. This district effect has its own geographic range of action: as a rule, it is hundreds of kilometers, but not thousands.

It is curious to see the transformation of the "single" project for the development of the Vankor field of Rosneft into the largest project "Vostok Oil", which provides for the infrastructural and organizational integration of the resource corridor of fields from Vankor to the Arctic Ocean: Vankor, Lodochnoye, Tagulskoye, Suzunskoye, and others, discovered "along the way" on this movement to the "ice sea". Also, if the oil from the Vankor cluster is transported to the south through the Vankor-Purpe oil pipeline with a length of 556 km to the main oil pipeline of Transneft, then the transformation of the Vankor mini-project into the Vostok Oil megaproject will lead to a change in the logistics scheme for oil export. Oil from all fields, except for Vankor, will be sent through a pipeline to the new sea terminal "Port of Bukhta Sever" on the Arctic Ocean.

The company plans to obtain the district effect (in reducing current and capital costs). Firstly, the Vankor field will be transformed into a base for the territory in terms of testing advanced technologies here. Secondly, the district effect is becoming even more powerful due to the integration of the Vankor and Payakhskaya group of fields and the formation of a single megaproject "Vostok Oil", which relies on the resource base of 13 fields in the Vankor cluster and the Payakhskaya group of fields and involves the construction of an oil pipeline between the Vankor and Payakhskaya group of fields. This will allow transporting products from all fields of a single industrial region through the oil loading terminal "Port Bukhta Sever" 40 km from the Dikson settlement to consumers in Europe and Asia via the Northern Sea Route (Oil Kapital, 2019).

The terminal will make it possible, at the first stage, to export annually up to 30 million tons of oil from the fields of the Payakha group by 2024. The construction of the second and third stages of the oil port will increase the transshipment volume up to 100 million tons. The implementation of the Vostok Oil project will create a new oil and gas province in the north of the Krasnoyarsk Territory, an airfield, 15 industrial towns, a railway, 3.6 thousand megawatts of power generation, and about 50 vessels of various classes. Our estimates have shown that the use of the district effect allows the company to reduce its costs in the development of a new vast oil and gas region in Eastern Siberia by 30–50 percent (Pilyasov & Polyachenko, 2021).

5.3.3 *Intercorporate Effect: Development of the Messoyakha Deposits and Other Projects*

Another effect of modern Arctic resource development can be called intercorporate. It arises as a result of the cooperation of several corporations in the development of a new resource facility. It is far from always that corporations manage to agree on the distribution of responsibilities or powers on the terms of interaction in the territories of new development.

Summarizing a dozen cases of conflicts and cooperation leads to the conclusion that the intercorporate effect often occurs when (1) there is a specialization of companies in their resource chain of added value (for example, gas, oil, condensate; in this case, even an equal weight category of partners does not bother to cooperate); (2) when it comes to a genuine technological challenge for both parties, and in order to overcome the simultaneous natural, technological, and financial challenges and risks (for example, in the operation of offshore facilities, competing companies in other territories are ready to cooperate here); and (3) in the case of one specialization, but different status of partners included in the alliance—when there is a senior company and a junior company—as the difference in status facilitates the achievement of partnership agreements.

The first case concerns the cooperation of Gazprom Neft and Rosneft in the project for the development of the Messoyakha fields: here Rosneft is responsible for the oil chain and Gazprom Neft for the gas. Cooperation is carried out within the framework of a general agreement between the companies made in 2006, which affects almost all aspects, except for sales: production, transportation, and processing of hydrocarbons, as well as informational, scientific, technical, and personnel interaction (Gazprom Neft, 2019). Another example is the joint venture between Gazprom Neft and Lukoil for the development of the Tazovskoye oil and gas condensate field, which was commissioned in June 2021: Gazprom Neft will be responsible for the gas chain, and Rosneft for the oil.

The second case is the intention to create a joint venture between Gazprom Neft and Novatek for the development of offshore projects (Kommersant, 2019a).

The third case concerns the logistic interaction of Rosneft and Neftegazholding in the integration of the development of the Vankor and Payakhskaya group of fields into a single Vostok Oil region (Kommersant, 2019b).

There is more detail about the first case that must be considered. The Vostochno-Messoyakhskoye and Zapadno-Messoyakhskoye fields are the northernmost of the developed continental fields in Russia, discovered in the 1980s, located in the Tazovsky region of the Yamal-Nenets Autonomous Okrug, 340 km north of Novy Urengoy. The Vostochno-Messoyakhskoye field has been in commercial operation since 2016. The West Messoyakhskoye field begins to develop in 2021.

The Vostochno-Messoyakhskoye field development project is a product of a successful intercorporate agreement. The license to develop the field is held by JSC Messoyakhaneftegaz, a joint venture between Gazprom Neft and Rosneft with equal shares in the project. Operational management of development is carried out by Gazprom Neft, which uses the Vostochno-Messoyakhskoye field as a testing ground for new technologies. For the development of the Vostochno-Messoyakhskoye field, all the necessary infrastructure has been created: an autonomous gas turbine power plant, an acceptance point, a shift camp along with supporting facilities. Transportation of products is provided by a local oil pipeline with a length of 98 km, connected to the main oil pipeline Zapolyarye-Purpe-Samotlor.

The digital twin of the Vostochno-Messoyakhskoye field was created by the joint venture between Gazprom Neft and Rosneft Oil Company Messoyakhaneftegaz. The digital model of the field includes several modules responsible for various stages of the production process: extraction, preparation, and transportation of raw materials, as well as maintaining pressure in the reservoirs. The digital system can be used to calculate the optimal parameters of the production process, taking into account the individual characteristics of the reservoirs and the capacity of the infrastructure. This makes it possible to improve the efficiency of working with hard-to-reach reserves at the northernmost field onshore in Russia in conditions of limited access to transport infrastructure.

Regarding the second case, Gazprom Neft and Novatek have set up a joint venture to develop offshore license areas, including the Severo-Wrangel area. The joint venture with Gazprom Neft will allow Novatek to gain access to offshore reserves, since currently only state-owned companies Rosneft and Gazprom have the right to develop offshore fields. It will provide the pooling of financial resources and industry expertise, as well as the logistic capabilities for the full range of exploration and development of offshore oil and gas fields.

In the actual process of developing new resource projects in the Russian Arctic, the platform effect of extreme localization, the district effect, and the intercorporate effect are often combined and reinforce each other. For example, from the "core" of the Prirazlomnaya platform, it is planned to drill dozens of wells in neighboring areas. The company wants to scale up the acquired competencies and skills of working in the freezing water area when developing neighboring production facilities. Thus, Prirazlomnaya is gradually becoming the nucleus of a new emerging offshore industrial district, from a whole cluster of projects and fields, with a district effect on the "neighbor's" conjugation of personnel, competencies, and infrastructure. Similarly, Gazprom Neft hopes to actualize the district effect as a result of the fact that the flagship Novoportovskoye field in the future may form a single cluster around itself together with other fields of the company in Yamal.

5.4 Bovanenkovskoye as a Greenfield-Brownfield Energy Project with Onshore Logistics

All energy projects implemented in the Russian Arctic can be classified into greenfield and brownfield: that is, respectively, projects "from scratch", or projects on infrastructural, technical, and organizational solutions of the old Soviet industrial model of oil and gas field development. All the previously described examples of projects—Novoportovskoye, Prirazlomnaya, Vankorskoye, Messoyakhskoye fields, LNG plants—belong to the "greenfield" category; that is, they are wholly innovative in terms of technical, organizational, and technological solutions. However, some resource corporations in Russia, most of all Gazprom, continue to use the old, Soviet development schemes in the development of new fields, relying on pipeline transport, the southern "land" export scheme (and not maritime logistics), the district effect with minimal localization of a new project in space, etc.

Among the energy projects that are based on development schemes dating back to the 1970s is the development by Gazprom of the Bovanenkovskoye field in Yamal. Despite the fact that the field is located only 40 km from the Kara Sea, its logistics uses the classic "southern" pipeline scheme for transporting the produced gas. The field is the largest in Yamal in terms of proven gas reserves. It was opened in 1971, and gas production has been conducted since 2012. According to Gazprom's estimates, production at the field will last for over 100 years (Gazprom, 2018). The development of the field is supported by three gas fields, which were gradually commissioned from 2012 to 2018. Employment is dominated by local residents, mainly from Nadym.

The project is interesting, in that it combines internal technological innovations with a very traditional logistics scheme of supply and export: a combination of brownfield and greenfield elements takes place. For the development of Bovanenkovskoye, for the first time in Russia, a unified production infrastructure is used to extract gas from the Cenomanian and

Aptian-Albian deposits, which makes it possible to increase the operational efficiency. Automation systems have been introduced at all fields to reduce the number of personnel involved.

Gas transportation from the field is provided by the traditional scheme with connection to the main gas pipeline. In 2012, the Bovanenkovo-Ukhta gas pipeline was commissioned from the field to the city of Ukhta in the Komi Republic, and in 2017, the Bovanenkovo-Ukhta-2 gas pipeline was commissioned. The Obskaya-Bovanenkovo railroad, currently the northernmost in the world and with a length of 572.2 km, was built for the material and technical supply of the deposit. The railway line is owned by Gazpromtrans, a subsidiary of PJSC Gazprom.

5.5 Syradasay Coking Coal Deposit as the Largest Non-oil and Gas Energy Project in the Russian Arctic

The development of the Syradasay coking coal deposit of unique grades is a complex project that includes mining, transport, and logistics blocks. The field is located 55 km from the Yenisei Gulf, 105 km south of the seaport of Dikson, in the Taimyr Dolgano-Nenets municipal district of Krasnoyarsk Krai.

The license for the development of the field belongs to Severnaya Zvezda LLC, which is part of the international private investment corporation AEON. In terms of the volume and quality of raw materials, the coal basin is one of the most promising in Russia.

At present, the Severnaya Zvezda LLC company is creating a complex for the extraction and processing of coal at the Syradasay deposit. The project includes the creation of a coal mine with a capacity of 5 million tons per year at the first stage and 10 million tons per year at the second, an enrichment plant, the construction of a dam, a sea coal port, an industrial road 60 km long to a new seaport, the longest covered conveyor in the Arctic for delivery of coal to the port, power plant on coal dust, etc. It is planned to build supporting infrastructure: an airfield, an explosives warehouse, a shift camp, etc.

The Syradasay project clearly shows how well-known Arctic deposits established long ago in Soviet times can, in new conditions, such as under the influence of a new marine logistics supply and marketing scheme (both in the traditional western and eastern, Asian directions, which were poorly used in Soviet times), suddenly acquire investment attractiveness.

At the start of the project, an interesting dispute arose between two logistics schemes for the implementation of the project: relying on the already existing port of Dikson as an input base for new development, or relying on the construction of a special new seaport. The initial version of the project envisaged the construction of a 120 km long railway from the deposit to Dikson to transport coal, where it was planned to transship coal to ships and, thus, to create a large Dikson coal mining center. In this case, it was necessary to carry out dredging works, as well as the creation of new port infrastructure facilities. On the other hand, the option of building a twice shorter (60 km) automobile road (temporary, seasonal) from the field to the Yenisei Gulf provides for the creation of a new seaport "Yenisei", up to 15.5 m deep, with a quay wall of 500 m and an industrial with a capacity of 5 million tons, connected by an industrial road with a quarry (Yenisei Siberia, 2019). It was this option that was accepted as the basic one.

The victory of the automobile scheme with the creation of a new port cannot be understood outside the institutional context (the transition from the development of Arctic resources by state structures to private, corporate ones). For government agencies, control issues were less important than for modern Russian resource companies. The new logistics scheme may even be more expensive than the railway scheme based on Dikson; however, only it provides the owner with full control over the situation, because Dikson has its own influential players, whose interests will have to be taken into account, whereas for the new port, the project is the owner's from start to finish.

For the construction of the infrastructure of the field, a temporary shift camp was organized, in which about 300 workers live. The construction of the road is one of the key infrastructure facilities of the field development project. The first stage of construction of the road—from the Yenisei seaport on the coast of the Yenisei Gulf to the site of the concentration plant—is almost completed, and the second stage of the road construction to the coal mine will start soon. The construction of the track is carried out using a special technology to minimize the impact on the environment; the track is covered with rocky soil to minimize disturbance of the soil and vegetation layer.

For the time being, the Dikson seaport is being used for the delivery of equipment for the construction of the field infrastructure; in the future, the Yenisei port will replace it in this function as well. The goods are dispatched from the port of Arkhangelsk. In winter, an ice berth is organized for unloading ships by clearing snow and freezing and strengthening ice using a special technology. This makes it possible to unload equipment directly onto the ice.

The fate of Arctic projects now directly depends on the extent to which they are provided with ice-class vessels capable of operating at least in the western sector of the Northern Sea Route without constant icebreaker support. It is here that the bottleneck in the logistics of Arctic projects has now emerged: for example, experts believe that the projects for the development of the Syradasay coal mine are not provided with ships (Grigoriev, 2019). To transport coal from the deposit, the Severnaya Zvezda mine has planned to design a new bulk carrier ArcticMax with a cargo capacity of 105 thousand tons. The bulk carrier will have ice class Arc5, its width will also allow to follow the icebreaker in the channel (Portnews, 2021a).

5.6 Minimizing the Carbon Footprint

The idea of modern energy projects in the Russian Arctic will be incomplete without noting their new common feature: the desire to introduce elements of green energy and low-carbon technological solutions. For example, when implementing the Vostok Oil megaproject, Rosneft plans to use wind power to reduce its environmental impact and its carbon footprint. For the construction of wind farms, Rosneft has entered into a cooperation agreement with the Danish company Vestas RUS, one of the largest wind turbine manufacturers in the world. Within the framework of the agreement, it is planned to study potential sites for the construction of wind generators, as well as the supply and maintenance of equipment. The use of wind power will reduce the carbon footprint by four times compared to other new oil projects. The volume of carbon dioxide emissions within the framework of the project will be about 12 kg per barrel (the average volume of emissions from the oil industry in the world is about 50 kg per barrel) (Oil Kapital, 2021).

To reduce carbon dioxide emissions, Novatek is considering a plan to build a wind farm in Sabetta. The station will be located next to the Yamal LNG plant and in the future it will be able to partially replace the electricity generation of the existing power plants using natural gas. The capacity of the projected wind farm will be up to 200 megawatts. The use of wind energy will allow Novatek to reduce emissions by 0.6 percent of the total.

The Syradasay energy project also provides for the construction of a wind park with a total capacity of 4 megawatts. Wind generators will provide electricity to consumers of the Yenisei Marine Terminal (Portnews, 2021b).

5.7 Conclusions

This review of the new features of the largest energy projects in the Russian Arctic indicates significant changes in comparison with the Soviet period of industrial development which peaked in the 1970s on the one hand; on the other hand, these projects have common features in comparison with the other large production projects of the present time: the inclusion of elements of green energy and green economy, technological innovation, "platform" compactness, the temporality of many of the technological schemes and organizational solutions used, etc.

Companies "peep" at each other's breakthrough technological solutions: Lukoil's breakthrough in maritime logistics through the new Arctic terminal Varandey became an incentive for innovative solutions decades later from Gazprom Neft in the development of the Novoportovskoye solution and Rosneft in the megaproject Vostok Oil. On the other hand, there are the efforts of many companies to create a corporate industrial area by linking licensed areas with each other and obtaining a district effect on the overall transport and communication infrastructure, the flow of knowledge and competencies from one corporate project to another, using specialists from one corporate market labor, etc. On the other hand, there is the use of the effect of "platform" localization, which makes it possible, on the basis of new technologies, to radically reduce the time of construction and commissioning of a new field.

On the one hand, many new energy projects are emerging based on the maritime logistics of transporting and marketing Arctic resources to the markets of Asia and Europe; on the other hand, the "southern" onshore pipeline scheme for the export of hydrocarbons is maintained inertly at some new projects. New energy projects in the Arctic are full of paradoxes: combining the features of the old and new economies; or encouraging hydrocarbon production and, at the same time, making efforts to minimize the carbon footprint through the growing use of renewable energy (primarily wind) in generation for these projects. The variety of features of new energy projects in the Russian Arctic, along with clearly marked new tendencies and trends, is their most important feature.

References

Gazprom. (2018). *The final gas production facility at the Bovanenkovskoye field and the Ukhta-Torzhok-2 gas pipeline were commissioned*. https://www.gazprom.ru/press/news/2018/december/article469802/ (in Russian).

Gazprom Neft. (2019). *Ability to negotiate: Gazprom Neft partnerships as a strategic tool*. https://www.gazprom-neft.ru/press-center/sibneft-online/archive/2019-september/3592575/ (in Russian).

Grigoriev, M. N. (2019). Thirty years later. *Oil and Gas Vertical, 14*, 50–56 (in Russian).

Kommersant. (2019a). *Shelf for two: Novatek may become a partner of Gazprom Neft in areas in the Arctic*. https://www.kommersant.ru/doc/3931718 (in Russian).

Kommersant. (2019b). *Oil admirals: Rosneft intends to load the Northern Sea Route with raw materials from its fields*. https://www.kommersant.ru/doc/3896485 (in Russian).

Kommersant. (2020). *Yamal LNG has been tested for cold brittleness*. https://www.kommersant.ru/doc/4226037 (in Russian).

Oil Kapital. (2019). *Rosneft and Neftegazholding will increase the cargo traffic of the NSR*. https://oilcapital.ru/news/markets/28-02-2019/rosneft-i-neftegazholding-narastyat-gruzopotok-smp-smi?ind=1892 (in Russian).

Oil Kapital. (2021). *Vestas and Rosneft to develop wind energy*. https://oilcapital.ru/news/companies/11-06-2021/sotrudnichat-v-oblasti-vetrogeneratsii-namereny-vestas-rus-i-rosneft (in Russian).

Pilyasov, A. N., & Polyachenko, A. E. (2021). District effect: Problematic issues of theory and practice. *Bulletin of Moscow University. Series 5: Geography* (4), 25–37 (in Russian).

Pilyasov, A. N., & Putilova, E. S. (2020a). New projects for the development of the Russian Arctic: Space matters! *Arctic and North, 38*(1), 21–43 (in Russian).

Pilyasov, A. N., & Putilova, E. S. (2020b). The modern resource project of the Arctic for the industrial policy of Russia: The pole of growth of the national economy or the "Cathedral in the Desert"? *North and the Market: The Formation of the Economic Order, 3*, 4–17 (in Russian).

Pilyasov, A. N., & Zamyatina, N. Y. (2019). Development of the north 2.0: Challenges of the formation of a new theory. *Arctic and North, 34*(1), 57–76 (in Russian).

Portnews. (2021a). *A new type of bulker needed for coal shipments from the Severnaya Zvezda terminal*. https://portnews.ru/news/316088/ (in Russian).

Portnews. (2021b). *Severnaya Zvezda will have an eco-friendly energy station*. https://portnews.ru/news/320007/ (in Russian).

TASS. (2016). *Yamal LNG contracted 96% of LNG produced on a long-term basis*. https://tass.ru/tek/2747164 (in Russian).

Yenisei Siberia. (2019). *West Taymyr cluster for production of coal concentrate from coking coal*. https://ensib.ru/en/kipproject/west-taymyr-cluster-for-production-of-coal-concentrate-from-coking-coal/ (in Russian).

Zamyatina, N. Y., & Pilyasov, A. N. (2018). A new theory of the development (space) of the Arctic and the North: Multi-scale interdisciplinary synthesis. *Arctic and North, 3*, 5–27 (in Russian).

Zamyatina, N. Y., & Pilyasov, A. N. (2019). Modern theory of development: The search for an integrating platform. *North and the Market: The Formation of the Economic Order, 64*(2), 16–28 (in Russian).

CHAPTER 6

Rethinking Russia's Energy Policy in the Arctic

Maria L. Lagutina and Valery N. Konyshev

6.1 Introduction

In the 2000s, attention to the Arctic began to grow, both in Russia and globally, mainly due to the increased interest in the region's resources. As a result, the resource development was declared one of the long-term strategic objectives of the Arctic policy of the Russian Federation, which was enshrined in strategic documents. According to 'The Basic Principals of The Russian Federation's State Policy in The Arctic Until 2020 And Beyond', published in 2008, the main goal of the state policy of the

M. L. Lagutina (✉)
World Politics Department, St. Petersburg State University, Saint Petersburg, Russia
e-mail: m.lagutina@spbu.ru

V. N. Konyshev
Theory and History of International Relations Department, St. Petersburg State University, Saint Petersburg, Russia
e-mail: konyshev06@mail.ru

V. I. Salygin (ed.), *Energy of the Russian Arctic*,
https://doi.org/10.1007/978-981-19-2817-8_6

Russian Federation in the Arctic in the field of socioeconomic development was to expand the resource base of the Arctic Zone of the Russian Federation (AZRF), which can largely meet the needs of Russia in hydrocarbon resources, aquatic biological resources, and other types of strategic raw materials. Thus, this document defined the Arctic as a resource base of the Russian Federation. Without any doubts, this approach was consistent with Russia's policy of those years, which had an 'energy dimension'. However, this approach was sharply criticized both by representatives of the Arctic regions of the Russian Arctic and by some Russian experts who pointed out that it is unacceptable and even dangerous to limit the development of the Arctic exclusively by resource extraction. According to them (Ryadova, 2020), this approach does not correspond to the unique potential of the region and the national interests of Russia in general, since the AZRF has not only a rich variety of resources but also a unique socio-cultural community. This narrow 'resource' approach could make the AZRF a peripheral territory of Russia with a high level of population outflow to other regions, meaning that, consequently, the security of Russia's northern borders may be threatened (Lagutina, 2021).

In the new edition of the 'The Basic Principles of The Russian Federation's State Policy in The Arctic Until 2035', published in March 2020, the Arctic is also defined as the main resource base for national economic growth, as it was in 'The Basic Principles – 2008', but it is no longer the first priority. Instead, this document primarily focuses on the sustainable socioeconomic development of the region, as well as on improving the living standards of the population of the Russian Arctic. It asks questions of whether it possible to speak about 'rethinking' of Russia's energy policy in the Arctic; of what this is; and of what factors influenced on the change of priorities.

The purpose of this chapter is to identify the main factors and results of the 'rethinking' of Russia's energy policy in the Arctic at the present stage.

6.2 General Characteristics of the Oil and Gas Potential of the Russia's Arctic

The presence of an 'energy' item on the agenda of Russian policy in the Arctic is understandable; the wealth of natural resources is still one of the AZRF's main competitive advantages and the most essential guarantee of its continuing exploration and development. The Russian Arctic's unique resource potential strengthens Russia's geopolitical standing in the world community and facilitates its full-scale integration into global markets.

The Arctic is divided into the shelf and the mainland. The mainland part of the AZRF has been developed for a long time (since the 1960s), while the development of the Arctic seas is difficult. The Arctic Zone of Russia produces 80 percent of its gas and 17 percent of its oil (Kremlin.ru, 2020), but its raw material potential is not fully exploited.

According to the estimates of the Chairman of the State Duma Committee on Ecology, Natural Resources, and Environmental Protection, D.N. Kobylkin, at the end of 2019, 'the oil reserves of the Russian Arctic Zone are estimated at 7.3 billion tons, condensate at 2.7 billion tons, natural gas at about 55 trillion cubic meters', and 'about 41 per cent of the region's oil and gas resources are located on the Arctic shelf' (Kobylkin, 2019: 8). It is important to note that at the same time, a significant portion of the resources is concentrated in the west of the Arctic shelf; the undiscovered reserves of the eastern part are mainly classified as assumed or conditional.

Gas. The State balance of mineral reserves in the Russian Arctic at the beginning of 2021 included 204 fields with technologically recoverable reserves of free gas, which amounted to 76.3 percent of the reserves in the Russian Federation. In 2019, free gas production within Russia's Arctic amounted to 87.4 percent of production in the whole of the Russian Federation. The Yamalo-Nenets Autonomous Okrug occupies the leading place in terms of reserves and production of free gas in the Russian Arctic and the Russian Federation as a whole, on the territory of which 151 fields with technological recoverable reserves of free gas are accounted for—this represents 75.0 percent of the reserves in the Russia's Arctic and 57.2 percent of reserves in the Russian Federation. A significant share of recoverable free gas reserves is concentrated on the shelf of the Arctic seas, where the State balance of reserves includes 21 fields with technological recoverable reserves of free gas, 22.3 percent of the reserves of the Russian Arctic, and 17.0 percent of the reserves of the Russian Federation, including 7 fields on the shelf of the Barents Sea and 14 fields on the shelf of the Kara Sea (VSEGEI, 2021).

Oil. The State balance of mineral reserves in the Russian Arctic at the beginning of 2021 accounted for 282 fields with an oil component with technologically recoverable oil reserves of 20.8 percent of the reserves in the Russian Federation. In 2019, oil production within the Russian Arctic amounted to 13.2 percent of production in the whole of the Russian

Federation. The leading place in terms of oil reserves and production is occupied by the Yamalo-Nenets Autonomous District (65.0 percent of reserves and 50.1 percent of production in the whole AZRF), while the Nenets Autonomous District (17.5 percent and 18.6 percent, respectively) takes the second place. The largest number of unique and large fields is concentrated in these regions: in the Yamalo-Nenets Autonomous Okrug, there are 3 unique (Russkoye, Vostochno-Messoyakhskoye, and Novoportovskoye—all oil and gas condensate [NGK]) and 21 large fields, while in the Nenets Autonomous Okrug there are 9 large fields (VSEGEI, 2021).

As of 2021, offshore oil production is carried out only at the Prirazlomnoye field in the Barents Sea, where 'Gazprom Neft Shelf' has been producing since 2013. The development of such fields as 'Pobeda' in the Kara Sea and 'Tsentralno-Olginskoye' in the Khatanga Bay of the Laptev Sea is not planned in the near future (Kobylkin, 2019: 8). The problem is that Russia is not technologically ready for the production of Arctic shelf oil, nor is it ready for its environmentally safe development.

Among the main oil and gas producing companies with licenses for fields on the Barents Sea shelf are 'Gazprom' and 'Rosneft'. For the fields on the Kara Sea shelf, the main such companies are 'Gazprom', 'Gazprom Dobycha Yamburg', 'RusGazAlliance', 'Novatek-Yurkharovneftegaz', 'Arctic LNG 3', and 'Rosneft' (VSEGEI, 2021).

In recent years, the pipeline transport has been actively developing in the Arctic. In 2017, the 'Bovanenkovo - Ukhta-2' gas pipeline began operation, and the 'Zapolyarye - Purpe - Samotlor' and 'Kuyumba - Taishet' oil pipelines were also commissioned. The main oil loading terminal in the Arctic is Varandey in the Pechora Sea, which is used for transshipment of produced oil onto tankers for subsequent export or transportation to Russian ports. The largest ports capable of receiving and shipping oil and oil products are Murmansk, Vitino, and Arkhangelsk. The port of Sabetta in the east of the Yamal Peninsula allows the export of liquefied natural gas (LNG).

Thus, today, the main production of hydrocarbon resources in the AZRF is onshore. According to experts' forecasts (Amiragyan, 2016), large-scale oil and natural gas production on the Arctic shelf of Russia is not planned until 2030–2035. At the same time, the best conditions for the development of oil and gas resources in the AZRF have been created in the coastal areas of the Barents and Kara Seas, the areas of which are characterized by good geological exploration, developed transport infrastructure, and more comfortable climatic conditions for economic activity.

6.3 Internal Factors of 'Rethinking' of Russia's Energy Policy in the Arctic

In Russian official documents ('Energeticheskaya strategiya Rossii na period do 2035 goda'; 'General'naya skhema razvitiya gazovoy otrasli na period do 2030'; 'Strategy for Developing the Russian Arctic Zone and Ensuring National Security until 2035'), the Arctic is defined as one of the most important regions for increasing the resource base, a strategic reserve for the development of the mineral resource base of Russia. Thus, according to the text of the 'Energy Strategy of the Russian Federation until 2035', it is planned to form an oil and gas mineral resource center, as well as an oil and gas chemical complex in the AZRF, and to expand the infrastructure for transporting energy resources. A new trend is clearly visible in the new editions of strategic documents defining the Arctic policy of modern Russia ('Osnovi gosudarstvennoy politiki Rossiyskoy Federatsii na period do 2035 goda'; 'Strategy for Developing the Russian Arctic Zone and Ensuring National Security until 2035'): 'Russia's view of the Arctic region's development has shifted from the expansive exploitation of just natural resources, with rather rapid infrastructure development, to a more conscious socio-economically centred approach' (Morgunova, 2021). Today, the development of hydrocarbons on land and on the shelf forms the basis of the strategy of socioeconomic development of the Russia's Arctic.

At the same time, Russian and international experts predict relatively insignificant volumes of oil and natural gas production on the shelf of the Russian Arctic in the long term, due to the 'high cost of developing offshore fields and the availability of alternative methods of increasing reserves and producing oil and gas onshore' (Amiragyan, 2016). In 2021, the Government of the Russian Federation approved the general schemes for the development of the gas and oil industries developed by the Ministry of Energy of the Russian Federation until 2035. In particular, according to the draft of the 'General Scheme for the Development of the Oil Industry until 2035', it is predicted that oil production on the shelf, including the shelf of the Arctic seas, will decrease by almost 2 times compared to the level of 2021 (State Duma Committee on Energy, 2021). As a result, in the next 15 years, the Russian Arctic shelf will play the role of an experimental platform for the creation and development of innovative technologies (Podlinova & Volobuyev, 2021). According to experts, 'until 2035, large-scale oil production on the Russian Arctic shelf will not start' (Poussenkova, 2019: 87); rather, in the near future, the offshore oilmen will mainly conduct geological exploration. The main reason for such changes, in addition to sanctions, the fall in oil prices, and the COVID-19 factor, is the lack of necessary technologies.

The development of the Arctic hydrocarbon resources of the Russian Arctic seems to be very promising, but at the same time, it is a complex area of the Russian Arctic policy, which is due to a number of factors that restrain the development and production of natural resources of the Russian Arctic at this stage.

Firstly, one should take into account the *natural and geological factors*: low temperatures, floating icebergs, year-round icing of the water area, seismic activity, the presence of permafrost, shallow bottom deposits of free gas, etc. This leads to poor knowledge of the geological structure of the Arctic Ocean, which seriously complicates the production of hydrocarbons in the Arctic. In addition, difficult natural and climatic conditions narrow the time opportunities for drilling and other offshore operations, and also create additional requirements for technical equipment and personnel.

Secondly, there are the *infrastructural factors*. An important condition for the development of hydrocarbon resources of the Russian Arctic is the provision of a system for transporting the volumes of oil and gas produced. The hydrocarbon transportation system includes the use of a tanker fleet and gas carriers, the traditional use of rail and road transport, and requires serious investments. At the same time, the Russian Arctic is characterized by a low density of transport and production infrastructure (Amiragyan, 2016). One of the features of the AZRF is that the main infrastructure facilities are located in the European part of the Arctic—to the west of the Yamal Peninsula. The main volumes of produced hydrocarbons in the Arctic fall on the European part and are transported to consumers via trunk pipeline systems (Russian Council of the Arctic Federation, n.d.). Along with the pipelines, the Northern Sea Route (NSR) plays an important role in the transportation of hydrocarbons. In 2018, President Putin set an ambitious goal: to reach 80 million tons of traffic by 2024 (National Projects of Russia, 2021). As far as the transportation of hydrocarbons along the NSR is concerned, according to the Ministry of Energy, by 2024 its volumes will grow from the current

18 million tons to 66 million tons (Kommersant, 2019). At the same time, at this stage, 96 percent of all cargoes of the NSR are mineral raw materials (Kommersant, 2019), and by 2024, this picture is unlikely to fundamentally change. As a result, in the absence of diversifying the NSR loading (there are no other non-oil and gas industrial projects in the region), it will become an incredibly costly project (Ampilov, 2020), requiring constant injections from the state budget. The problem of the unprofitability of a number of infrastructure facilities due to their high cost is one of the most pressing in the Russian Arctic. For example, as part of the 'Pechora LNG' project, it was planned to build and put into operation overground pipelines with a length of 395 km, but 'Rosneft' assessed this project in 2018 as unpromising and withdrew from it.

Another important internal problem is related to the paradoxical imbalance and irrationality of the energy supply system of the AZRF itself: on the one hand, the region is a strategic resource base of Russia, and on the other, fuel is the object of the 'northern delivery': 'oil produced in the Arctic is exported, processed 'on the mainland' and returned to the North in the form of motor fuel in a small fraction and at a huge cost. Gas is used extremely little: locally in oil and gas fields and for adjacent settlements' (Tolstukhova, 2021). In this regard, there is another ongoing urgent task of integrating the Arctic energy system into the all-Russian one.

Thirdly, there is a set of *environmental factors*, mainly due to anthropogenic activity in the Russian Arctic. The process of field development is accompanied by a large number of emissions into the atmosphere and discharges into the marine environment, which creates environmental risks for the economic system, the environment, and the safety of the Russian Arctic. Oil and gas activities are one of the main sources of 'greenhouse' gases that determine the process of climate change (Mitko, 2017). The heavy load of pipelines and their deterioration require reconstruction, since the emergency potential on the pipelines increases the possibility of an environmental catastrophe. Oil transportation by tankers is also accompanied by the danger of spills during loading and unloading, and bunkering operations at the terminals. Unfortunately, it should be noted that there is a lack of experience in eliminating the consequences of oil spills in extreme Arctic conditions. In the case of the AZRF, a serious danger is also posed by the risks of stumbling upon radioactive waste buried in the Kara Sea.

Technological factors are determined Russia's lacking of its own technologies for the development of hydrocarbon resources of the Arctic shelf.

Without any doubts, the development of the Arctic is a serious technological challenge for Russia. Unfortunately, at this stage, Russia is highly dependent on the supply of imported equipment and technologies for its oil and gas complex as a whole. At the same time, according to experts, today Russia has a certain independence in downstream technologies (stages after production: refining, petrochemicals, and transportation), having on average 40 percent of its equipment; in the upstream areas (including and before production: directional drilling, production and enhancement of wells, LNG, offshore production), the foreign presence is more noticeable, amounting to 80 percent (Monokin, 2019). Since 2014, Russian companies have switched to a policy of import substitution, but the real effect from it will take some time. FSUE 'Sevmashpredpriyatie' and JSC 'Severstal' are among the main Russian equipment manufacturers, but so far, their capabilities are limited and cannot meet the increased demand for their products. The world leaders in technologies for offshore fields are companies from countries that have imposed sanctions against Russia: the United States (FMC technologies, Cameron, GE Vetco), Norway (Aker solutions ASA, Equinor ASA), and Germany (Siemens AG, MAN SE), while Asian partners (China, South Korea, Singapore, Japan), which are being relied on today, do not have technologies designed to work in the harsh conditions of the Arctic (Kosivecz et al., 2018). Thus, the lack of the necessary equipment significantly complicates and slows down the development of the Arctic shelf of the Russian Arctic.

Nevertheless, in practice, the development of the Arctic shelf is still ongoing, even despite the lack of domestic technologies and equipment due to the sanction's restrictions. In the period 2012–2019, 5 fields were discovered on the Arctic shelf ('Pobeda', 'Tsentralno-Olginskoe', 'Severo-Obskoe', 'Named A.V. Dinkov', and 'Nyarmeyskoe') (Kobylkin, 2019: 9). The state takes various measures to provide economic incentives for the development of new offshore hydrocarbon deposits: for example, in March 2020, Putin signed a federal law providing for amendments to the Tax Code of the Russian Federation, which will stimulate the implementation of new projects for the extraction of hydrocarbons in the Russian Arctic (Federal'nyy zakon ot 18.03.2020 № 65-FZ). Besides that, in August 2020, a package of federal laws on state support for entrepreneurial activity in the Arctic came into force (Federal'nyy zakon ot 13 iyulya 2020 g. № 193-FZ), and a system of preferences was launched.

6.4 External Factors of the 'Rethinking' of Russia's Energy Policy in the Arctic

Among the external geopolitical and geoeconomic factors, the greatest limiting influence on the prospects for the development of hydrocarbon resources of the Arctic shelf was exerted by the fall in international oil prices (according to experts, oil production in the Russian Arctic will be profitable only at a price above $80–90 per barrel [Podlinova & Volobuyev, 2021]) and the introduction of Western sanctions against the Russian oil and gas industry in 2014.

The imposed 'sanctions' of the United States and the EU's 'restrictive measures' against civil and military officials and heads of large companies affected the oil and gas sector, including in the AZRF. Two bans were formed for citizens and companies: a financial ban (a ban on the conclusion of transactions and services in securities and short-term debts with Russian counterparties, a restriction on investments and foreign financial institutions in Russian projects for the production of crude oil and the laying of export pipelines) and a ban on the supply or (re)export of goods, technologies, and services for the exploration, development, and production of deep-sea, arctic offshore, and shale projects with the potential to produce oil (Congressional Research Service, 2021; European Council, 2021). The list of banned companies includes 'Rosneft', 'Lukoil', 'Gazprom', 'Novatek', and 'Surgutneftegas', as well as their mining, processing, regional subsidiaries. A number of Western companies were forced to refuse to work in the AZRF, or to adapt to the new working conditions. For example, ExxonMobil, which has long and effectively cooperated with 'Rosneft', was forced to leave 9 out of 10 joint projects after lengthy litigation and fines. Italian Eni S.p.A. also suspended cooperation with 'Rosneft' in the Barents Sea. In its turn, the British-Dutch Royal Dutch Shell withdrew from almost all joint ventures with 'Gazprom' in October 2014, but continued to work on the projects of Sakhalin Energy (operator of the Sakhalin-2 oil and gas project, since 1994) and Salym Petroleum Development N.V. (operator of the Salym group of oil fields, since 1996) in those areas that were not subject to sanctions (e.g., within the framework of the second project, shale production was replaced by the development of mature fields). The restrictions introduced covered both components of the implementation of Arctic projects: upstream (exploration, development/production) and downstream (processing, transportation, sale) (Monokin, 2019). Nevertheless, the work continues, and the impact of US sanctions and EU restrictive measures was mainly reflected in the extension of the project implementation deadlines, in some cases in the change of directions of cooperation with Western companies and the reorientation of Russian companies to suppliers from the East. Cooperation with such Chinese companies as CNPC, Sinopec, the Japanese Sodeco consortium, and the Indian state oil company ONGC has already started. However, the experts note that 'the peculiarity of the sanctions against the Russian oil and gas industry is their delayed and accumulated damage: they threaten not current, but future, development much more seriously' (Monokin, 2019), the greatest impact will be felt in the long term, after about 2030.

The *2020 crisis* also made serious adjustments, when the OPEC and non-OPEC countries entered into an agreement to cut production until early May 2022, as a result of which oil production began to decline in Russia from April 2020. The COVID-19 pandemic also played a negative role.

Today, the energy dimension of Russia's Arctic policy cannot be studied without taking into account the global trends in the development of the world energy system associated with the transition to a new technological structure, due to the 'environmental situation and the transition to energy-saving and resource-saving technologies in the energy sector and other sectors of the economy' (Safonova, 2020). First of all, this refers to the so-called *energy transition*, launched by the Paris Agreement of 2015, which involves the abandonment of such traditional fuels as oil, gas, and coal, as well as the transition to 'green' energy sources. Greenhouse gas reduction policy is becoming the main instrument of economic development in many countries around the world. It is interesting to note that the key role here was played by the climatic factor, which in the 2000s was associated with the prospects of turning the Arctic into an ice-free region and, as a result, a promising region for the production of energy resources. However, by the beginning of the third decade of the twenty-first century, it became obvious that this approach was wrong, and the climatic factor contributed to a radical revision of the energy development paradigm, and thus qualitative changes in the global economy.

Within the framework of the global 'energy transition', regional initiatives have begun to appear, which in the future will also have global implications. Thus, at the end of 2019, the EU announced the so-called Green Deal—a new EU strategy aimed at implementing 'ambitious environmental, climate and energy policies with the ultimate goal of stimulating sustainable development', and a 'roadmap for the transition to a clean circular economy' (WECOOP, 2020). The EU's Green Deal, focused on achieving carbon neutrality by 2050, will also have serious geopolitical implications for the EU's key trade, economic, and energy partners, including Russia, one of the largest exporters of fossil fuels: 'decarbonizing of the EU and phasing out the use of fossil fuels can lead to a significant reduction in Russia's revenues in the energy market' (Sukhoverkov, 2020). Undoubtedly, this will also affect the Russian Arctic. Citing the severity of the consequences of climate change for the Arctic, in the fall of 2021, the European Commission proposed introducing a global moratorium on the production of hydrocarbon fuels in the Arctic (European Commission, 2021), which caused a violent reaction from a number of Arctic countries, including European ones (e.g., Norwegian Prime Minister Jonas Gahr Store criticized this approach [Milne, 2021]). In early December 2021, it became known that, under pressure from environmental activists, the French authorities prohibited banks from investing in the Russian megaproject 'Arctic LNG-2' in Yamal, despite the fact that the French company Total Energies participates in the LNG plant project (PRO-Arctic, 2021). The main motive on the part of activists was fears of serious damage to the Arctic ecosystem.

As a result, countries such as Russia are faced with the task of integrating into the next technological order, for which it is necessary to actively develop technologies for oil extraction from hard-to-reach and depleted fields, as well as alternative energy.

6.5 Are There Any Prospects for Alternative Energy in the Russian Arctic?

Taking into account the factors and tendencies noted above, the development agenda of the energy complex of the AZRF brings up the issue of the development of alternative energy sources, or 'green' energy. For a long time, this topic was viewed as irrelevant in relation to the Russian Arctic, rich in hydrocarbon resources; however, today, the situation is beginning to change. First of all, this is due to the low cost of alternative energy in comparison with traditional types of energy: 'now the cost of electricity in the regions of the Far North is in the region of ₽22–237 kilowatt per hour, which is 5–55 times higher than in the rest of Russia, the cost of electricity production at solar power plants averages ₽9.5 kilowatt per hour, wind turbines ₽6.3 kilowatt per hour, new combined-cycle plants ₽3.6 kilowatt per hour' (Mayorov, 2020). The Arctic began to be considered as a promising region from the point of view of the introduction of alternative energy, despite the fact that, according to some experts, 'at present, the state of renewable energy in the Arctic region is fairly minimal' (Boone, 2021).

Experts believe that the most promising alternative energy source in the Arctic is wind, and the simplest and cheapest-to-use renewable energy installations are solar plants: they barely require any special maintenance. At this stage, a number of wind power plants already operate in the Russian Arctic: the experimental wind power station in Labytnangi and the Polaris project in the Yamalo-Nenets Autonomous Okrug, the Anadyr wind power station at the Observatsii cape of the Anadyr region in the Chukotka Autonomous Okrug, and the wind park and the experimental wind power station 'Bykov Mys' in the village of Bykov Mys (Mayorov, 2020). Solar power plants successfully operate in the Yamalo-Nenets Autonomous Okrug. Since the AZRF is also home to big Russian rivers (Lena, Ob, etc.), with many inhabited localities by the riverside, it would make sense to use hydropower as well—to build modern hydropower stations that do not require the construction of levees or dams. Today, in the Murmansk region, there are 17 hydroelectric power plants; in Yakutia, there are 2; and in the Arkhangelsk region, there is 1 (Mayorov, 2020). Biofuels and geothermal energy are considered promising for the Russian Arctic.

However, given the region's climate and the costliness of these technologies, mixed-technology solutions (combining wind, solar energy, and diesel generators) appear as the most promising renewable energy solution for the AZRF. The subject of alternative energy, meanwhile, is nothing new for the AZRF, although presently these alternatives are approached non-systemically and promoted locally: 'There are no federal projects in the Arctic as yet, but there are regional projects promoting renewable energy systems. Say, solar panels are installed in Yakutia and wind-and-diesel generators are used in the Nenetsky Okrug – such projects do exist, but still only locally' (TASS, 2017). Small power plants using combinations of diesel engines and alternative energy units are already up and running in small villages belonging to the Pomors people in the Murmansk Oblast, supplying electricity to locals 24/7. For instance, the first hybrid power installation was set in place in Pyalitsa village at the end of 2014; then, hybrid wind-solar-diesel utilities were mounted in the villages Chavanga and Tetrino; and in the spring of 2016, a similar utility was launched in the village of Chapoma.

Undoubtedly, it is still unlikely that, in the near future, alternative energy will be able to completely replace traditional energy, but the combination of traditional and renewable energy sources in the Arctic seems to be quite real and in demand.

6.6 Conclusions

To conclude, the huge reserves of hydrocarbons, which are concentrated on the territory of the Arctic Zone, became the reason for the so-called Arctic optimism (Poussenkova, 2019: 87) regarding the energy prospects of the region in the early 2000s, which was reflected in the priority of Russia's national interests in the Arctic in 'Basic Principles – 2008', and then in 'Basic Principles – 2020'. However, after the events of 2014 and the fall in global oil prices, and acknowledging the fact that many of the explored fields are not mined due to the inaccessibility and high cost of its development, adjustments were made: a period of 'Arctic realism' began in Russia, when oilmen started to reduce their plans for oil production on the Arctic shelf.

Moreover, as noted earlier, the definition of the Arctic exclusively as a 'resource base' raises some criticism, since the understanding of such a rich and strategically important region is very narrow and leads to an underestimation of its complex importance. In addition, it creates the preconditions for the development of an exclusively energy sector of the Arctic economy, giving it a raw material character. In recent years, such a narrow raw material approach to the Arctic, fortunately, is undergoing changes, under the influence of the internal and external factors discussed in this chapter; more and more have begun to talk about the need for the sustainable development of the region, which, in particular, implies taking into account the need to ensure environmental security of all energy projects in the Arctic, the development of 'green' energy.

At the same time, the presence of a number of difficulties leads to an understanding of the need for an integrated approach to the development of the Russian Arctic, which implies the development of transport and social infrastructure in the region, an increase in the standard of living of the population, and the provision of environmental safety in the Arctic. An effective state policy for the development of the Russian

Arctic is required, which creates favorable conditions for the formation of a competitive innovative economy, taking into account the internal and external factors of the development of the region. Obviously, the AZRF needs to change the development model of the energy complex from an extensive-operational model to an innovative one, built on the principles of sustainable development.

Acknowledgements This research was supported by a Jean Monnet Centre of Excellence grant (Erasmus+ Programme of the European Union) 'The EU's Green Deal and Russia: Raising Awareness, Looking for Cooperation Possibilities' (project number 101047820); by RFBR and FMSH 'Building an optimal model of a human security system in the Arctic Zone of the Russian Federation' (project number 20-514-22001).

References

Amiragyan, A. (2016). Oil and gas in the Russian Arctic. *Fuel and Energy Complex of Russia, 9*, 34–39. http://ac.gov.ru/files/content/10406/neft-i-gaz-v-rossijskoj-arktike-pdf.pdf (in Russian).

Ampilov, Y. (2020). A new look at the concept of energy development of the Russian Arctic. *Gas Business, 3*, 2–9.

Boone, C. (2021, July 5). *Renewable energy in the Russian Arctic and beyond: Can a region dominated by oil and gas contribute to Russia's Paris agreement targets?* RIAC. https://russiancouncil.ru/en/blogs/carter-boone/renewable-energy-in-the-russian-arctic-and-beyond-can-a-region-dominat/

Congressional Research Service. (2021). *U.S. sanctions on Russia: An overview*. Updated September 1, 2021. https://sgp.fas.org/crs/row/IF10779.pdf

European Commission. (2021). *Questions and answers on the EU's Arctic strategy*. https://ec.europa.eu/commission/presscorner/detail/en/qanda_21_5164

European Council. (2021). *EU sanctions against Russia over Ukraine*. https://www.consilium.europa.eu/en/infographics/eu-sanctions-against-russia-over-ukraine/

Federal'nyy zakon ot 18.03.2020 № 65-FZ "O vnesenii izmeneniy v chast' vtoruyu Nalogovogo kodeksa Rossiyskoy Federatsii" [Federal Law of 18.03.2020 No. 65-FZ 'On amendments to part two of the tax code of the Russian federation']. http://publication.pravo.gov.ru/document/view/0001202003180041 (in Russian).

Federal'nyy zakon ot 13 iyulya 2020 N 193-FZ "O gosudarstvennoy podderzhke predprinimatel'skoy deyatel'nosti v Arkticheskoy zone Rossiyskoy Federatsii" [Federal Law of July 13, 2020 N 193-FZ "On state support

for entrepreneurial activity in the Arctic Zone of the Russian Federation"]. https://rg.ru/2020/07/16/193-fz-ob-arkticheskoy-zone-dok.html (in Russian).
Kobylkin, D. N. (2019). Resursy arkticheskogo shelfa – eto nash strategicheskiy zapas [The resources of the Arctic shelf are our strategic reserve]. *Energy Policy, 3*(141), 6–10 (in Russian).
Kommersant. (2019). *Difficult northern sea route*. https://www.kommersant.ru/doc/3937692 (in Russian).
Kosivecz, A. A., Maxmutov, T. A., & Polosina, D. S. (2018). US and EU sanctions against Russia: Implications for Arctic projects. *Arkticheskie Vedomosti, 2*(25), 12–24 (in Russian).
Kremlin.ru. (2020). *Strategy for developing the Russian Arctic Zone and ensuring national security until 2035, dated October 26, 2020*. http://en.kremlin.ru/acts/news/64274
Lagutina, M. (2021). Russia's Arctic policies: Concepts, domestic and international priorities. *Polar Journal, 11*(1), 1–18. https://doi.org/10.1080/2154896x.2021.1932262
Mayorov, M. (2020, June 30). Alternative energy in the Arctic. *Go Arctic*. https://goarctic.ru/work/alternativnaya-energetika-v-arktike/
Milne, R. (2021, October 25). Drilling shutdown would mean end of green transition, Norway PM warns. *The Financial Times*. https://www.ft.com/content/cbf5bc66-660f-46da-a4f7-4f26eb752298
Mitko, A. V. (2017). *Arctic hydrocarbon routes*. Neftegaz.Ru. https://magazine.neftegaz.ru/articles/transportirovka/549195-marshruty-arkticheskikh-uglevododorodov/ (in Russian).
Monokin, E. N. (2019). Impact of US & EU economic sanctions on Arctic projects of Russian Federation. *Scientific Notes of V. I. Vernadsky Crimean Federal University. Juridical Science, 4*(72), 451–464. https://cyberleninka.ru/article/n/vliyanie-ekonomicheskih-sanktsiy-ssha-i-es-na-arkticheskie-proekty-rossiyskoy-federatsii (in Russian).
Morgunova, M. (2021). Oil and natural gas exploitation in the Russian Arctic. *ISPI*. https://www.ispionline.it/en/pubblicazione/oil-and-natural-gas-exploitation-russian-arctic-29947
National Projects of Russia. (2021). *In 2024, it is planned to transport 80 million tons of Cargo along the northern sea route*. https://национальныепроекты.рф/news/v-2024-godu-po-sevmorputi-planiruetsya-perevezti-80-mln-tonn-gruzov (in Russian).
Podlinova, A., & Volobuyev, A. (2021, April 11). Russia will postpone shelf development for at least 15 years. *Vedomosti*. https://www.vedomosti.ru/business/articles/2021/04/11/865485-rossiya-otlozhit (in Russian).

Poussenkova, N. N. (2019). Arctic offshore oil in Russia: Optimism, pessimism, realism. *Outlines of Global Transformations: Politics, Economics, Law, 12*(5), 86–108.

PRO-Arctic. (2021, December 2). *French banks were banned from financing Arctic LNG-2.* https://pro-arctic.ru/02/12/2021/news/45118 (in Russian).

Russian Council of the Arctic Federation. (n.d.). *Hydrocarbon transport systems in the Arctic.* https://russiancouncil.ru/arctic-petroleum-transport

Ryadova, M. (2020). Russian Arctic: Science is more important than resources. *Russia in Global Politics.* https://globalaffairs.ru/articles/arktika-nauka-vazhnee/ (in Russian).

Safonova, T. Y. (2020). Prospects for Russian oil and gas production in the Arctic: From collapse to development. *Kreativnaya ekonomika, 14*(10), 2569–2590. https://creativeconomy.ru/lib/111085 (in Russian).

State Duma Committee on Energy. (2021). *General'naya skhema razvitiya neftyanoy otrasli na period do 2035 goda [General scheme for the development of the oil industry for the period up to 2035], dated April 14, 2021.* http://komitet2-13.km.duma.gov.ru/rabota/rekomendacii-po-itogam-meropriyatij/item/26205621/ (in Russian).

Sukhoverkov, K. (2020, December 11). Is EU Green Deal a threat for Russia? *RIAC.* https://russiancouncil.ru/analytics-and-comments/columns/europeanpolicy/zelyenaya-sdelka-es-ugroza-dlya-rossii/ (in Russian).

TASS. (2017, July 2). *Experts: Widespread use of alternative energy is possible in the Arctic.* http://tass.ru/v-strane/4382568 (in Russian).

Tolstukhova, N. (2021, June 22). Development of the Russian shelf. *Morskiye Vesti Rossii.* http://www.morvesti.ru/analitika/1691/90216/ (in Russian).

VSEGEI (Russian Geological Research Institute). (2021). *Spravka o sostoyanii i perspektivakh ispol'zovaniya mineral'no-syr'yevoy bazy Arkticheskoy zony RF na 15.03.2021. [Certificate on the state and prospects of using the mineral resource base of the Arctic zone of the Russian Federation as of 03/15/2021].* The certificate was prepared by FSBI "VSEGEI" as part of the implementation of the State Assignment of the Federal Agency for Subsoil Use dated 01/14/2021 No. 049-00016-21-00. https://www.rosnedra.gov.ru/data/fast/files/202104/45bb8bcc7b844220954744c0149a86f4.pdf (in Russian).

WECOOP (European Union—Central Asia Water, Environment and Climate Change Cooperation). (2020). *EU Green Deal.* https://wecoop.eu/glossary/green-deal/

CHAPTER 7

China's Participation in Energy Cooperation with Russia in the Arctic

Yana V. Leksyutina

7.1 Introduction

The past decade has seen a historical transition in Russia's gas sector, from its production base in Western Siberia to the new giant gas cluster at the Yamal and Gydan Peninsulas in the Russian Arctic. Moreover, a new oil cluster is expected to emerge in Russian Arctic's Krasnoyarsk region when Rosneft's "Vostok oil" mega-project is realized. Russia's aspirations to maintain its status as a major energy exporter in the world, to become a global supplier of natural gas, and particularly to become a major player in the LNG world market are closely connected with developing Russian Arctic hydrocarbon reserves. Nevertheless, the development of Arctic natural resources faces multiple technical difficulties (conditioned by low temperatures, severe weather conditions, etc.) and logistical challenges (the lack of transport infrastructure, special ice-class vessels being

Y. V. Leksyutina (✉)
School of International Relations, St. Petersburg State University, Saint Petersburg, Russia
e-mail: lexyana@yandex.ru

V. I. Salygin (ed.), *Energy of the Russian Arctic*,
https://doi.org/10.1007/978-981-19-2817-8_7

required, etc.), the operational costs of resource exploration, extraction, and transportation are relatively high, the required investment is massive, and the environmental risks are to be seriously taken into account while implementing energy projects in the Arctic. To develop its Arctic resource base, Russia needs to cooperate with foreign partners which are capable of providing capital and technical assistance, special equipment, and relevant technologies and services. In general, foreign investors are very cautious in participating in natural resource development in the Arctic and are not rushing to extract Arctic resources, instead preferring to take a wait-and-see approach. Transatlantic sanctions imposed on the Russian energy and financial sectors since 2014 have caused additional difficulties and challenges to the energy development of the Russian Arctic.

Against such a background, Beijing's enthusiasm in joining the development of the Russian Arctic has received a warm welcome from Moscow. In contrast to various and loud speculations of many Western scholars and politicians for many years, arguing that the Arctic is to become the area of strong Russia–China competition and contradictions, so far, the Arctic has become one of the most successful dimensions of Russia–China commercial cooperation. Currently, China is the largest investor and a creditor in Russian Arctic energy projects, with a focus on LNG projects. China is the world's largest energy consumer, it has a steadily rising demand for hydrocarbons, its dependence on energy imports is increasing, and the Russian Arctic has become a new choice for China's energy resource imports. Moreover, for Beijing, participation in Russian Arctic energy projects is not exclusively driven by commercial profits or the push to ensure reliable energy supplies, but is also driven by political considerations—a desire to promote China's status as an important Arctic stakeholder, and to solidify comprehensive strategic relations with Russia by adding a new pillar of cooperation.

7.2 Russia's Arctic Hydrocarbon Development: Major Foreign Partners and China's Role

Russia's hydrocarbon exploration activities in the Arctic have always been closely connected with its foreign partners' involvement in energy projects. Foreign companies have been acting either as providers of financial capital, suppliers of advanced technologies and equipment, or sources of unique competences or services.

Until roughly 2013, in developing energy partnerships in the Arctic, Moscow has prioritized Western companies, which were seen as valuable technological partners with exclusive competences. In 2005, Russia's Lukoil and American ConocoPhillips established the Naryanmarneftegaz joint venture to develop resources in the northwest Russian Arctic. In 2008, Russia's state-run company Gazprom signed an agreement with Norwegian Statoil and French Total to develop the Shtokman field in the Barents Sea. In 2011, Russia's Rosneft signed a strategic cooperation agreement with American ExxonMobil to develop the potentially vast but largely untapped oil and gas reserves on the Russian Arctic shelf (in the Kara Sea). In 2012, Rosneft signed agreements with Statoil and Italian ENI establishing joint ventures for the development of the Arctic shelf (in the Barents and the Okhotsk seas, and in the Barents Sea correspondingly) (Sørensen & Klimenko, 2017: 13). In 2013, Gazprom and British-Dutch Shell inked a memorandum setting the principles of cooperation within hydrocarbon exploration and development in Russia's Arctic shelf and a section of deep-water shelf abroad.

Some non-Western countries, like China and Japan, also pursued active Arctic policies and announced their interest to take part in Russia's Arctic energy development. However, prior to 2013, Russia has resisted non-Arctic countries' engagement in Arctic issues and cooperation in the Arctic development. Russia viewed non-Arctic countries as revisionist per se, willing to expand their role in the Arctic governance, to challenge sovereign interests of the Arctic nations (including Russia), or to transfer the NSR under international regulation (Leksyutina, 2021: 140). A breakthrough in Moscow's approach toward Asian non-Arctic states happened in 2013, when five of them eventually acquired an observer status in the Arctic Council under the conditions of their recognition of the Arctic states' sovereignty and jurisdiction within their Arctic areas and the existing legal frameworks. Co-opting non-Arctic states—rather than excluding them from Arctic affairs—has been viewed in Moscow as much better option. The need for foreign technologies, financing, and investment acted as a powerful incentive for Moscow to seek cooperative rather than confrontational type of relations with non-Arctic actors (Konyshev et al., 2017: 6). After Asian non-Arctic countries acquired Arctic Council observer status, it became crystal clear that Moscow indeed planned to diversify its partnerships in the Arctic development to also include Asian countries.

Crucially, the years of 2012–2014 played a pivotal role in breaking Western companies' dominance in Russia's Arctic hydrocarbon projects. After 2012–2014, when Moscow lifted unofficial restrictions on attracting Chinese investment in Russia, Chinese investors joined a number of large projects in the Russian Arctic. Previously, Chinese investors encountered various barriers, particularly in oil, gas, other natural resources, as well as construction and infrastructure projects across the Russian Far East, Siberia, and the Arctic. In 2013, Rosneft and Chinese CNPC have agreed on joint oil exploration projects on the Arctic shelf in the Barents and Pechora Sea. In September 2013, Novatek (a private Russian gas company) signed a contract with CNPC for the sale of a 20 percent stake in Yamal LNG (an extra-large LNG project in the Arctic).

Incentives for Moscow to have greater cooperation with non-Western countries in the Arctic hydrocarbon projects were further reinforced in 2014, when, reacting to the 2014 Ukrainian crises and responding to the Russian government's actions and policies toward Ukraine, the United States and the European Union (EU) introduced multi-vector restrictive measures against Russia, including sectoral sanctions targeting the energy and the financial sectors in Russia. The restrictions on Russia's access to Western capital markets and advanced technological equipment for oil and gas exploration and production have been imposed.

Transatlantic financial sanctions affected major publicly owned Russian companies in the energy sector, such as Gazprom Neft, Rosneft, and Transneft. In particular, the ban on the circulation of new bonds and shares of listed companies was tightened: the maximum possible duration of these instruments was reduced from 90 to 30 days in case of the EU measures (according to the 2014 EU sanctions) and to 60 days in case of the US restrictive measures (according to the 2017 US sanctions). The US restricted access to the US capital market for such energy companies as Gazprom Neft, Rosneft, Transneft, and Novatek, as well as their subsidiaries; meanwhile, the EU similarly restricted access to the EU capital market for Gazprom Neft, Rosneft, and Transneft (Novatek was not listed in the EU sanction list). Financial restriction was not applied to Lukoil and Surgutneftegas, but these two energy companies also faced the problem of worsening terms for attracting financial capital. Moreover, as an indirect result of the sanctions, leading rating agencies, like Moody's, Standard & Poor's, and Fitch, lowered the credit ratings of Russian energy companies, which has led to higher costs of borrowing in the Asian markets as well (Mitrova et al., 2018: 7). In sum, the financial sanctions limited foreign borrowing opportunities for Russian energy companies.

On top of that, the imposition of sanctions resulted in the US' ban on the export of equipment for specialized hydrocarbon exploration to Russia, and the EU's stopping of providing Russia with services and technologies for deep-water oil exploration and production, Arctic oil exploration and production, and the development of shale oil deposits. Restrictions also apply on providing technical help, intermediary services, or financing in respect of these technologies. The US sanctions have been tougher as these touched both the oil and gas industry, while the EU measures targeted only Russian oil industry (not gas industry, which would be harmful to Europe, which is a large importer of Russian gas) while the EU shielded the Russian gas companies, including Gazprom and Novatek, from the sanctions.

The imposed transatlantic sanctions created problems and challenges for the Russian oil and gas industry, which has traditionally greatly relied on Western technologies. Such dependency on Western technologies is especially noteworthy for offshore drilling rigs (including self-elevating floating drilling rigs), pump-and-compressor and downhole equipment, power generation equipment, and software (Makhmutov et al., 2018). For instance, at the Prirazlomnoye field—operated by Gazprom Neft Shelf, and currently the only active production site on the Russian Arctic shelf—many Western companies acted as partners (contractors, suppliers, service providers) for developing this field. Such service companies, like Baker Hughes, Halliburton, and Schlumberger, were employed in the drilling at the Prirazlomnoye field. The drilling rig was supplied and installed by American Indrill International. Foreign contractors performed 50 percent of the total workload in servicing the systems operating at the Prirazlomnaya platform, and in the provision of services accompanying drilling (Mitrova et al., 2018: 16). This is unsurprising, given the overall situation: the majority of drilling rigs used on Russian oilfields are imported, spare parts for drilling rigs are produced by Western companies, and Western companies also provide technical maintenance services.

Following the introduction of transatlantic sanctions and due to the drop in international oil prices, many previously planned joint Arctic energy projects between Russian and Western companies were put on hold or frozen. The sanctions have most negatively affected the Arctic offshore projects intended by Rosneft. Rosneft had to stop geological exploration of the Arctic shelf, declaring that it would not be able to resume drilling in the Kara Sea because its partners, ExxonMobil and Norwegian North Atlantic Drilling,[1] were obliged to comply with the transatlantic sanctions (Sørensen & Klimenko, 2017: 21–22). The joint project between Rosneft and ENI in the Barents Sea was also put on hold. Novatek, which appeared on the US sanction list and thus was deprived of an opportunity to borrow capital in the US financial markets, faced temporary problems finding money for its Yamal LNG project.

Under the new conditions of sanction regime and limited access to Western technologies and services, Moscow had to start either developing Russian indigenous technologies and equipment, which required considerable time, finance, and efforts, or looking after alternative technological partners, capable of bringing advanced technologies to Russian Arctic projects, with those partners mostly coming from Asia (Trenin, 2020).

As a result, Russian energy companies started searching for new technological and financial partners for Arctic energy development. Russian high-level officials continuously have been voicing their warm welcome for Asian countries to join Russian Arctic development. In 2015, the Minister of Natural Resources, Sergey Donskoy, voiced his support for a Chinese partnership in Arctic energy resource development if it would bring technology and investment. In November 2015, the head of Rosneft, Igor Sechin, offered Japanese companies the opportunity to become a technological partner as a supplier of ships and equipment for the development of the resources of the Russian shelf, including the Arctic. In July 2016, the Minister of Foreign Affairs, Sergey Lavrov, said that Chinese investors were welcome in the Arctic. In September 2016, Deputy Prime Minister of Russia, Dmitry Rogozin, invited India to participate in Russia's Arctic offshore projects (Sørensen & Klimenko, 2017: 21–22). In 2016 Special Representative for International Cooperation in the Arctic and Antarctic, Artur Chilingarov, underlined Russia's interest in further partnership with Singapore in the Arctic, and specifically in Singaporeans' drilling platform technologies for the Arctic shelf.

[1] According to the agreement with Rosneft, the North Atlantic Drilling was to be engaged in drilling on the shelf (with providing of six offshore drilling rigs for the long-term lease), but due to the sanctions it eventually deferred its participation in the Kara Sea project.

In September 2019, Indian Prime Minister Narendra Modi was invited as the chief guest to Russia's Eastern Economic Forum at Vladivostok, where Vladimir Putin welcomed India to participate in Far Eastern LNG and Arctic LNG 2 projects.

One of the side effects of the transatlantic sanctions against Russian energy sector has been the broader opening of the Russian market of oilfield services to Asian companies, specifically to Chinese ones. For example, Rosneft opted to use a Chinese platform of China Offshore company in the stage of exploratory drilling at the Sea of Okhotsk shelf. In 2016, Rosneft signed a memorandum with China's Shandong Kerui Petroleum Equipment on strategic cooperation in the area of oilfield services.

Nevertheless, no matter what kind of difficulties and challenges Russia faces resulting from transatlantic sanctions against the Russian energy sector, Moscow has very ambitious plans related to the development of Arctic hydrocarbon reserves and specifically natural gas resources. The exploration of vast resources of natural gas available in the Arctic falls perfectly within Russia's aspirations to become a global supplier of natural gas and specifically a major player in the LNG world market, along with Qatar, Australia, and the United States. In June 2020, the Russian government adopted the "Energy Strategy to 2035" policy document, which prioritizes the diversification of energy exports and aims to significantly increase the production and export of liquefied natural gas (LNG), particularly those produced in the Arctic region (Energy Strategy, 2020: 49). "Energy Strategy to 2035" also sets a goal to establish oil and gas resource bases in Russia's Arctic, providing (among other things) the development of the continental shelf within the Barents, Kara, Pechora, and Okhotsk seas (Energy Strategy, 2020: 50).

A targeted export market for Russia's energy resources is the Asia–Pacific region. According to the "Energy Strategy to 2035," the share of the Asia–Pacific region in the total volume of exports of Russian energy resources should increase from the 2018 baseline of 27 percent to no less than 40 percent by 2024, and 50 percent by 2035 (Energy Strategy, 2020: 75). Specifically for natural gas, the Strategy aims to increase annual LNG exports from the baseline of 26.9 billion cubic meters in 2018 to 59.8–65.1 in 2024 and to 108–189 billion cubic meters in 2035 (Energy Strategy, 2020: 89).

LNG is viewed as key to maintaining Russia's competitive position in the global gas market and providing opportunities for economic development domestically. Since Russia opened its first LNG facility in 2009 (the Sakhalin 2 LNG project), the production and export of LNG have become one of the priority areas in Russia's energy policy . As of the beginning of 2021, there are three operational LNG facilities in Russia—Sakhalin 2 LNG, Yamal LNG, and Vysotsk LNG. There are a number of facilities under construction, like Novatek's Arctic LNG 2 and Gazprom's Portovaya LNG, as well as those that are planned: Ob LNG (in the Arctic region), Baltic LNG, Far East LNG, Arctic LNG 1, and Arctic LNG 3 (in the Arctic region), Vladivostok LNG, Yakutsk LNG and Kara LNG (in the Arctic region), etc. In March 2021, the Russian government approved a Long-term Liquefied Natural Gas Production Development Program, which envisages ambitious plans for the development of LNG production. In 2031–2035, LNG production is expected to reach 80–140 million tons per year (Russian Government, 2021: 41). These plans would make Russia one of the top three or four LNG producers in the world, and enhance Russia's position as the world's largest natural gas exporter. Importantly, a lion's share of LNG produced in Russia is expected to be produced in the Russian Arctic. Most of Russia's LNG projects implemented both at present and in the future are situated in the Arctic. According to the Russian government plans, in 2035, LNG production in the Russian Arctic is to reach 91 million tons (Strategy for Developing the Russian Arctic Zone, 2020: 41), compared to 80–140 million tons planned for 2035 for all LNG production in Russia.

As a matter of fact, Russia's ambitious plans for LNG development derive from the expected growth in world demand for LNG. The past several years have seen mounting global attention to environmental protection issues and a push for the use of green and renewable energy sources in an attempt to mitigate climate change. However, the widespread utilization of renewable energy sources will take time and will not substitute traditional fossil fuel at least in the next 15–20 years (Vasiliev, 2021). Natural gas—being the cleanest fossil fuel—seems to be the last among fossil fuels to retreat from the scene (Vasiliev, 2021). The environmental protection policies are expected to promote (among other things) a rise in the world's demand for natural gas, by far the cleanest fossil fuel. The growth in natural gas consumption is expected in many regions of the world with the highest growth presumably in the Asia–Pacific, including China. LNG has high chances to become Russia's major export product "made in the Arctic."

7.3 China's Interests in the Russian Arctic's Energy Reserves

China is the world's largest energy consumer with 24 percent of the global energy consumption in 2020 (Enerdata, 2021) and energy producer in the world. According to various estimates, China's energy demand is expected to steadily increase, with much of this demand covered by energy imports.

In its energy consumption, China has traditionally been heavily reliant on coal, which has led to the grave air pollution situation in China. Driven by environmental considerations, in recent years China has undertaken various measures to diversify its energy consumption by replacing coal and oil use with cleaner burning fuels. However, in 2019, coal still accounted for around 58 percent of China's total energy consumption, petroleum and other liquids for 20 percent, hydroelectric sources for 8 percent, natural gas for 8 percent, nuclear power for 2 percent, and other renewable sources for nearly 5 percent (U.S. Energy Information Administration, 2020: 2).

For the past several years, and increasingly against the global trend of transition to green development, China has been paying a lot of attention to building a clean, low-carbon, and efficient domestic energy system. At the 75th United Nations General Assembly in September 2020, Xi Jinping pledged that China will adopt more vigorous policies and measures so to have carbon dioxide emissions peak before 2030 and to achieve carbon neutrality before 2060. To build a clean and low-carbon domestic energy system, Beijing has set a goal to accelerate the utilization of natural gas and increase the proportion of natural gas in primary energy consumption with coal being gradually replaced by low-carbon and clean energy natural gas (Sun Kai & Ma Yanhong, 2018: 1).

The acceleration of coal-to-gas switching—in the fight against air pollution and tackling climate change—has led to the rapid rise in natural gas demand in China. In the past decade, natural gas demand in China has quadrupled. At present, China is the world's sixth-largest natural gas producer, the third-largest consumer (behind the United States and Russia), and the second-largest importer. China is one of the fastest-growing natural gas markets in the world. According to the US Energy Information Administration's estimates, in 2050, China will consume nearly three times as much natural gas as it did in 2018 (US International Trade Administration, 2021).

Despite steadily rising, China's natural gas production has not been able to cope with the huge increase in demand, and natural gas imports are on the rise. To fill the widening gap between its domestic natural gas production and demand, China has increased natural gas pipeline imports and LNG imports.

China imports natural gas via three pipeline systems: the Turkmenistan-China (known as the Central Asia-China Gas Pipeline), the Myanmar-China, and the Russia–China (called the Power of Siberia) natural gas pipeline systems. In 2020, about 20 percent of China's natural gas imports came from Turkmenistan. The Turkmenistan-China pipeline can transport 55 billion cubic meters annually. The Myanmar-China gas pipeline can deliver 12 billion cubic meters per year, but only 4.1 billion cubic meters of gas were transported in 2019 (US Office of the Secretary of Defense, 2021: 138). The Power of Siberia pipeline, which became operational in December 2019, is projected to reach an annual capacity of 38 billion cubic meters annually by 2025. The Power of Siberia pipeline is the first natural gas pipeline to deliver Russia's natural gas exports to China. For the past several years, China has been considering the construction of either Line D of the Central Asia-China Gas Pipeline[2] (Line D—the fourth pipeline of the Central Asia-China Pipeline system—is to run from Turkmenistan through Kyrgyzstan and Tajikistan to China), or a natural gas pipeline that could deliver natural gas to China from Russia via Mongolia.

In 2019, all in all, China's pipeline imports accounted for 38 percent of the total, and LNG imports accounted for 62 percent (U.S. Energy Information Administration, 2020: 8). China started to import LNG in 2006, and now China is the second-largest LNG importer globally (after Japan), with 68.91 metric tons or a 19.3 percent world market share (GIIGNL, 2021: 6). In 2020, the largest sources for Chinese LNG imports were Australia (29.67 metric tons), Qatar (10.72), Malaysia (6.38), Indonesia (5.37), Russia (4.92), and the United States (3.21) (GIIGNL, 2021: 32–33).

For at least the past two or three decades, as the Chinese economy demonstrated rapid growth and China's energy demand boomed, China's energy policies have been driven by an imperative to ensure China's

[2] The Central Asia—China Gas Pipeline currently combines three lines—A, B, and C—which originate in Turkmenistan and run parallel to each other through Uzbekistan and Kazakhstan to China.

energy security and the diversification of energy import sources and import transportation options. During the past few years, China has diversified its LNG supply sources, but Australia, with which Beijing has had strong contradictions since 2020, takes a lead. Russia's share in China's LNG imports is still comparatively low—only 7 percent. Against the background of China's rapidly growing natural gas demand and increasing dependence of energy consumption on foreign countries, Russia's LNG Arctic projects seem very attractive to China. Due to close strategic relations between Russia and China, labeled by some experts as quasi-alliance relations, Russia is seen in Beijing as a reliable supplier of LNG. Russia's Arctic LNG projects' natural gas exports to China can ease the pressure of domestic market demand growth and ensure the sustainable growth of Chinese economy.

7.4 Russia–China Cooperation in LNG Projects in the Arctic

Most of the Russia's LNG projects carried out both at present and in the future, including the largest, are localized in Russia's Arctic region. Currently, in the Russian Arctic, there is the fully operational Yamal LNG plant—a project of Russian privately owned natural gas company Novatek—which is comprised of three LNG trains of 5.5 million tons per year each and one LNG train of 900 thousand tons per year, utilizing the hydrocarbon resources of the South Tambey field on the Yamal Peninsula. The Yamal LNG plant launched LNG production in 2017, and now its total capacity reaches at least 17.4 million tons natural gas per year. From the project's commencement through to March 2021, Yamal LNG has shipped 50 million tons of LNG and dispatched 685 cargos of LNG to Europe and Asia. In 2020, the Yamal LNG plant produced 18.8 million tons of LNG, accounting for more than 5 percent of the global LNG market (Yamal LNG, 2021). It is worth mentioning that the Yamal LNG project is among the global plants with the lowest carbon footprint, with 0.26 tons of carbon dioxide per ton of LNG produced, against an industry average of about 0.4 tons (Shiryaevskaya & Khrennikova, 2021).

Another LNG production-related project of Novatek is the Arctic LNG 2 plant, with projected total LNG capacity of the three trains at 19.8 million tons per year and about 1.6 million tons per year of stable gas condensate. The resource base for Arctic LNG 2 is the Utrenneye field, which is located in the Gydan Peninsula approximately 70 km across the Ob Bay from Yamal LNG plant. The Arctic LNG 2 project reached its final investment decision in September 2019 and is currently under construction. Novatek expects the first train to be completed in 2023, the second and third trains in 2024 and 2025, respectively.

Both the Yamal LNG and the Arctic LNG 2 projects are located in Russia's Yamal-Nenets Autonomous Region, which is the world's largest natural gas producing area and accounts for approximately 80 percent of Russia's natural gas production and about 15 percent of the world's gas production. In both projects, China is the largest foreign investor. In 2013, the Chinese CNPC acquired a 20 percent stake in the Yamal LNG project, and in December 2015, the Chinese Silk Road Fund bought another 9.9 percent in the Yamal LNG project (French Total has 20 percent, and Russian Novatek keeps 50.1 percent stake). In the Arctic LNG 2 project, China acquired a 20 percent stake (the CNPC and the CNOOC secured a 10 percent share each), while Novatek keeps 60 percent, Total holds 10 percent stake, and Japan's consortium of Mitsui & Co and Japan Oil, Gas and Metals National Corporation has a 10 percent stake.

The Yamal LNG project has been a real breakthrough for China into the Russian energy sector and the development of Arctic reserves. Although China was on the list of the Yamal LNG project's investors back in 2013, it managed to acquire an even larger stake at the project than it expected initially. In 2014, due to transatlantic sanctions, Novatek faced severe financial problems and had to look for new investors and creditors. Novatek got subject to US sectoral sanctions that restricted its access to long-term debt financing, and its large shareholder, Gennady Timchenko, was included in the Specially Designated Nationals list. On top of that, Japanese companies, which (following the imposition of sanctions against Russia in 2014) were invited by Novatek to join the Yamal LNG project, opted not to get connected with the Yamal LNG project as shareholders. As a result, China acquired an additional 9.9 percent share in the Yamal LNG project and became a large creditor to the project. In December 2015, the Chinese Silk Road Fund signed an agreement with Novatek for the provision of a 15-year loan of around EUR 730 million for financing the Yamal LNG project. Additionally, in 2016, the Export–Import Bank of China and the China Development Bank opened two 15-year credit lines for the total amount of EUR 9.3 billion and RMB 9.8 billion to finance the project. According to some estimates, China has totally provided up to 60 percent of the capital for the Yamal LNG project (Sørensen & Klimenko, 2017: 32–33).

In the Yamal LNG project, China has also provided vast technical support and was awarded multiple billion-dollar contracts for providing various equipment and services. During the building of the plant, Chinese enterprises have undertaken 85 percent of the construction of all modules, produced one of the four polar drilling rigs for the project, exported products more than $10 billion, and received shipping contracts under the project worth at $8.5 billion (Yamal LNG Project Completed and Put into Operation, 2018). In particular, China's CNOOC has received a contract from Novatek worth about $1.6 billion to build core modules for the liquefication process on the Yamal LNG project. For China, it was the pioneering experience to export LNG core process modules to the outside world.

Another Chinese company, Shanghai-based Wison Offshore and Marine Ltd (WOM), received a large contract for the fabrication of industrial modules for the Arctic LNG 2 project. According to Zhang Shouyue, vice-president of WOM's Zhoushan manufacturing base, "the delivery of modules for the liquefied natural gas project fills a gap in the building of China's marine engineering prowess, indicating the nation has joined the world's top club in the construction of high-end marine engineering products and large and complex module construction" (Offshore Energy, 2019). As a matter of fact, China's participation in Russia's LNG projects in the Arctic brought new opportunities for China in production capacity and equipment manufacturing: China accumulated experience and the developed technologies needed for Arctic resource development, and also, for the construction and management of large-scale LNG projects, China has explored a new model for Chinese companies to carry out oil and gas cooperation overseas.

Overall, through joining Novatek's LNG projects, China expects to obtain returns in terms of investment income, promote its technological progress, expand energy and Arctic cooperation with Russia, and also to ensure LNG supply to China. Being an investor in the Yamal LNG, in 2014, Chinese CNPC signed a binding contract with Novatek, subscribing for 3 million tons of LNG per year in the Yamal LNG project for a period of 20 years. In February 2021, Novatek and Shenergy Group signed a 15-year LNG sales and purchase agreement for the cumulative supply of more than 3 million tons of LNG produced from the Arctic LNG 2 project to China. Additionally, in January 2022, Novatek and

China's Zhejiang Energy Gas Group signed a long-term LNG sale and purchase agreement, which stipulates the supply of up to 1 million tons of LNG per year from the Arctic LNG 2 project for a term of 15 years to China.

7.5 Conclusions

Since 2013, China has been an active participant in Russia's hydrocarbon development in the Arctic. Given China's ever-rising energy consumption and demand in hydrocarbon imports, the green development drive alongside Beijing's corresponding energy policy of coal-to-gas switching due to environmental considerations, and a focus on increasing the share of natural gas in its energy consumption, China's primary interests in hydrocarbon development in the Russian Arctic falls within LNG projects, while Russia's Arctic oil projects have not inspired Chinese investors' enthusiasm so far. Currently, China is the largest foreign investor in Russia's largest Arctic LNG projects—the Yamal LNG and Arctic LNG 2—with plans to participate in subsequent Russian Arctic LNG projects to be executed by Novatek.

LNG cooperation in the Arctic has become an area where Russia's and China's expectations and interests fully converged and complemented each other. China has turned out to be a large financial contributor to Russia's LNG projects (both as investor and creditor, which is increasingly important against transatlantic financial and sectoral sanctions against Russia), a provider of technical support, and a contractor for various equipment and services for LNG projects, which mean a lot to Russia's aspirations to maintain its status as a major energy exporter in the world. In its turn, China—through participation in Russia's LNG projects—receives commercial benefits as a large shareholder of relevant projects, ensures reliable supply of competitively priced LNG, and diversifies its LNG supply sources away from traditional (albeit not wholly reliable) partners (during the past several years China has had strong contradictions with Australia and the United States). By taking part in Russia's Arctic LNG projects as a contractor and technological partner, China develops its Arctic competences and the technologies needed for the Arctic resource development and for the construction and management of large-scale LNG projects. Participation in Russia's Arctic LNG projects is also seen by Beijing as solidifying Chinese positions in Arctic affairs, as well as its status of a large and important stakeholder in the Arctic.

The Yamal LNG and the Arctic LNG 2 projects are the first large-scale overseas projects implemented within China's flagship Belt and Road Initiative. It is the meeting point of the conjugation of China's Belt and Road Initiative and Russia's Eurasian Economic Union strategy. Basically, these two LNG projects have become the largest platforms for Russia–China economic and trade cooperation, a significant pillar solidifying Russia–China strategic comprehensive partnership.

References

CNPC. (2018). Yamal LNG project completed and put into operation. *CNPC Special Report*. https://www.cnpc.com.cn/en/csr2014enhms/201807/771c0549950f4420ad6de3972fe27073/files/88c8e515fb734e6984ef3599ab2255c2.pdf

Enerdata. (2021). *Global energy statistical yearbook 2021*. https://yearbook.enerdata.net/total-energy/world-consumption-statistics.html

GIIGNL. (2021). *GIIGNL Annual Report 2021 Edition*. https://giignl.org/wp-content/uploads/2021/11/GIIGNL_Annual_Report_November2021.pdf

Konyshev, V., Sergunin, A., & Subbotin, S. (2017). Russia's Arctic strategies in the context of the Ukrainian crisis. *The Polar Journal*. https://doi.org/10.1080/2154896X.2017.1335107

Leksyutina, Y. (2021). Russia's cooperation with Asian observers to the Arctic Council. *The Polar Journal, 11*(1), 136–159. https://doi.org/10.1080/2154896X.2021.1892833

Makhmutov, T., Polosina, D., & Kosivets, A. (2018). *Sanctions activity of the US and the EU in regard to Russia: Consequences for Arctic projects*. https://russiancouncil.ru/en/analytics-and-comments/analytics/sanctions-activity-of-the-us-and-the-eu-in-regard-to-russia-consequences-for-arctic-projects/

Ministry of Energy of the Russian Federation. (2020). *Energy strategy of the Russian Federation to 2035*. https://minenergo.gov.ru/node/1026 (in Russian).

Mitrova, T., Grushevenko, E., & Malov, A. (2018). *The future of oil production in Russia: Life under sanctions*. https://energy.skolkovo.ru/downloads/documents/senec/research04-en.pdf

Novatek. (2021). *Yamal LNG reaches fifty million tons milestone*. https://www.Novatek.ru/common/upload/doc/Yamal_LNG_50_mln_Eng.pdf

Offshore Energy. (2019). *Wison offshore & marine starts work on Arctic LNG 2 project*. https://www.offshore-energy.biz/wison-offshore-marine-starts-work-on-arctic-lng-2-project/

Russian Government. (2021). *Long-term liquefied natural gas production development program*. http://static.government.ru/media/files/l6depkb3cdktgzxbb6sdfc2npepad7se.pdf (in Russian).

Shiryaevskaya, A., & Khrennikova, D. (2021). *Russia's biggest LNG producer joins race to make fuel greener*. https://www.bloomberg.com/news/articles/2021-03-11/russia-s-biggest-lng-producer-joins-race-to-make-fuel-greener

Sørensen, C., & Klimenko, E. (2017). *Emerging Chinese–Russian cooperation in the Arctic: Possibilities and constraints* (SIPRI Policy Paper, 46). https://www.sipri.org/sites/default/files/2017-06/emerging-chinese-russian-cooperation-arctic.pdf

Strategy for Developing the Russian Arctic Zone and Ensuring National Security until 2035. (2020). http://publication.pravo.gov.ru/document/view/0001202010260033 (in Russian).

Sun, Kai, Ma Yanhong [孙 凯, 马艳红. "冰上丝绸之路"背景下的中俄北极能源合作–以亚马尔LNG项目为例]. (2018). China–Russia Arctic energy cooperation in the context of the "Polar Silk Road"—A case study of Yamal LNG project. *Journal of Ocean University of China (Social Sciences), 6*, 1–6. https://doi.org/10.16497/j.cnki.1672-335x.2018.06.001 (in Chinese).

Trenin, D. (2020). *Russia and China in the Arctic: Cooperation, competition, and consequences*. https://carnegie.ru/commentary/81407

U.S. Energy Information Administration. (2020). *Country analysis executive summary: China*. https://www.eia.gov/international/content/analysis/countries_long/china/china.pdf

US International Trade Administration. (2021). *China—Country commercial guide*. https://www.trade.gov/country-commercial-guides/china-energy

US Office of the Secretary of Defense. (2021). *Annual report to congress: Military and security developments involving the People's Republic of China*. https://media.defense.gov/2021/nov/03/2002885874/-1/-1/0/2021-cmpr-final.pdf

Vasiliev, A. (2021). *The Russian Arctic/LNG production strategy: Implications of the global decarbonization drive*. https://www.arcticcircle.org/journal/the-russian-arcticlng-production-strategy-implications

CHAPTER 8

Activities of Major Oil and Gas Companies in the Arctic Zone of the Russian Federation

Maria L. Lagutina and Valery N. Konyshev

8.1 Introduction

One of the key objectives of the Russian Arctic policy is the socio-economic development of the region, which is reflected in the Basic Principles of The Russian Federation's State Policy in The Arctic Until 2035, published in 2020. In order to accelerate the economic development of the Arctic Zone of the Russian Federation (AZRF), the document provides for such economic measures as an increase in private investment (while maintaining state control over the implementation of projects) in key energy projects in the AZRF, and the development of the Northern

M. L. Lagutina (✉)
World Politics Department, St. Petersburg State University,
Saint Petersburg, Russia
e-mail: m.lagutina@spbu.ru

V. N. Konyshev
Theory and History of International Relations Department,
St. Petersburg State University, Saint Petersburg, Russia
e-mail: konyshev06@mail.ru

V. I. Salygin (ed.), *Energy of the Russian Arctic*,
https://doi.org/10.1007/978-981-19-2817-8_8

Sea Route (NSR) in order to export oil and gas to foreign markets. In the implementation of the tasks of the socio-economic development in the region, and along with the state, an active role is now played by business, such as large oil and gas producing and processing companies.

Today, several large Russian energy companies already operate in the Russian Arctic. It is important to underline that the onshore part of the AZRF has been developed since the 1960s, whereas the development of the Arctic seas is a relatively new and difficult direction in the activity of Russian oil and gas companies: this is "because of the rich onshore resource base, offshore development was for a long time a marginal activity, despite exploration indicating huge offshore resources" (Moe, 2020: 128). The legislation of the Russian Federation strictly limited the composition of possible licensees for the Arctic shelf (Law 21.02.92 [2019]): since 2008, only Russian legal entities with more than 50 percent state participation and 5 years of experience in shelf development can be licensees. These requirements are fully met by two Russian companies: Gazprom PJSC and Rosneft Oil Company PJSC, which have received exclusive licenses for offshore operations and a mandatory work program (MWP).

It is obvious that the study and development of the resources of the Arctic requires significant investments, which Russian companies alone cannot always cover. Therefore, the creation of joint ventures with both Russian and foreign companies has become one of the most effective models for working in the Russian Arctic. Such partnerships make it possible to combine both financial capabilities and material, human, and logistical resources of the participants. Today, in the Russian Arctic, along with the main Russian subsoil users on the shelf (Gazprom and Rosneft), such Russian private companies as Novatek and Lukoil, as well as a number of other companies, are engaged in the development of mainland fields, as are a number of foreign companies participating in the joint study and development of oil and gas areas in the Arctic, among them the French company Total, China National Petroleum Corporation (CNPC), China National Offshore Petroleum Corporation (CNOOC), etc.

The purpose of this chapter is to determine the main directions and specific features of the activities of large oil and gas producing and processing companies in the AZRF.

8.2 Key Oil and Gas Projects in the AZRF: Great Expectations and Realities

At this stage, the main Russian oil and gas projects in the Arctic are concentrated onshore. At the same time, according to experts, there are not that many existing projects in the Russian Arctic in relation to explored deposits, which is due to difficult mining conditions primarily, as well as the lack of domestic mining technologies (Tolstukhova, 2021) and the underdevelopment of infrastructure. As far as the development of the Russian Arctic shelf is concerned, the expectations of both Russian and foreign companies in the early 2010s are not justified here. Still, the companies have many plans and a sufficient number of projects in development, including offshore ones.

Due to the climatic and technological features of the development of hydrocarbons in the Arctic, as well as according to Russian legislation, the largest Russian energy companies—such as Gazprom and Rosneft, whose controlling stakes are owned by the state—can conduct production on the Arctic shelf. As of July 2020, Gazprom held 30 licenses for fields located on the Russian Arctic shelf (Skolkovo, 2020: 9). Today, Gazprom is the only Russian company engaged in hydrocarbon production on the Russian Arctic shelf (in the Prirazlomnoye field). In the case of Rosneft, as of June 2021, the company owned 28 licensed areas: in the Western Arctic—the Barents, Pechora, and Kara Seas (19 areas)—and in the Eastern Arctic—the Laptev Sea, East Siberian and Chukchi Seas (9 areas) (Rosneft, n.d.). Rosneft has offshore projects, but so far, their implementation is considered only as future projects. Nevertheless, these are the two largest monopolists on the Russian Arctic shelf, to some extent implementing their Arctic strategies.

According to experts, from the very beginning, the Gazprom and Rosneft companies adhered to several different strategies in the development of the Arctic shelf: Rosneft took on large, poorly explored areas in areas remote from the supply center; Gazprom chose small areas, with greater exploration and close proximity to existing deposits (Buzovskiy & Konoplyanik, 2016: 18). Practice also shows that Rosneft was more focused on cooperation with foreign partners than Gazprom, with almost all Rosneft projects in the Western Arctic having been associated with Western companies (ExxonMobil, Statoil, ENI), this was largely due to the lack of the company's own technologies for deep-water exploration of the Arctic shelf. The Gazprom company has technologies for working on the coastal and shallow water shelves; the Prirazlomnaya platform is in operation. According to the information on the official website of the Rosneft company, in 2020, the most difficult stage of drilling the northernmost shallow wells in the history of the development of the Russian shelf was successfully completed. However, in the face of low oil prices and the imposed sanctions, both companies undoubtedly came under serious pressure and were forced to make serious adjustments to their plans and strategies.

Gazprom. In 2017, Gazprom approved the Program for the Development of Hydrocarbons on the Russian Shelf until 2040, under which Gazprom planned to start natural gas production at a number of fields located on the Russian continental shelf, including the Arctic: for example, it was planned to start gas production at the Shtokman gas condensate field on the shelf of the Barents Sea in 2028; gas production at the Leningradskoye field on the shelf of the Kara Sea may begin either in 2034 or in 2040. The most interesting update is the development of *the Shtokman field*, which was frozen in 2012. The Shtokman gas condensate field was discovered back in 1988 and is located in the central part of the shelf of the Russian sector of the Barents Sea, about 600 km northeast of Murmansk. In terms of explored reserves of natural gas, the Shtokman field is currently one of the largest in the world. Geological reserves of the field amount to 3.9 trillion cubic meters of gas and about 56 million tons of gas condensate (Shtokman, n.d.). Initially, the United States was considered the main consumer of gas from this field, but their shale revolution has seriously changed the balance of power on the energy map of the world. The fall in oil prices internationally once again put off the prospect of developing Shtokman for an indefinite period. According to experts, the gas produced at the Shtokman field could not cost less than $450–500 per thousand cubic meters, while the market price for it in 2012 was $300–320, and by 2021 it fell to $107 (according to the New York Mercantile Exchange as of April 29, 2021) (Tolstukhova, 2021). However, in February 2021, once again, one of the leaders of Gazprom, Vladimir Vavilov, said that Gazprom plans to return to the development of Shtokman in 2029.

As for oil, its production in the AZRF has been implemented since December 2013 at only one project—*the Prirazlomnoye field* in the Pechora Sea (Prirazlomnoye mestorozhdeniye). The deposit was discovered in 1989. The Prirazlomnoye field is currently the only operating project in Russia for the production of hydrocarbons on the shelf of the Russian Arctic, and the production is carried out by Gazprom Neft. Reserves at the field are estimated at 70 million tons (Gazprom, n.d.-b.). The most important stage in the development of the Prirazlomnoye project was November 2015, when the millionth ton of the first Russian Arctic oil was produced at the field. A new grade of oil produced at Prirazlomnoye has been named ARCO (Arctic Oil). It first entered the global market in April 2014. 3.27 million metric tons of oil produced at the field in 2020 (Gazprom, n.d.-b.).

In 2020, Gazprom Neft's Arctic production accounted for 31 percent of the company's total hydrocarbons (Savosin, 2021). A special place in the Arctic strategy of Gazprom Neft is occupied by Yamal, where the company is implementing such projects as Novy Port and Messoyakha, while Bovanenkovo, Kharasavey, and Urengoy are being considered in the future. According to forecasts, they will add another 10 percent to the company's total production (Alekseyev, 2021).

The company's largest Arctic asset is *the Novoportovskoye field* on the Yamal Peninsula, which produced 14.87 million tons of oil equivalent in 2020 (Alekseyev, 2021). This is the largest developed oil and gas condensate field in Yamal, discovered back in 1964. However, due to the lack of transport infrastructure and complex geology, for a long time full-scale development could not begin at the field. The project entered its active phase in 2010, when Gazprom decided to transfer it to Gazprom Neft, which managed to gain unique experience in the harsh conditions of the Arctic and test new oil production technologies. A new grade of oil is produced here—Novy Port, which is delivered by sea to European countries. For the export of oil, a sea route was developed, and a unique offshore export terminal "Gate of the Arctic" was built. To work on the Novy Port project, the company created its own fleet: seven tankers and two high-tech diesel-electric icebreakers (Gazprom Neft, n.d.). Gazprom Neft plans to create a powerful hydrocarbon production cluster here. The company's portfolio of new assets includes the Yuzhno-Kamennomyssky, Yuzhno-Novoportovsky, and Surovy license areas.

Another major project in the YNAO is *Messoyakha*. This is a joint project between Gazprom Neft and Rosneft and is the northernmost onshore oil field in Russia. Production is currently underway at the Vostochno-Messoyakhskoye field on the Gydan Peninsula of the YaNAO, which produced 6.2 million tons of oil equivalent in 2020 (Yakovlev, 2021). Exploration work continues at Zapadno-Messoyakhskoye. The project uses a number of unique technologies (for example, fish bones).

Rosneft is the largest holder of licenses for areas of the Russian continental shelf. From the very beginning, the company has been focused on cooperation with foreign partners on the shelf. The start of large-scale practical activities of Rosneft in the development of the Arctic shelf fell on the period after 2012, when the company began field work in the Kara and Pechora Seas, including with the participation of foreign partners.

Today, Rosneft is implementing projects both in the Western and Eastern Arctic. In the Western Arctic, according to the official website of Rosneft (Rosneft, n.d.), the company owns 19 licensed blocks: 7 blocks in the Barents Sea (Fedynsky, Central Barents, Perseevsky, Albanovsky, Varneksky, Zapadno-Prinovozemelsky, and Gusinozemelsky); 8 sites in the Pechora Sea (Russky, Yuzhno-Russky, Yuzhno-Prinovozemelsky, Zapadno-Matveevsky, North Pomorsky-1, North Pomorsky-2, Pomorsky, and Medynsko-Varandeysky); and 4 sites in the Kara Sea (Vostochno-Prinovozemelsky-1, 2, 3, and Severo-Kara). Seven fields were discovered on the territory of the sites (Pobeda, named after Marshal Zhukov, named after Marshal Rokossovsky in the Kara Sea, Severo-Gulyaevskoye, Medynskoye-Sea, Varandey-Sea, and Pomorskoye in the Pechora Sea). The total recoverable reserves of these fields as of January 1, 2021 are: oil + condensate/333 million tons, gas/1.9 trillion m3 (Rosneft, n.d.). It is important to note that at this stage, contrary to experts' forecasts, Rosneft is fulfilling its license obligations ahead of schedule in the areas of the shelf of the seas of the Western Arctic.

At the moment, Rosneft holds licenses for 9 blocks on the continental shelf of the seas of the Eastern Arctic: 5 blocks in the Laptev Sea (Ust-Oleneksky, Ust-Lensky, Anisinsko-Novosibirsky, Khatangsky, and Pritaimyrsky); 1 site in the East Siberian Sea (East Siberian-1); 3 sites in the Chukchi Sea (Severo-Vrangelevsky-1, Severo-Vrangelevsky-2, and South Chukchi).

An important event in 2017 was the first ever drilling of the northernmost exploratory well on the Russian shelf of the Laptev Sea in Khatanga, which allowed Rosneft to be considered a pioneer of fields on the shelf of the Eastern Arctic. The first well in the Eastern Arctic was *Tsentralno-Olginskaya-1* with recoverable reserves of more than 80 million tons of oil (Rosneft, n.d.).

Despite the difficulties that arose after 2014 and the subsequent adjustments in the company's strategy in the Russian Arctic, the management of Rosneft does not plan to abandon investments in the Arctic and its development, as it considers the development of the continental shelf of the

Arctic seas as a strategic direction for the development of offshore projects (Rosneft, n.d.). In addition, Rosneft plans to develop shipbuilding and industrial clusters. To develop the Arctic shelf and the NSR, Rosneft is building the Aframax vessels, Arc4 ice-class tankers, Arc7 ice-class LNG tankers, and a nuclear-powered icebreaker.

Rosneft also has a major project to develop a group of oil fields on the Taimyr Peninsula and the north of the Krasnoyarsk Territory—*Vostok Oil*. The Vostok Oil project includes 52 license blocks in the north of the Krasnoyarsk Territory and in the Yamalo-Nenets Autonomous Okrug, where 13 oil and gas fields are located, including the Vankor field (includes the Vankor, Suzunskoye, Tagulskoye, and Lodochnoye fields) and the Payakhsky cluster (Rosneft, 2021). This is the largest project for the development of oil and gas fields, requiring huge financial resources, and therefore it is important to attract large international companies as investors. However, in the context of a new trend—the transition to "carbon neutrality"—the largest Western companies refrained from participating in the project. Still, foreign corporations received shares in this project: Singaporean Trafigura, Swiss-Dutch Vitol, and Singaporean oil transportation company Mercantile & Maritime. The most important advantage of the Vostok Oil project is the possibility of supplying raw materials from the fields in two directions at once—to European and to Asian markets. In 2020, Rosneft began full-scale development of the project's fields. At this stage, the main task is to create the necessary infrastructure. One of the key facilities will be the largest seaport in the Arctic Zone—Bay Sever. The launch of oil production is planned for 2024 (TASS, 2021b). Vostok Oil production growth potential: 25 million tons in 2024, 50 million tons in 2027, 115 million tons in 2030 (Interfax, 2020). In the future, Vostok Oil should provide a significant share of the cargo traffic along the NSR. Nevertheless, the issue of the project's profitability at the current world oil price is being raised in expert circles: the Vostok Oil project is considered marginal at an oil price of $35 per barrel (Kuznetsov, 2020a). However, the company's management intends to implement the project no matter what.

The prospects for the profitability of Arctic oil and gas projects have caused great controversy in expert circles, not only against the backdrop of falling world oil prices and sanctions, but also in the context of the economic crisis caused by COVID-19. As a result, at the beginning of 2020, the State Duma adopted a law on tax incentives for oil and gas projects in the Arctic (Federal Law dated March 18, 2020 No. 65-FZ), which provides for the provision of tax incentives to companies engaged in geological exploration and production of hydrocarbons in certain AZRF areas. The goal is to create more favorable conditions at the national level for the development of companies' Arctic projects. Thus, for the Vankor cluster, Rosneft is granted a deduction for the mineral extraction tax, subject to investment in infrastructure development. Such a special position of two state-owned companies, as well as the lack of real competition in the oil and gas market of the Russian Arctic, only intensifies disputes and discussions in expert circles, contributing to the emergence of new initiatives aimed at expanding the range of companies involved in the economic development of the Russian Arctic.

The trend over recent years is the expansion of the liquefied natural gas (LNG) market. 3 of the 7 new gas liquefaction plants planned in Russia will be located directly in the Arctic (Arctic LNG, Arctic LNG-2, Arctic LNG-3). The largest player here is another Russian company, *Novatek*, which produces gas, oil, and gas condensate on the Yamal Peninsula. Novatek has been implementing its projects in the Arctic for over 20 years. At this stage, the company is more focused on the domestic market, but in the future will also focus on the development of LNG supplies abroad (Yukhtenko, 2021). Today, Novatek's activities in the Russian Arctic are associated with two projects: the first private LNG plant, Yamal LNG, launched in 2018, and the construction of a second plant, Arctic LNG-2.

Yamal LNG is the northernmost LNG plant in the world and is being sold at the resource base of the Yuzhno-Tambeyskoye field on the Yamal Peninsula. The plant project aroused interest among foreign investors even at the initial stage; however, the events of 2014 and the placement of Novatek on the sanctions lists made serious adjustments: the bet on Western investors did not materialize, and Chinese banks and enterprises supported it. However, today, the project's shareholders include the world's largest energy company—Total Energy—as well as Chinese energy companies—CNPC and the Silk Road Fund. The Yamal LNG project received, among other things, serious state support in the form of tax incentives. The implementation of the project was not only a success for Novatek, which proved that it is capable of implementing such large projects, but also gave impetus to Russian–Chinese cooperation in the energy sector in the Arctic.

The resource base of the *Arctic LNG-2* project is the Utrenneye field located on the Gydan Peninsula in the YaNAO. Despite the fact that the

Arctic LNG-2 project is still under construction (its launch is scheduled for 2023), a number of foreign companies have also shown interest in it: in addition to the French Total, Chinese CNPC, and CNOOC, the Japanese companies Mitsui and Jogmec have been added. The success of the Yamal LNG project also contributed to the interest of foreign companies. However, in this case, one can see the desire of the Russian side for a greater diversification of partners. It is known that negotiations were also held with Saudi Aramco and the Korean Kogas, but it was not possible to agree. In early 2021, it became known that Novatek's main gas buyers were the project investors: Total, CNOOC, CNPC, the Mitsui & Co consortium, and JOGMEC will receive approximately 2 million tons of LNG annually (Spivak & Gabuyev, 2021).

Another Novatek project is the launch in 2027 of the gas chemical project *Ob GKH* (Yamalo-Nenets Autonomous Okrug) (TASS, 2021a). This project replaced the idea of building the Ob LNG, which was supposed to be based on the Russian gas liquefaction technology, the Arctic Cascade, but the investment decision was not accepted. The Yamal LNG project was built by Novatek on Western equipment and liquefaction technology from the American APCI; the future Arctic LNG-2 is planned to be implemented on the liquefaction technology of the German Linde; and a project on which it was planned to test Russian gas liquefaction equipment and its own "Arctic Cascade" technology was closed. Thus, the problem of dependence of LNG production on foreign technologies has not yet been resolved. At the same time, at this stage, LNG is the main type of product produced in the Russia's Arctic and the basis of cargo traffic along the NSR.

Lukoil is another Russian private company that operates on the territory of three subjects of the Russian Arctic: in the Yamalo-Nenets Autonomous Okrug, the Pyakyakhinskoye and Nakhodkinskoye fields are being developed, which are the basis of gas production; in the Nenets Autonomous Okrug, oil and natural gas are being produced; in the village of Varandey, Russia's northernmost terminal for the shipment of oil and oil products (Bezopasnost' Arktiki) is located. For a long time, Lukoil sought the right to explore and produce hydrocarbons on the Arctic shelf, but state-owned monopoly companies did not allow private Russian companies that could seriously compete. Instead of the Russian shelf, paradoxically, Lukoil became the first Russian company to receive a qualification permit to operate on the Norwegian continental shelf as an operator.

As noted earlier, until recently, according to the legislation, only Rosneft and Gazprom could develop the Arctic shelf. This caused great controversy and even criticism of the two state monopolists. According to critics, including Yury Trutnev, Deputy Prime Minister for the Development of the Far East and the Arctic, the lack of competition has led to a rather slow development of projects on the shelf: "since 2012, the government has issued 69 licenses for the exploration and production of oil and gas, obliging companies to drill 86 wells for 20 years. However, by the summer of 2019, only five wells had been drilled" (Ampilov, 2020). For the state, the development of the Arctic shelf is a strategically important task, not only from the standpoint of the economy, but also politics—even the image of the state on the world stage. From the positions of the companies themselves, it is more reasonable at this stage to pay attention to onshore projects, where sufficient experience has already been accumulated and the necessary technologies are available, while retaining its offshore licenses for later. Here there is a discrepancy between the interests of companies and the state in the Arctic: "the government, however, wanted Arctic offshore development for political reasons" (Moe, 2020: 128). As a result, at the end of 2019, the Russian Ministry for the Development of the Far East and the Arctic posted a bill proposing the creation of the state corporation "Rosshelf" to work in the Arctic to coordinate offshore projects. The document envisaged a new system for granting subsoil use rights on the Russian continental shelf in the Arctic and the Far East, as well as the introduction of a new system of one-time payments for such use. In addition, it provides for the admission of private investors that meet the established requirements to participate in projects for the development of deposits on the terms of joining consortiums with the participation of a state corporation (Kommersant, 2020). However, the initiative did not find wide support, in particular from the Ministry of Energy of Russia. However, it is interesting that in the summer of 2021, Novatek created a joint venture with Gazprom Neft to develop the Severo-Vrangelevsky license area on the Arctic shelf, in the waters of the East Siberian and Chukchi Seas (Galperin, 2021).

8.3 Features of Cooperation Between Russian and Foreign Oil and Gas Companies in the AZRF

In the early 1990s, there was an acute need for foreign capital in the Russian oil and gas sector. This was due to the fact that Russian companies were just emerging and did not have sufficient funds for independent development of large projects (Volgina, 2009: 155).

The Federal Law "On Subsoil" of 1992 determined the owner of the resources, determined the procedure for their development (later the bill was finalized and supplemented) (O nedrakh, 1992). In accordance with the Law and its Regulations, foreign investors were granted the right to explore and develop mineral resources. In the mid-1990s, the development of a regulatory framework for attracting foreign investment to the country began. In a number of laws of the Russian Federation, such as "On the Continental Shelf of the Russian Federation" of 1995 or "On Production Sharing Agreements" of 1995, it was fixed by the provision that, in exchange for granting the right to foreign states to explore and develop mineral deposits (in particular, hydrocarbons), the state received license and tax payments (Federal Law 30.11.95; Federal Law 30.12.95), but there were no norms defining how this should be implemented in practice.

As for the AZRF, the legislation of the Russian Federation (more than other Arctic countries) limits the composition of possible licensees for the shelf, including the Arctic: as noted earlier, since 2008, they can only be Russian legal entities with more than 50 percent state participation and 5 years of experience in the development of the shelf (O nedrakh, 1992). From 2020, it is planned to expand the access of private investors to the Arctic shelf, including foreign ones, while maintaining state control.

Today, corporations use several tools to ensure their presence in foreign markets: foreign participation in projects is implemented on the basis of a strategic cooperation agreement, under a production sharing agreement, or through a joint venture. In all forms of joint participation, both foreign and Russian companies operate through subsidiaries (hereinafter referred to as "daughters") (Monokin, 2019). It is important to emphasize that foreign companies are more involved in projects related to exploration and production (the so-called upstream), while in all of them the share of a Russian company is from 50 percent.

The largest foreign companies involved in the Russian oil and gas sector in the Arctic are British Petroleum (since 1990), Royal Dutch Shell (since 1911), Exxon Mobil (since the end of 1980), ENI Group (since 1950), Total (since 1992), and Statoil Hydro (since the 1990s) (Volgina, 2009: 155–156), etc. The peak of interest in the resources of the Russian Arctic at the present stage fell on the period 2011–2014, when "strategically important cooperation agreements were signed between Rosneft and ExxonMobil in the Kara Sea in 2011, and with ENI and Equinor in the Barents Sea in 2012. In 2013, Rosneft and ExxonMobil set up joint ventures to conduct Arctic offshore research and exploration" (Brodt, 2021: 239). Western companies were seen as the main source of new technologies needed by Russia for the successful development of the Russia's Arctic. There was optimism on the part of Western companies regarding the prospects for such cooperation, given the then high oil prices. However, after the events of 2014—the imposition of sanctions against Russia, and also as a result of the fall in world oil prices—a number of Western companies were forced to leave or freeze their participation in Russian Arctic oil and gas projects: in 2014, "ExxonMobil was required to abandon the drilling campaign in the Kara Sea before the scheduled end of season. Equally, the cooperation with Statoil and Eni, which had not yet properly started in situ, was more or less frozen" (Moe, 2020: 130). US and EU sanctions prohibit foreign companies from participating in the exploration and production of oil from the Arctic shelf in Russia. However, Western companies have not lost interest in the Russian Arctic and, rather, have taken a wait-and-see attitude. This is evidenced by the statements of the heads of large Western companies. Thus, according to Cederic Cremers head of Shell, "despite the economic uncertainties and pressure on our industry, Russia remains a country of strategic focus for Shell. We welcome the expansion of our cooperation with Gazprom Neft, our reliable and highly professional long-term partner" (Griffin, 2020).

The imposition of sanctions has undoubtedly deprived Russia of a number of valuable technologies, the development of which will take extra time and money. Russia can conduct geological exploration itself, but industrial drilling on the shelf is a special technology that is not yet fully available to Russian industry. Offshore production technologies are being developed in China, but Chinese companies have no experience in the Arctic, which means they have no experience in managing these risks, especially environmental ones.

Nevertheless, despite the sanctions and the difficult geopolitical situation, the presence of Western companies in the Russian Arctic remains, albeit in a very modest amount.

It is interesting to note the position of the UK, which joined the anti-Russian sanctions, but continues to work closely with Russian companies, including on Arctic initiatives: British Petroleum owns 19.75 percent of Rosneft shares, and in 2017 the companies launched a number of joint projects in the Russian Arctic (Todorov & Lyzhin, 2019). As Russian expert Eremina notes as an example, BP currently also owns a 49 percent stake in Yermak Neftegaz Company, formed in 2016, which is conducting onshore exploration in the West Siberian and Yenisey-Khatanga basins (Eremina, 2021). In the summer of 2020, the Russian company Gazprom Neft and the British Shell set up a joint venture to explore for and develop hydrocarbons on the Gydan peninsula in the AZRF (at the Leskinsky and Pukhutsyayakhsky license blocks) (Griffin, 2020). Unable to participate in Arctic offshore projects, Western companies retain their participation in onshore projects in the Russian Arctic. At the same time, Western companies are aware of the high risks of oil and gas projects in the Russian Arctic, giving preference to projects outside the AZRF.

Today, the French company *Total* continues its work in the AZRF. One of the projects in which the French company is involved is the Kharyaga Production Sharing Agreement (PSA). Since 2016, the shares of participants have been distributed as follows: 40 percent for the Zarubezhneft group of companies (operator), 30 percent for Statoil Kharyaga AS, 20 percent for Total Exploration Development Russia, and 10 percent for Nenets Oil Company JSC (Skolkovo, 2020). Total most actively works in partnership with the Russian company Novatek on such projects as: Termokarstovoye field (since 2009) in the YaNAO, Yamal LNG, and Arctic LNG-2. In 2011, a strategic alliance was created between Total and Novatek. Today, Total's share in the capital of the Russian company is 19.4 percent (Skolkovo, 2020). At the same time, Total is a supporter of the goals of the Paris Agreement, which in practice means an increase in the share of gas as the most environmentally friendly fossil energy resource from the current 50–60 percent by 2035 alongside a higher share of low-carbon projects (Skolkovo, 2020).

It is important to note that foreign Western companies and their participation in Russian projects have played a particular role in the development of the oil and gas industry in Russia in general, and in the Arctic in particular. It was they who largely contributed to the introduction of modern production technologies and enterprise management, thereby helping to form a modern model for the development of the Russian Arctic. Russian companies partially adopted the experience of foreign corporations in the field of hydrocarbon production and processing, and in improving the efficiency of management and financial activities; interaction with foreign companies contributed to, among other things, the introduction of the principles of social and environmental responsibility.

Undoubtedly, the presence and withdrawal of foreign companies from the Russian Arctic is determined today by the so-called “geopolitical risks”, which most clearly manifested themselves after 2014. As a result, in recent years, there has been an increase in interest in Russia’s Arctic projects from the Asian countries that are experiencing a rapid upsurge: the most interested being China, Japan, Singapore, etc.

As far as the countries of Asia is concerned, for Russia, cooperation with them in the Arctic is a fairly new direction. Until recently, “non-Arctic Asian countries were seen in Russia as pursuing a revisionist agenda, wishing to expand the role of non-Arctic countries in the management of the Arctic, to revise the legal status of the Arctic” (Leksyutina, 2021: 34), so the attitude toward their interest in the Arctic was wary. However, over time, the need for investment and advanced technologies outweighed fears and concerns; the consequences of the events of 2014 only accelerated the reorientation of Russian companies toward cooperation with Asian countries, including in the Arctic.

The largest eastern shareholder in Russian projects in the Arctic is China, represented mainly by China National Petroleum Corporation (CNPC). In 2013, the Chinese company CNPC acquired a 20 percent stake in the Yamal LNG project, and in December 2015, the Chinese Silk Road Fund bought another 9.9 percent in this project; in the Arctic LNG-2 project, CNPC acquired a 10 percent stake. The Chinese side took an active part in all stages of the implementation of the Yamal LNG project. Moreover, as Russian expert Leksyutina states: “the Russian company Novatek has adopted a typical Chinese financing model with ‘embedded conditions’, where financing is conditional on the requirements to attract Chinese contractors and labor during the implementation of the relevant project, as well as the purchase of Chinese materials and equipment” (Leksyutina, 2021: 37). Obviously, the Russian company took such a step in the face of an acute shortage of its own financial resources for the implementation of the project, as well as the loss of promising Western investors due to the imposed sanctions. Novatek received loans from the Chinese banks Export–Import Bank of China (EIBC) and China Development Bank (CDB) on very favorable terms (Spivak & Gabuyev, 2021), as a result of which the company managed to solve all the financial problems of the Yamal project LNG and successfully implement it.

The second Chinese company present in the Russian Arctic is China National Offshore Oil Corporation (CNOOC, a subsidiary of CNPC), which in 2018 received the first LNG cargo from the Yamal LNG plant, and then signed an agreement with the Russian company Novatek under which CNOOC Ltd. acquires 10 percent in the Arctic LNG-2 project. It is obvious that while this is the beginning of the journey, Russian–Chinese cooperation in the Arctic has great potential, and there is mutual interest in such cooperation: "in the Arctic, Russia needs China to realize many of its goals for infrastructure development and resource extraction. China is eager to access the Arctic's economic potential and enhance its technological prowess by partnering with Russia on key projects" (Stronski & Ng, 2018). China has become Russia's largest foreign partner in LNG production. Of course, there are a number of obstacles on this path, including insufficiently developed infrastructure in the Russian Arctic, mistrust between Russian and Chinese business circles, the Chinese strategy of "embedded conditions", etc. Undoubtedly, one of the fears of the Russian side is complete dependence on the Chinese partner, in connection with which Russian companies are striving for greater diversification of partners in their Arctic projects, attracting companies from other Asian countries: Singapore, Japan, India, etc.

On the Russian side, work continues to create the most comfortable conditions for the participation of foreign companies in Russian Arctic projects. However, a number of initiatives cause dissatisfaction on the part of large monopolists on the Arctic shelf of the Russian Arctic. Thus, in June 2020, the Ministry for the Development of the Far East and the Arctic prepared a draft law on amending certain legislative acts in terms of establishing the specifics of regulating subsoil use relations related to the development of minerals in areas of the continental shelf of the Russian Federation in the Arctic and Pacific Oceans, which establishes a new model of investors' work on the Arctic and Far Eastern shelf of Russia (Neftegaz.ru, 2020). In particular, the project provides for the admission of all interested investors to participate in projects for the development of resources on the continental shelf in the Arctic. The activities of such investors will be carried out on the basis of agreements on a consortium for the joint use of rights to subsoil plots. At the same time, among the mandatory requirements for such an agreement, it assumes that the agent's share in the agreement is at least 25.1 percent (Ministry of the Development of the Russian Far East, 2020). The purpose of such an initiative is to intensify investment and entrepreneurial activity in the region, and, consequently, to involve the Arctic resource base in the economic turnover.

8.4 Social and Environmental Responsibility of Oil and Gas Companies in the AZRF

A feature of the AZRF—as well as the entire Arctic as a whole—is not only its rich resource potential, but also a unique ecosystem. Undoubtedly, the economic activity of oil and gas companies in the Arctic has a direct impact on the development of the region as a whole and the population living there, both positive and negative. Thus, the positive consequences include: an increase in the number of jobs in the Russian Arctic; growth in the incomes of local contractors and suppliers of goods and services for the needs of the ongoing project; regular revenues to the regional budget; improvement of the infrastructure of the region (for example, construction of housing, sports complexes, construction of roads, bridges), etc. However, one should not forget that the main goal of any business is, first of all, to make a profit; therefore, in practice, the interests of companies often do not coincide with the interests of the state and society. Possible negative consequences of the activities of large companies include: an increase in the burden on the environment, including a high risk of accidental pollution (for example, the oil spill from the Kharyaga-Usinsk oil pipeline in 1994 in the Timan-Pechora oil and gas province); differentiation of the population by income, which, in turn, can cause social tension, etc. As Russian expert Filimonova notes: "the process of subsoil use is inherently the cause of a negative impact on the environment. Each stage of field development, from search and exploration to production and transportation of hydrocarbon raw materials, is accompanied by environmental damage – landscape changes, pollution of water bodies, soil and atmospheric air" (Kuznetsov, 2020b). The concepts of social and environmental responsibility are a kind of compromise between the interests of business, government, and society.

Obviously, active economic activity in the Arctic should be focused not only on maximizing the benefits for the state and companies conducting this activity in the region, but also on minimizing environmental damage, maintaining the ecological balance in the region, and also having minimal negative consequences for those living in the region. People, on the other hand, contribute to the improvement of the conditions and quality of life of the population, the development of transport and social infrastructure, etc. Environmental protection and improving the living standards of the population are the priorities of Russia's state policy in the Arctic. "Green", socially responsible business is a global trend, and the requirements for compliance with environmental standards are laid down in the Paris Agreements. Thus, in addition to the tasks of improving the technologies for exploration, production, and supply of extracted energy resources, companies that are active in the region also face tasks of a social and environmental nature. Gradually, it becomes an indispensable condition for doing business in the Arctic.

Today, Russian oil and gas companies already have experience in implementing various environmental and social projects in their arsenal: from preventive measures (remote sensing of the Earth and environmental monitoring) and projects to preserve biological diversity to social guarantees for employees of their companies, local communities, indigenous peoples of the Arctic. Some companies include the UN sustainable development goals in their development strategies and identify a set of priority goals (for example, Gazprom, Gazprom Neft, Novatek) (Morgunova & Kovalenko, 2021: 32). It has become a norm for large Russian companies to publish annual environmental and social reports. For example, Gazprom has been implementing its environmental policy since 1995, as a result of which the company became "the first Russian oil and gas company to declare its voluntary responsibility in the field of environmental protection" (Gazprom, 2020). In 2020, Gazprom Group's environmental protection expenditures amounted to ₽49.12 billion (Gazprom, n.d.-c.). As for Rosneft, in 2010 the company became a member of the UN Global Compact in the field of social responsibility, and in 2018 it published the first report on its contribution to the implementation of the 17 UN Sustainable Development Goals (SDGs). According to 2020 data, Rosneft has become one of the leaders in environmental management in Russia (Rapsi, 2021). As part of the implementation of its environmental responsibility, Rosneft not only provides so-called "green" investments, but also seeks to reduce greenhouse gas emissions. Rosneft also has an active scientific environmental activity in its arsenal: for the past ten years, the company has been conducting large-scale scientific expeditions (for example, the Program for the Study of Key Species of Arctic Ecosystems, with the support of the Ministry of Nature of the Russian Federation, 2019–2022), during which the company has conducted a study of hydrometeorological, ice, engineering-geological and environmental conditions in licensed areas in the Arctic (Rosneft, n.d.). Rosneft has an Arctic Research and Development Center in Moscow, which operates jointly with ExxonMobil.

Today, the priority of the environmental component over the social one can be seen in the policy of Russian oil and gas companies. The disadvantage of corporate principles is that the social policy is not sufficiently developed in comparison with the environmental one. Nevertheless, it cannot be said that nothing is being done in the social sphere. The main areas in the field of social corporate responsibility of Russian companies operating in the Russian Arctic are: financial support for citizens; improvement of working conditions and work with personnel; investment or construction of social facilities, infrastructure facilities; sponsoring mass events in the field of culture and sports; support for education; preservation of the culture and traditions of the indigenous peoples of the Arctic, etc.

For example, many companies operating in the Arctic, such as Novatek, Gazprom, and Lukoil, are engaged in social support for the population of the region, including the indigenous population, and allocate funds to support their employees. For example, in 2020, Novatek provided financial support to the District Association of Indigenous Minorities of the North "Yamal for Descendants!" and its regional branches; Novatek regularly finances events under the People's Program of the Indigenous Minorities of the North in the Yamalo-Nenets Autonomous Okrug, etc. (Novatek, n.d.). The Lukoil company annually allocates funds to representatives of the indigenous peoples of the North under agreements with reindeer herding farms, as a result of which roads are built and maintained, snowmobiles are bought; the project "Red Chum" ("Naryana me") is aimed at providing health services to the representatives of the indigenous peoples of the North. Lukoil actively supports sports, cultural, and educational projects in the region. For example, the company sponsors "the Buran-Day races", the traditional winter national games "Kanin Mebeta" ("Belt of Friendship"), the classical music festival "Days of High Music in Arctic", the Interregional Boxing Tournament "Arctic Lights", a scholarship program for schoolchildren, the modernization of a school in the village of Krasnoye, and the construction of sports grounds with safety surfaces and a skate park in Naryan-Mar, etc. (Yakusheva & Turov, 2020). Gazprom often supports sports and various cultural projects in the regions where it implements its projects. For example, Gazprom annually supports many public projects, assisting in the development of culture, science, education, and the promotion of a healthy lifestyle (for example, "the Fakel Festival", "Russian Seasons in Germany", "German Classics in Russia", "Siberian Odyssey", Competition "Literary torch", "Music without borders. Russia – Italy", etc.) (Gazprom, n.d.-a.). Today, almost all Russian companies are implementing projects in all areas of social policy.

Undoubtedly, Russian oil and gas companies make a certain contribution to the social development of the Russian Arctic, but in this, according to experts, a certain duality is manifested: "on the one hand, the financial burden on the budget decreases in the region, but on the other hand, the region becomes dependent on the activities of companies" (Yakusheva & Turov, 2020). In this context, it becomes important to find a balance between the interests of the state and society alongside the interests of companies, which requires companies to be transparent about the goals and objectives of their social policy.

8.5 Conclusions

Summing up, the development of the AZRF is one of the key priorities of Russia's policy, which determines the leading role of the state in the development of the region. At the same time, today, on the one hand, a new model of the development of the AZRF is being formed, different from the Soviet one, in which the main role was played mainly by the state, focused on the formation of "large regional territorial production complexes that linked into a single system < ... > production from unique large deposits, road construction, energy supply, social infrastructure and others" (Pilyasov & Putilova, 2020: 28); the model of the development of the modern Arctic involves the participation of not only the state, but also large companies focused on a more local, compact development of the Russian Arctic, which is manifested in the creation of platforms and economic "islands" (Pilyasov & Putilova, 2020: 28), the use of new technological solutions, and new management approaches with minimization of material costs. On the other hand, the leading oil and gas companies that have received privileged positions in the Russian Arctic are the state-owned companies Rosneft and Gazprom, an arrangement which is criticized in certain circles. However, over the past few years, there has been an active discussion in Russian political circles about the possibility of providing a wider range of participants with access to the Arctic shelf. The leading position in the Russian LNG market has already been occupied by the private company Novatek, and the policy of attracting foreign partners and investors continues. Under these conditions, it is important to avoid prioritizing political and geopolitical ambitions over economic and social ones.

Undoubtedly, the fall in global oil prices, as well as the sanctions and restrictive measures introduced after 2014, made certain changes in relations between Russian and Western companies, contributed somewhat to the reorientation of Russian companies to suppliers from the East, and made serious adjustments to the plans of the companies themselves. Most of the ongoing Russian oil and gas projects in the Arctic are onshore, while only one project (Prirazlomnoye) is being implemented on the shelf, and the situation is unlikely to change in the near future. Under the pressure of geopolitical and geo-economic factors, it was objectively necessary to revise the priorities and energy policy as a whole: "some ambitions had to be totally scrapped (Arctic offshore gas), some have become highly uncertain and will be scaled down radically (Arctic offshore oil), and some are pending and dependent on special concessions (several Arctic onshore oil and other mineral projects)" (Moe, 2020: 135). The most promising under these conditions are LNG projects.

However, the work continues, and the role of oil and gas companies is not limited to economic activities. Today, Russian oil and gas companies in the AZRF are quite active in the field of social policy (social corporate responsibility), as well as environmental protection in the Arctic (social environmental responsibility). In many ways, this was facilitated by the experience of cooperation with Western partners, who have been implementing such practices in their activities for quite a long time. Moreover, often the extraction of hydrocarbons is associated with the development of other related industries, for example, with the development of shipbuilding and research projects.

Acknowledgements The reported study was funded by RFBR and FMSH, project number 20-514-22001 "Building an optimal model of a human security system in the Arctic Zone of the Russian Federation".

References

Alekseyev, A. (2021, May). Yamal strategy: The most important Arctic projects of Gazprom Neft. *Sibirskaya Neft*, 181. https://www.gazprom-neft.ru/press-center/sibneft-online/archive/2021-may/5410679/ (in Russian).

Ampilov, Y. (2020). The limitless potential of the Arctic shelf. *Neftegaz, 5*(101), 13–17. https://oilcapital.ru/article/general/23-08-2019/arkticheskiy-shelf-kak-pr (in Russian).

Brodt, L. (2021). The development of arctic offshore oil and gas resources in Russia: Energy policy updates and new activities by companies. In L. Heininen, H. Exner-Pirot, & J. Barnes (Eds.), *Arctic yearbook 2021—Defining and mapping the Arctic: Sovereignties, policies, and perceptions* (pp. 234–248). Arctic Portal.

Buzovskiy, V. V., & Konoplyanik, A. A. (2016). Factor analysis of Russian Arctic offshore development strategies by domestic companies. *Gazovaya Promyshlennost, 12*(746), 16–23 (in Russian).

Eremina, N. (2021). UK–Russia relations in the Arctic: From damage limitation to a cooperative agenda? *The Polar Journal, 11*(1), 57–74. https://doi.org/10.1080/2154896x.2021.1932275

Federal Law "On the Continental Shelf of the Russian Federation" of November 30, 1995. No. 187-FZ (with amendments and additions) [Federal'nyy zakon «O kontinental'nom shel'fe RF» ot 30.11.1995. № 187-FZ (s izmeneniyami i dopolneniyami)]. *Garant*. http://base.garant.ru/10108686/ (in Russian).

Federal Law "On Production Sharing Agreements" of December 30, 1995. No. 225-FZ [Federal'nyy zakon «O soglasheniyakh o razdele produktsii» ot 30.12.1995. №225-FZ]. *Consultant-plus*. http://base.consultant.ru/cons/cgi/online.cgi?req=doc;base=law;n=117376;fld=134;dst=4294967295;from=51067-0 (in Russian).

Federal Law No. 65-FZ dated March 18, 2020 "On Amendments to Part Two of the Tax Code of the Russian Federation" [Federal'nyy zakon ot 18.03.2020 № 65-FZ «O vnesenii izmeneniy v chast' vtoruyu Nalogovogo kodeksa Rossiyskoy Federatsii»]. pravo.gov.ru. http://publication.pravo.gov.ru/document/view/0001202003180041 (in Russian).

Galperin, A. (2021, July 8). Gazprom Neft and Novatek created an enterprise for development in the Arctic. *Praime*. https://1prime.ru/oil/20210708/834145920.html (in Russian).

Gazprom. (n.d.-a.). *Help with affairs*. https://www.gazprom.ru/social/ (in Russian).

Gazprom. (n.d.-b.). *Prirazlomnoye field*. http://www.gazprom.ru/about/production/projects/deposits/pnm/ (in Russian).

Gazprom. (n.d.-c.). *Environmental management system*. https://www.gazprom.ru/nature/ems/ (in Russian).

Gazprom. (2020). *Environmental report of PJSC Gazprom for 2020*. https://www.gazprom.ru/f/posts/57/982072/gazprom-environmental-report-2020-ru.pdf (in Russian).

Gazprom Neft. (n.d.). *Novoportovskoye field*. https://www.gazprom-neft.ru/company/major-projects/new-port/ (in Russian).

Griffin, R. (2020, July 21). Gazprom Neft, shell to set up new Arctic hydrocarbons JV. *S&P Global*. https://www.spglobal.com/platts/ru/market-insights/latest-news/oil/072120-gazprom-neft-shell-to-set-up-new-arctic-hydrocarbons-jv

Interfax. (2020, May 12). *Rosneft launched a new Vostok Oil project in Taimyr. Overview*. https://www.interfax.ru/business/708350 (in Russian).

Kommersant. (2020). *LUKOIL does not yet plan to invest in the development of the Arctic shelf*. https://www.kommersant.ru/doc/4227659 (in Russian).

Kuznetsov, S. (2020a, May 15). *Arctic projects: Why and when?* Oilcapital.ru. https://oilcapital.ru/article/general/15-05-2020a/arkticheskie-proekty-zachem-i-kogda (in Russian).

Kuznetsov, S. (2020b, June 9). *Her majesty ecology or Arctic paradoxes*. Oilcapital.ru. https://oilcapital.ru/article/general/22-06-2020b/eyo-velichestvo-ekologiya-ili-arkticheskie-paradoksy (in Russian).

Law of the Russian Federation from 21.02.1992 # 2395-1 "On Subsoil" (edited from 2.8.2019) [Zakon Rossijskoj Federacii ot 21.02.1992 № 2395-1 «O nedrakh» (v red. ot 2.8.2019)]. http://pravo.gov.ru/proxy/ips/?docbody=&nd=102014778&rdk=61 (in Russian).

Leksyutina, Y. V. (2021). Russia's cooperation with northeast Asian countries in the commercial development of the Arctic. *Far Eastern Affairs, 1*, 33–48. https://doi.org/10.31857/S013128120013932-6

Ministry of Development of the Russian Far East. (2020, June 9). *The ministry for the development of the Far East proposes to introduce a number of amendments to the legislation for the work of investors on the shelf of the Arctic and the Far East*. https://minvr.gov.ru/press-center/news/25445/ (in Russian).

Moe, A. (2020). Russia and the development of Arctic energy resources. In K. Spohr, D. S. Hamilton, & J. Moyer (Eds.), *The Arctic and world order* (pp. 119–142). Brookings Institution Press.

Monokin, E. N. (2019). Impact of US & EU economic sanctions on Arctic projects of Russian Federation. *Scientific Notes of V. I. Vernadsky Crimean Federal University. Juridical Science, 4*(72), 451–464. https://cyberleninka.ru/article/n/vliyanie-ekonomicheskih-sanktsiy-ssha-i-es-na-arkticheskie-proekty-rossiyskoy-federatsii (in Russian).

Morgunova, M., & Kovalenko, A. (2021). Energy innovations in the Arctic. *Energy Policy, 4*(158), 30–44. https://energypolicy.ru/wp-content/uploads/2021/04/№4158-2021.pdf (in Russian).

Neftegaz.ru. (2020, June 9). *The ministry for the development of the Russian Far East has proposed issuing licenses to foreign companies for mining on the Arctic shelf for 30 years*. https://neftegaz.ru/news/shelf/553176-minvostok razvitiya-predlozhilo-vydavat-inostrannym-kompaniyam-litsenzii-na-dobychu-na-shelfe-arktiki/ (in Russian).

Novatek. (n.d.). *Interaction with society*. https://www.Novatek.ru/ru/develo pment/responsibility/ (in Russian).

Pilyasov, A. N., & Putilova, Y. S. (2020). New projects for the development of the Russian Arctic: Space matters! *Arctic and North, 38*, 21–43. https://doi.org/10.37482/issn2221-2698.2020.38.21 (in Russian).

Rapsi. (2021). *Ecology is no longer a secret: New trends, real leaders*. http://rap sinews.ru/incident_publication/20210316/306852299.html (in Russian).

Rosneft. (n.d.). *Shelf projects*. https://www.rosneft.ru/business/upstream/off shore/ (in Russian).

Rosneft. (2021, October 1). *Rosneft has begun construction of a river berthing infrastructure for the Vostok Oil project*. https://www.rosneft.ru/press/news/item/207873/ (in Russian).

Savosin, D. (2021, March 18). *Share of Arctic Oil in Gazprom Neft's total production exceeds 30%*. Neftegaz.RU. https://neftegaz.ru/news/dobycha/671658-dolya-arkticheskoy-nefti-v-obshchem-obeme-dobychi-gazprom-nefti-prevysila-30/ (in Russian).

Shtokman. (n.d.). *Shtokman gas condensate field*. http://www.shtokman.ru/pro ject/gasfield/ (in Russian).

Skolkovo. (2020). *The activities of large oil and gas companies in the Arctic Zone of Russia*. https://energy.skolkovo.ru/downloads/documents/senec/research/skolkovo_enec_ru_arctic_vol2.pdf (in Russian).

Spivak, V., & Gabuyev, A. (2021, December 27). *Ice age: Energy cooperation between Russia and China in the Arctic*. https://carnegie.ru/2021/12/27/ru-pub-86088 (in Russian).

Stronski, P., & Ng, N. (2018). *Cooperation and competition: Russia and China in Central Asia, the Russian Far East, and the Arctic*. Carnegie Endowment for International Peace. https://carnegieendowment.org/2018/02/28/cooperation-and-competition-russia-and-china-in-central-asia-russian-far-east-and-arctic-pub-75673

TASS. (2021a, October 15). *Novatek plans to launch the Ob GCC in 2027*. https://tass.ru/ekonomika/12676067 (in Russian).

TASS. (2021b, November 12). *Rosneft will launch Vostok Oil in 2024 with the shipment of up to 30 million tons of oil by sea*. https://tass.ru/ekonomika/129 09141 (in Russian).

Todorov, A. A., & Lyzhin, D. N. (2019). Interests of Great Britain in the Arctic. *Arctic and North, 36*, 84–95. http://www.arcticandnorth.ru/upload/iblock/2d3/05_todorov_lyzhin.pdf (in Russian).

Tolstukhova, N. (2021, June 22). Development of the Russian shelf. *Morskiye Vesti Rossii*. http://www.morvesti.ru/analitika/1691/90216/ (in Russian).
Volgina, N. A. (2009). Foreign TNCs in the oil and gas sector of Russia. *Bulletin of the Tomsk State University, 329*, 155–160.
Yakovlev, V. (2021, March 18). The share of Arctic Oil in Gazprom Neft's total production exceeded 30%. *Gazprom Neft*. https://www.gazprom-neft.ru/press-center/news/dolya_arkticheskoy_nefti_v_obshchem_obeme_dobychi_gazprom_nefti_prevysila_30/ (in Russian).
Yakusheva, U. Y., & Turov, Y. V. (2020, December 21). Interaction between oil and gas companies and interested communities in the Arctic. *GOArctic*. https://goarctic.ru/work/vzaimodeystvie-neftegazovykh-kompaniy-i-zainteresovannykh-soobshchestv-v-arktike/ (in Russian).
Yukhtenko, K. (2021, November 1). *Conquest of the Arctic: What should an investor bet on?* Investing.com. https://ru.investing.com/analysis/article-200287157 (in Russian).

CHAPTER 9

The Exploration of Arctic Energy Resources and the Competitiveness of Big Russian Oil and Gas Companies

Natalia Yu. Konina

9.1 Introduction

The modern global energy market is primarily characterized by the erosion of sectoral boundaries and increased attention to environmental problems, particularly the growing role of alternative energy. According to Western agencies, the planet is approaching the level of oil consumption before the pandemic—100 million barrels per day. Furthermore, civil aviation has not yet started flights due to new pandemic waves in many countries. When civil aviation resumes its work, it should provide about 80 percent more oil consumption to the total volume. Several years ago, the prevailing point of view was that the world had passed peak coal consumption, with coal being the fuel of the nineteenth century.

N. Yu. Konina (✉)
MGIMO University, Moscow, Russia
e-mail: nkonina777@gmail.com

V. I. Salygin (ed.), *Energy of the Russian Arctic*,
https://doi.org/10.1007/978-981-19-2817-8_9

However, coal consumption is actually increasing despite the announced energy transition, growing in Europe and Asia.

There is no doubt that the global economy is moving to new energy sources in the long term, but the terms and time of that transition in each country are unclear. Renewable energy sources (including biofuels) have shown record growth in energy consumption (3.2 EJ) over the past year (Baiman, 2021).

Given the new trends, the position of oil in the energy balance remains controversial, with some researchers taking a contrary view, suggesting the impossibility of large-scale displacement of hydrocarbon resources soon (Aalto, 2016; Konina, 2018b). The International Energy Agency (IEA) predicted that after 2025 oil demand is expected to peak right after 2025 at 97 million barrels per day. The demand for natural gas would also peak after 2025 and then plateau to 3850 billion cubic meters in 2050—just below current levels (International Energy Agency, 2021).

As of Autumn 2021, oil remains the dominant fuel globally. However, its share in global primary energy consumption continues to decline due to restructuring the international energy sector, reaching 33.1 percent in 2019 (Degtyaryova, 2019). New sources of oil and gas can significantly change the balance of power in the global energy market.

The most critical tasks for the largest oil and gas companies are diversifying activities, minimizing economies of scale, decentralizing management, and raising funds to develop new technologies. At the same time, the development of Arctic resources is very much in doubt.

Despite the current uncertainty, Russian oil and gas companies aim to implement their oil and gas development programs in the Arctic, in order to remain competitive, based on resources amounting to about 13 billion tons of oil and 86 trillion cubic meters of gas (Larchenko et al., 2019). This competitive strategy is somewhat risky because the majority of those Russian oil and gas resources are not adequately explored and are located in highly severe climate conditions, including the Arctic shelf.

An extensive and growing literature examines the current and future role of hydrocarbon resources in primary energy consumption. The growing importance of renewable energy sources makes issues related to oil demand and production increasingly uncertain. The development of Arctic oil and gas resources seemed to be very important and aroused great interest before the global financial and economic crisis of 2008–2009. Development of the global energy markets since then has confirmed that importance.

Russian researchers have already published many works on the prospects for developing Arctic oil and gas resources (Chanysheva & Ilinov, 2021; Gielen et al., 2019; Kontorovich et al., 2019; Laverov et al., 2016; Mitrova et al., 2018; Poussenkova, 2019; Razmanova & Steblyanskaya, 2020; Trofimov, 2019).

These publications identified both the positives and the problems and risks of oil and gas development in the Arctic, considering the technical, environmental, geological, and economic circumstances (Carayannis et al., 2020; Ilinova et al., 2020; Kontorovich et al., 2019; Morgunova, 2020; Muslimov, 2021).

Despite several existing works concerning various aspects of oil and gas development on the Arctic shelf, there are no studies devoted to assessing the key factors affecting the prospects for oil and gas production in the Arctic in the context of energy transition and the relationship between the development of Arctic resources and the competitiveness of the leading oil and gas companies in Russia.

9.2 Methodology and Statement of the Research Problem

Global warming and associated climate change have an increasing impact on the global economy amid the technological and digital revolution. The authors base their research on the vision that oil and gas will continue to play a significant role in the global energy market till 2040 as a minimum, with further growth in their importance as extra precious raw materials for chemical and other necessary products.

Research goals and objectives are connected with the investigation of Russian companies' competitiveness through the main phases of the value chain.

The purpose of the study is to empirically and analytically test the hypothesis about the transitional nature of the current stage of the development of oil and gas resources in the Arctic and to determine possible directions for increasing the competitiveness of Russian oil and gas companies.

9.3 Results

The strategic need for managing oil and gas companies at the present stage of energy market development is to maintain their competitiveness, develop competitive strategies, and assess their effectiveness. Thus, significant structural changes in the oil and gas industry affect the basis for applying competitive techniques by companies, which are developed taking into account their competitive advantages as factors of maintaining competitiveness.

It is becoming increasingly difficult for all companies to compete in the global digitally transforming market (Konina, 2021). For this reason, the question of determining the main factors affecting the competitiveness of oil and gas companies is becoming especially relevant.

Competitiveness may be understood as a firm's ability to offer products and services that meet the quality standards of the local and international markets at competitive prices and provide adequate returns on the resources employed or consumed in producing them. The competitiveness of oil and gas companies at the present stage is determined by the influence of various factors, considered from their interrelation and mutual influence. Only with this in mind, it becomes possible to give the most objective assessment of competitiveness to develop an effective strategy to improve it.

Oil and gas companies need to consider the degree of industry globalization, particularly the implementation of economies of scale, the international division of labor, mergers, and acquisitions, outsourcing, and other activities.

In the current conditions of the development of the energy market, the competitive capabilities of companies are not limited to the domestic market only. Oil and gas companies in the aggregate solve specific problems that differ in the final result. Still, in one way or another, in conditions of increased market competition, they strive to form sustainable competitive advantages as factors of their further development.

Ensuring the competitiveness of transnational companies in the current situation of the Fourth Industrial Revolution is associated with both external and internal factors (Konina, 2021).

The competitive advantages of companies in the oil and gas industry are, as a rule, built on several factors reflecting the influence of both external and internal environment:

- market factor (location and expansion of sales markets, the level of world prices and the dynamics of their development, other market features);
- resource-production factor (availability of modern technologies and innovations, conditions of access to natural resources, their quality);
- financial factor (level of financial stability of the company, access to financial resources, capital resources);
- management factor (reputation and image of the company in the market, the level of qualifications of personnel, corporate governance, ESG).

The assessment of the very competitiveness of oil and gas companies, which is the basis of their competitive strategies, is necessarily carried out, taking into account all the competitive advantages that ensure a specific position in the market.

At the end of the twentieth century, American and European oil and gas companies such as ExxonMobil, Royal Dutch Shell, BP, Total, and Chevron occupied the top role in global oil production and refining. At the beginning of the twenty-first century, there was a noticeable change in the balance of power in favor of national companies from emerging economies, such as, for example, Petrobras, PetroChina, or Saudi Aramco. In the face of technological change, oil and gas companies face an acute issue of competitiveness management.

In current market conditions, competitive advantages based only on a raw material base are unreliable due to constant fluctuations in raw material prices. Consequently, proper investment policy, innovations, and technological leadership are becoming essential factors in the growth of competitiveness (Sapir & Karachev, 2020). In the oil and gas industry, complimentary access to resources and new technologies are among the main factors ensuring competitiveness. Technology essence is primarily determined by the specifics of the industry and the competitive position of the company itself.

To maintain the competitiveness of modern production processes, oil and gas companies strive to constantly provide a significant amount of investment in developing new technologies. American and European oil and gas companies are leading in terms of R&D funding. ExxonMobil and Total can be distinguished in this regard. For example, ExxonMobil spent \$1.02 billion on research and development in 2020, \$1.214

billion in 2019, and $1.116 billion in 2018. Research and development expenditure in the oil and gas industry is not as high as in other major industries. For example, the chemical industry spends around three percent of its revenue for R&D purposes. Pharmaceutical companies pay even more—often between 15 and 20 percent. For comparison, ExxonMobil's operating revenues were over 170 times higher than its R&D spending. The amounts of R&D spending by the most significant Russian Company—Gazprom—in 2016–2019 amounted to ₽6.3 billion in 2016, ₽8.2 billion in 2017, ₽9.0 billion in 2018, and ₽12.1 billion in 2019, ranging equivalent from about $100 to $150 million per year (Morgunova, 2020).

However, exploration expenses are often counted separately—usually greater than R&D expenditures in oil refining, petrochemicals, and renewable energy.

In the oil and gas industry, to acquire sustainable competitive advantages, companies, as a rule, use vertical integration due to the unique specifics of production processes, the improvement of which plays a role in the growth of their competitiveness. Depending on the specialization, oil firms adopted specific operating models: vertically integrated, oil-producing and oil refining companies at the turn of the century.

Thus, vertically integrated companies are engaged in activities that involve participation in all stages—from exploration to sales. These companies include the largest oil and gas companies (BP, Total, Royal Dutch Shell, ExxonMobil). In the global energy market, such companies can achieve synergy by pooling their resource capabilities. In this case, they can retain the main competitive advantage—high-performance efficiency in implementing complex and large-scale projects. The second group of companies is engaged only in oil production (for example, Saudi Aramco). The third group is oil refining and marketing petroleum products (Marathon Petroleum, Phillips 66, PBF Energy, Valero Energy).

The global oil market is dominated by several prominent players, including national oil companies and the most significant transnational oil companies, such as ExxonMobil, ChevronTexaco, BP, Shell, Total, ConocoPhillips, which are traditionally called "oil majors." Transnational oil and gas companies have historically competitive advantages: a leading position in the oil and gas industry, opportunities to diversify their investment portfolio by working with a wide range of international investors, the global size of operations, and a universal, efficient business model. The energy transition and profound technological changes significantly influence the competitive positions of transnational oil and gas companies. The development of oil and gas fields and the extraction of natural resources are still the basis of the activities of these oil companies, and this activity continues to bring them their main profit. European companies such as Total, BP, and Shell can be considered the leaders of the energy transition. These companies provide open reporting of their activities in renewable energy sources, form long-term strategies in the field of green energy, and use new technologies for capturing and utilizing CO_2. The business models of American companies (for example, ExxonMobil and Chevron) appear less adapted to the new climate agenda than European competitors, as the main activities of American oil and gas companies are focused on the development of fossil energy sources.

In the past few years, there has been a clear trend toward a change in the business models of international oil and gas companies, which are forced to invest heavily in developing projects in alternative energy and technologies to reduce emissions. For example, Britain's BP has partnered with Chevron, Petrobras, and Suncor to establish a carbon capture and storage facility. An alliance was formed between Shell and solar energy producer British Solar Renewables—in 2018, the companies signed a long-term agreement to buy electricity. Moreover, Shell has funded the Indian Company Husk Power Systems, which provides renewable energy to rural communities and businesses. Total acquired a majority stake in SunPower, an American solar energy company, in 2011; in Lampiris, a Belgian company, in 2016; and in GreenFlex, a French energy company, in 2017. In return, in 2017, BP acquired 43 percent of the solar energy developer Lightsource, marking its return to solar power after the closure of its previous solar business in 2011. European companies are showing greater readiness for the energy transition. They are more actively introducing renewable energy sources into their production and financial portfolios, setting goals directly related to climate requirements, and investing in low-carbon production.

The most advanced company on this path energy transition can be considered the French TNC Total. Until recently, Total's activity in ecological fuels was focused on biofuels, but over the past few years, the company has been able to reach a qualitatively new level, including by increasing the share of strategic investments in renewable energy sources. Total was one of the first among the majors to cut oil production by 30 percent and increase the share of gas to 50 percent by 2030. In addition, the company plans to increase electricity generation by a third exclusively through LNG and renewable energy sources and to increase investments in the renewable energy sector up to $3 billion per year, which will amount to about 20 percent of the company's capital investments. One of the most significant factors in this leadership is that Total is developing an integrated business model from production to trading renewable energy sources while developing a client base in parallel. In addition, Total has sold or converted several of its refineries into zero platforms over the past few years while increasing production and sales of biofuels. The Total Group also plans to invest $1 billion in battery production and charging stations for electric vehicles, with stations exceeding 150,000 in five years.

Competitiveness in the oil and gas industry is ensured by active investing to achieve technological leadership and diversifying activities to meet the growing demands of consumers (Sapir & Karachev, 2020).

The competitiveness of an oil and gas company is usually understood as their stable position in the market, at best achieving some kind of superiority over their competitors. In the case of oil and gas companies, one of the essential tasks is working out a proper competitive strategy based on the most apt forecast of the development of the world energy market, particularly the projected growth in demand, taking into account the growth of renewable energy sources.

The competitiveness of a company, in other words, is its ability to use its core strengths and compensate for weaknesses in critical areas of competition (Konina, 2018a). Exploration competitiveness is essential and is determined by such factors as wells drilled, success rate, well costs, seismic data availability, high-level competencies, and knowledge sharing of the talented professionals.

The most prominent companies in the oil and gas industry are characterized by high competitiveness, which is explained by their resistance to unfavorable factors in the external environment. They also reduce their costs due to economies of scale, and minimize commercial risks due to diversification, including geographic ones (Degtyaryova, 2019; Ilinova et al., 2020).

In the current unstable environment, the management of companies in the oil and gas industry is forced to change business models and strategies. At the present stage, these issues are receiving increased attention.

The competitiveness of a firm is a complex dynamic category that characterizes the totality of the features and the capabilities of the firm that provide it with a victory in the competitive struggle with other agents. Competitiveness is the totality of the properties and characteristics of the firm, taking into account the framework of dynamically changing environmental conditions to surpass other firms and to meet all the requirements imposed by consumers at the level of economic sectors, country, or world market (Konina, 2018a).

The specificity of developing competitive strategies for oil and gas companies is complicated because the oil and gas industry is represented by high-tech methods of oil and gas production and processing and relatively stable distribution channels. Competitive development strategies of companies in the oil and gas industry are usually developed for 10–15 years due to the long-term focus of their investments.

Cost leadership strategies served as the core of the competitive advantage of the majority of oil and gas companies. Cost leadership is based on economies of scale, as well as profitable access to resources. Another approach is the differentiation strategy, when oil and gas companies develop competitive strategies based on the uniqueness of the products they offer. Traditionally, the competitiveness of oil and gas companies was based on vast natural resources, vertical integration of production, and supply, combined with dominance in the global energy market to achieve economies of scale (Mitrova et al., 2018).

Companies can no longer adhere to the old competitive strategies in current market conditions negatively affected by the volatility of global prices on traditional energy resources and stricter requirements for their extraction and processing due to climate change. In the energy transition to a low-carbon future, oil and gas companies focus on increasing productivity and applying new technologies in a long time, building a profitable portfolio of assets to ensure sustainability and economic security (Baiman, 2021).

The major international oil and gas companies have appreciated the prospects to improve their reserves and competitiveness by exploring Arctic energy resources. According to the 2008 US Geological Survey, 90 billion barrels of oil and 1669 trillion cubic meters of natural gas are concentrated beyond the Arctic Circle: 13 and 30 percent of all undiscovered reserves in the world, respectively (Carayannis et al., 2020). These reserves are becoming available thanks to the melting of the Arctic ice caused by global warming. One of the first to enter the Arctic shelf was the Anglo-Dutch Royal Dutch Shell, the greatest oil and gas company in Europe in terms of capitalization. Shell is well-acquainted with the Arctic: it began looking for oil in Alaska back in 1917. By the mid-1980s the company has drilled 15 wells on the Chukchi Sea and the Beaufort

Sea shelf off the coast of North America. But in the early 1990s, Shell abandoned the project: it was unprofitable to produce oil in such difficult conditions at low prices, and the company switched to the Gulf of Mexico.

Global warming and energy transit have a significant impact on the competitive strategies of the biggest oil and gas companies. Price uncertainty and unstable demand result in some promising projects in the Arctic may come to a standstill. An example of this is Gazprom's refusal (at least for the time being) to develop the giant gas Shtokman field, 600 kilometers from the Russian coast. Gazprom's partners—France's Total and Norway's Statoil—agreed that the costs were too high and available technology did not permit them to operate in severe Arctic unfavorable conditions (Ilinova et al., 2020). Ice remains the biggest obstacle for a big part of Arctic oil and gas sea projects. Global warming creates a threat of icebergs; some of these can exceed 1 million tons. Companies should dig deep trenches at the bottom of the sea for pipelines on the shelf to prevent passing icebergs from hitting bottom.

Special requirements are imposed on equipment for eliminating oil leaks, which significantly increases the cost of production and price competitive pressure. Developing oil fields on the Arctic shelf is advisable only if the recoverable reserves in one area are at least 500 million–1 billion barrels. Of the more than 400 hydrocarbon fields discovered in Alaska, only 60 met this requirement. Of these, only 12 were oil. The rest were gas (Larchenko et al., 2019). It is not surprising that many companies prefer to work in warmer regions. In addition, horizontal drilling and hydraulic fracturing technologies can extract natural gas from shale formations without investing billions of dollars.

Russian oil and gas companies are playing a significant role in the Russian economy. In 2020, oil and gas condensate production in Russia was lower, due to a decrease in global demand for hydrocarbons in the context of COVID-19 compared to 2019, and amounted to 512.8 million tons. Production of natural and associated petroleum gas (APG) amounted to 692.9 billion cubic meters (Muslimov, 2021).

The Russian Energy Strategy until 2035, approved by the Government of the Russian Federation, provides for an increase in oil and gas condensate production in Russia under the upper scenario to 560 million tons in 2024 and to 555 million tons by 2035. In the lower plan, production will grow to 556 million tons in 2024 and will decline to 490 million tons in

2035 due to the energy transition in the global energy sector. Gas production may grow to 820 billion cubic meters in 2024 and up to 1000.7 billion cubic meters in 2035 under the upper scenario, and 795.1 billion cubic meters and 859.7 billion cubic meters, respectively, under the lower scenario. Oil reserves in Russia will be sufficient for 58 years, but only 19 of these will be profitable (Merkulov, 2020). Production volumes on the Arctic shelf are predicted to increase by 2035 to the level of 31–35 million tons of oil. LNG exports should reach the level of 100–120 million tons by 2030 (Lazhentsev, 2021).

Most of the oil and gas resources are located in the Russian Arctic. The total recoverable resources of the Russian Arctic include 106 billion tons of oil equivalent, and gas reserves at 69.5 trillion cubic meters. Around 60 large hydrocarbon deposits are located beyond the Arctic Circle, 43 of which are Russian (Katysheva & Tsvetkova, 2017).

In 2017, 1.7 million kilometers squared were issued on the Russian Arctic shelf (24.5 percent of the total area); 122 licenses are in force in 12 seas, with Rosneft having 51 licenses, and Gazprom having 40 licenses (Kontorovich et al., 2019).

To a considerable extent, the future competitiveness of several big Russian state and private oil and gas companies—PJSC "Gazprom," PJSC "Gazprom Neft," PJSC "NK" Rosneft" and PAO "Novatek"—is connected with big Arctic projects.

PJSC Gazprom holds 40 licenses for fields located on the Russian Arctic shelf. The share of Arctic fields of PJSC Gazprom Neft has reached 27 percent in production. The company is implementing three major projects under the Arctic Time program: the Novoportovskoye, Prirazlomnoye, and Vostochno-Messoyakhskoye fields (Morgunova, 2020).

Rosneft owns 51 licensed areas on the Arctic shelf with unlimited resources of 34.6 billion tons of oil equivalent: 19 projects in the Western Arctic (Barents, Pechora, and Kara seas), and 9 projects in the Eastern Arctic (Laptev Sea, East Siberian, and the Chukchi Sea).

Rosneft, the biggest Russian oil company, has considerably increased its competitiveness by speedily developing considerable Arctic resources. The activities of Rosneft mainly concentrate on upstream operations, including production and distribution of oil and gas; downstream operations are not efficient enough. The company's assets are located in 23 countries, but most of them, of course, are located in Russia. The share of Rosneft in global crude oil production is 6 percent, more than the share of Canada.

The company receives about 92 percent of its revenue from the sale of oil and petroleum products. The share of gas and petrochemicals in revenue is still not significant. The key sales regions are Russia, Asia (mainly China), and Europe. The company's relatively weak refining is fully offset by the highly efficient exploration and production segment (being, as stated above, more than the share of Canada). Rosneft's production cost (excluding the tax burden) is the lowest among the international oil and gas majors. Rosneft is one of the most ardent supporters of investments in exploration and production. Back in the first half of 2020, when oil prices went negative, Rosneft continued to invest in oil exploration and production, noting that a sudden energy transition and a lack of investment in green energy could lead to a shortage of supply in the market. Rosneft estimated the resources of Vostok Oil at 6 billion tons of oil. Rosneft plans to achieve 100 million tons of oil production at the project by 2030 (Chanysheva & Ilinov, 2021; Morgunova, 2020).

Rosneft plans to start drilling and equipping pilot sites with wells on the Vostok Oil project in 2022, in particular drilling and preparing pilot sites with high-tech wells of various designs, including multilateral and horizontal wells with multistage hydraulic fracturing. The critical tasks are finding the best development system, assessing water cut levels and flow rate decline, and water injection schemes to correctly determine the effectiveness of the water flooding system.

Vostok Oil is a Rosneft project that combines several fields: the Vankor cluster, the Payakhskoye field, and the West Irkinsky area. Reserves include: Suzunskoye field (51 million tons of oil, 37 billion cubic meters of gas); Tagulskoye field (296 million tons of oil and condensate, 251 billion cubic meters of gas); Lodochnoe field (89 million tons of oil and condensate), 81 billion cubic meters of gas); and the Payakhskoye field in the north of the Krasnoyarsk Territory (1.2 billion tons of oil). Production volume is 26 million tons/year of oil.

PAO Novatek is one of Russia's largest independent natural gas producers, ranking third in proven gas reserves among public companies globally. At the end of 2019, the company's share accounted for more than 10 percent of the total Russian natural gas production, including several production and export of liquefied natural gas (LNG) projects (Poussenkova, 2019). Novatek's competitiveness is dynamically growing by the realization of several LNG projects. Yamal LNG is the first major LNG project in the country, built by Novatek in partnership with French Total (20 percent), Chinese CNPC (20 percent), and the Silk

Road Fund (9.9 percent) (Zagorskii, 2019). Novatek is already building another plant—Arctic LNG-2. The plant's capacity is about 18 million tons of LNG per year. The plant, with a capacity of 19.8 million tons, is expected to be launched in three phases in 2023–2025; in addition, Obsky LNG was announced (5 million tons, launch in 2022–2023), and in the longer term, Arctic LNG-1 will be launched (19.8 million tons per year) (Chanysheva & Ilinov, 2021).

A combination of factors determines the growth of competitiveness of Russian oil and gas firms. Those factors are the availability of resources, minimizing production costs, technological development level, the company's level of preparedness for market and prices fluctuations, government policy, and the level of exploration of available resources.

The level of innovations and technological development, as well human resource professionalism, is critical. The story of technological development influences the break-even point of Arctic projects and determines their sustainability in the highly turbulent energy sector. This indicator is under control, unlike the unmanaged oil price indicator. Oil prices are an essential factor influencing the profitability of Arctic exploration projects. The pandemic and global warming have significantly corrected the previous optimistic forecasts for the dynamics of oil prices. Before the pandemic, the International Energy Agency (IEA) predicted a sharp rise in oil prices to $113 per barrel by 2035 (IEA, 2021). The International Monetary Fund expects the level of oil prices at $50 per barrel by 2035, and certainly no more than $64 per barrel (Baiman, 2021).

There is a direct relationship between the profitability of Arctic projects and the technological and innovations level achieved. Thanks to technological progress, Norwegian oil and gas companies have reached a break-even point of $20 per barrel for deep-water offshore projects in the Arctic (oil fields Johan Sverdrup and Johan Kastberg) (Ilinova et al., 2020). For Russian offshore projects, a break-even point was calculated from $50 to $70 per barrel. Thus, R&D costs are considered a criterion for assessing the level of technical progress, although sometimes the amount of R&D costs is not summed up with the actual effect. Equinor (the world leader in developing the Arctic shelf) invested approximately $307 million, $315 million, and $300 million in R&D in 2017, 2018, and 2019, respectively (Morgunova, 2020). In recent years, the amount of R&D funding by Russian oil and gas companies is increasing: in 2019, for Gazprom Neft it reached about ₽3.3 billion (about $52 million), and for Rosneft it reached ₽30 billion (about $476 million) (Muslimov, 2021).

State support is one of the tools to ensure an acceptable level of the profitability of Arctic projects. The level of support shows the degree of strengthening fiscal and monetary measures to ensure the profitability of offshore projects (tax incentives, investment multiplier for exploration, provision of affordable project financing), including a preferential tax regime for income tax and mineral extraction tax.

Oil and gas companies' readiness level (in addition to the R&D investments) assumes that to foster the growth of competitiveness, companies need to develop a proactive measures system, including structuring corporate Arctic funds and venture funds, participation in international consortia, and focus on innovation. Exploration level is an indicator that depends both on active measures (the policy of the state and subsoil users) and on macroeconomic conditions. In Russia, subsoil users finance about 90 percent of geological exploration work. Thirty-three new exploration wells are anticipated to be drilled on the Arctic shelf in 2020–2025 (12 in 2020–2022, 17 in 2023–2024, and 4 in 2025) (Lazhentsev, 2021; Trofimov, 2019). At the same time, the successful implementation of those exploration plans can ensure the development of about 187 million tons of oil reserves (with the prospect of growth to 400 million tons by 2035) and 1.7 trillion cubic meters of gas.

Without Vostok Oil, annual oil production in the Arctic by 2030 may reach 15.5 million tons (plus 4 million tons in the Dolginskoye field). Several planned large projects on the Arctic shelf can increase oil and gas production in the Arctic. The increase is predicated on the fact that all planned projects are successfully implemented on time, including the Silurian and Lower Devonian deposits of the Prirazlomnoye field (0.8 million tons) and the Yuzhnoobskoye (0.8 million tons) and North-West (6 million tons) oil fields licensed areas (Laverov et al., 2016). Oil prices are highly volatile, and this macroeconomic criterion is difficult to predict, but they can determine the interest in Arctic projects, including oil exploration and production.

9.4 Discussion

Today it is necessary to work out direct mechanisms for Russian exporters to enter international markets, primarily to the demands of those countries where the climate plan is already developed with their products, entirely or partially cleared of the carbon footprint. According to various scenarios, for the next 3, 15, or 30 years, Russia can supply from 20 to 90 percent of export products of oil, oil products, gas, or coal, which are partially or entirely cleaned of their carbon footprint. Russia can almost certainly form a market, while at the same time being its most significant producer and exporter.

The implementation of Arctic oil and gas projects has potent multiplier effects. Up to 80 percent of work in implementing offshore projects falls on suppliers, which are transport service companies and companies in the construction sector, since the development of the shelf requires a significant volume of supplies: metal structures, equipment, overalls, food, and scientific research (Laverov et al., 2016).

The Arctic needs highly qualified specialists. Experts exploring gas reserves in the Arctic are faced with complex tasks to create modern, independent, and unique scientific solutions and technologies. Offshore development is an extremely high-tech and science-intensive process. The presence of a developed coastal infrastructure is a factor in supplying projects in the Arctic. Today, a modern integrated supply base is a modern port that carries out storage, transshipment, and handling of goods around the clock.

9.5 Conclusions

Based on the research conducted, the following conclusions can be drawn. Firstly, the future competitiveness of Russian big oil and gas firms is connected to a significant extent with the profitability of Arctic projects, which depends on oil prices and the achieved level of technology and innovation development.

Secondly, the prospects for the development of the Arctic big oil and gas onshore resources are sound, and their development will increase the competitiveness of Russian oil and gas companies in promising energy products. Projects for the further development of oil and gas resources in the Arctic should take into account the tasks of creating innovative climate-neutral products without a carbon footprint. This task is challenging and complex.

References

Aalto, P. (2016). Modernization of the Russian energy sector: Constraints on utilizing Arctic offshore oil resources. *Europe-Asia Studies, 68*(1), 38–63. https://www.tandfonline.com/doi/full/10.1080/09668136.2015.1113509

Baiman, R. (2021). In support of a renewable energy and materials economy: A Global Green New Deal that includes Arctic sea ice triage and carbon cycle restoration. *Review of Radical Political Economics, 53*(4), 611–622.

Carayannis, E., Ilinova, A., & Chanysheva, A. (2020). Russian Arctic offshore oil and gas projects: Methodological framework for evaluating their prospects. *Journal of the Knowledge Economy, 11*, 1403–1429. https://doi.org/10.1007/s13132-019-00602-7

Chanysheva, A., & Ilinov, A. (2021). The future of Russian Arctic oil and gas projects: Problems of assessing the prospects. *Journal of Marine Science and Engineering, 9*(5), 528. https://doi.org/10.3390/jmse9050528

Degtyaryova, E. V. (2019). Economic effects of the arctic continental shelf development. *IOP Conference Series: Earth and Environmental Science, 302*(1), 012136. https://doi.org/10.1088/1755-1315/302/1/012136

Gielen, D., Boshell, F., Saygin, D., Bazilian, M. D., Wagner, N., & Gorini, R. (2019). The role of renewable energy in the global energy transformation. *Energy Strategy Reviews, 24*, 38–50. https://doi.org/10.1016/j.esr.2019.01.006

Ilinova, A., Solovyova, V., & Yudin, S. (2020). Scenario-based forecasting of Russian Arctic energy shelf development. *Energy Reports, 6*, 1349–1355. https://doi.org/10.1016/j.egyr.2020.11.022

International Energy Agency. (2021). *World energy outlook 2021*. https://www.iea.org/topics/world-energy-outlook

Katysheva, E., & Tsvetkova, A. (2017). The future of oil and gas fields development on the arctic shelf of Russia. *International Multidisciplinary Scientific GeoConference Surveying Geology and Mining Ecology Management (SGEM), 17*(53), 935–940.

Konina, N. (2018a). Evolution of the largest multinational german companies. *Sovremennaya Evropa, 2*(81), 49–59. https://doi.org/10.15211/soveurope220184959 (in Russian).

Konina, N. (2018b). Issues of modern geo-economics. *Mezhdunarodnye Protsessy, 16*(2), 186–196. https://doi.org/10.17994/it.2018.16.2.53.11 (in Russian).

Konina, N. (2021). Introduction: At the dawn of the fourth industrial revolution—Problems and prospects. In N. Konina (Ed.), *Digital strategies in a global market*. Palgrave Macmillan. https://doi.org/10.1007/978-3-030-58267-8_1

Kontorovich, A. E., Burshtein, L. M., Livshits, V. R., & Ryzhkova, S. V. (2019). Main directions of development of the oil complex of Russia in the first half of the twenty-first century. *Herald of the Russian Academy of Sciences, 89*(6), 558–566. https://doi.org/10.1134/s101933161906008x

Larchenko, L. V., Gladkiy, Y. N., & Sukhorukov, V. D. (2019). Resources for sustainable development of Russian Arctic territories of raw orientation. *IOP Conference Series: Earth and Environmental Science, 302*(1), 012121. https://doi.org/10.1088/1755-1315/302/1/012121

Laverov, N. P., Bogoyavlensky, V. I., & Bogoyavlensky, I. V. (2016). Fundamental aspects of rational development of oil and gas resources in the Arctic and Russian shelf: Strategy, prospects and problems. *Arctic: Ecology and Economics, 2*(22), 4–13 (in Russian).

Lazhentsev, V. N. (2021). The Arctic and the North: A Russian spatial development context. *Economy of Region, 16*(3), 737–754. https://doi.org/10.17059/ekon.reg.2021-3-1 (in Russian).

Merkulov, V. I. (2020). Analysis of Russian Arctic LNG projects and their development prospects. *IOP Conference Series: Materials Science and Engineering, 940*(1), 012114. https://doi.org/10.1088/1757-899x/940/1/012114

Mitrova, A., Grushevenko, E., & Malov, A. (2018). Prospects for Russian oil production: Life under sanctions. *SEneC Skolkovo Foundation*. https://energy.skolkovo.ru/downloads/documents/senec/research04-ru.pdf (in Russian).

Morgunova, M. (2020). Why is the exploitation of Arctic offshore oil and natural gas resources ongoing? A multi-level perspective on the cases of Norway and Russia. *The Polar Journal, 10*(1), 64–81. https://doi.org/10.1080/2154896X.2020.1757823

Muslimov, R. K. (2021). On a new paradigm for the development of the oil and gas complex in Russia. *Neftyanoe Khozyaystvo-Oil Industry, 3*, 8–13. https://doi.org/10.24887/0028-2448-2021-3-8-13 (in Russian).

Poussenkova, N. N. (2019). Arctic offshore oil in Russia: Optimism, pessimism and realism. *Outlines of Global Transformations: Politics, Economics, Law, 12*(5), 86–108. https://doi.org/10.23932/2542-0240-2019-12-5-86-108 (in Russian).

Razmanova, S., & Steblyanskaya, A. (2020). Arctic LNG cluster: New opportunities or new treats? *IOP Conference Series: Earth and Environmental Science, 539*(1), 012165. https://doi.org/10.1088/1755-1315/539/1/012165

Sapir, E. V., & Karachev, I. A. (2020). Challenges of a new investment policy: Investment promotion and protection. *Finance: Theory and Practice, 24*(3), 118–131. https://doi.org/10.26794/2587-5671-2020-24-3-118-131 (in Russian).

Trofimov, S. E. (2019). Arctic shelf as a strategic region of development of the Russian economy. *Voprosy Ekonomiki, 7*, 147–160. https://doi.org/10.32609/0042-8736-2019-7-147-160 (in Russian).

Zagorskii, A. (2019). China accepts rules in the Arctic. *Mirovaya Ekonomika i Mezhdunarodnye Otnosheniya, 63*(7), 76–83. https://doi.org/10.20542/0131-2227-2019-63-7-76-83 (in Russian).

CHAPTER 10

Expert Systems for Risk Management in the Energy Complex of the Arctic: A Lesson for Russian Oil and Gas Companies

Vadim O. Evseev and Vladimir S. Osipov

10.1 Introduction

According to experts, most projects in Russia and with elements abroad run over budget. In nine out of ten transport infrastructure projects, costs are underestimated, which leads to cost overruns. The real cost of all types of projects is on average 28 percent higher than the estimated

V. O. Evseev
Plekhanov Russian University of Economics, Moscow, Russia
e-mail: manrus@mail.ru

Russian Academy of National Economics and Public Administration, Moscow, Russia

V. S. Osipov (✉)
MGIMO University, Moscow, Russia
e-mail: vs.ossipov@gmail.com

Lomonosov Moscow State University, Moscow, Russia

V. I. Salygin (ed.), *Energy of the Russian Arctic*,
https://doi.org/10.1007/978-981-19-2817-8_10

cost and even higher in the Arctic Zone. Underestimation of costs and cost overruns are present in 20 countries on five continents and are a global phenomenon; as a rule, this is due to underestimation of the risks of the construction process when making managerial and organizational decisions.

In the era of digitalization, the capabilities of artificial intelligence are increasingly being used to solve practical problems, which, as a rule, are the most complex algorithms of computational procedures for the management of professional knowledge (Guterres, 2021; Kasatkin et al., 2021; Kravchenko & Salygin, 2015; Osipov, 2020; Popova, 2020; Stepanov, 2021; Vasiliev & Pechatnova, 2020; Yankovskaya et al., 2021).

In modern conditions, *knowledge management* has become the most important and integral part of the management strategy of the most successful Russian and foreign companies. Employees' knowledge and skills are a valuable asset of the company. According to a survey conducted among managers of leading companies, the main part of corporate knowledge exists in an informal form—it is knowledge in the heads of employees and specialists (42 percent) as well as electronic documentation (26 percent).

Among the expected effects of the introduction of a knowledge management system are time savings when working with information, increased employee productivity, improved quality of decisions and efficiency of business processes, optimized operating costs and increased efficiency of capital expenditures, improvement of the innovation management system, and an increase in the value of intangible assets of the company.

Expert systems are an integral component of the knowledge management system, and one of the directions of creating artificial intelligence systems is computer programs of information and analytical orientation. In general, their structure is shown in Fig. 10.1. In terms of the quality and effectiveness of the solution, expert systems should not be inferior to the solutions of a human expert.

Artificial intelligence, in the knowledge management system, can be used to realize new human capabilities in all spheres of activity, including: freeing a person from monotonous work by automatically creating software; decision-making support; automation of dangerous types of work; and support of communication between people.

Expert systems are a field of research in the area of artificial intelligence on the creation of computing systems capable of making decisions

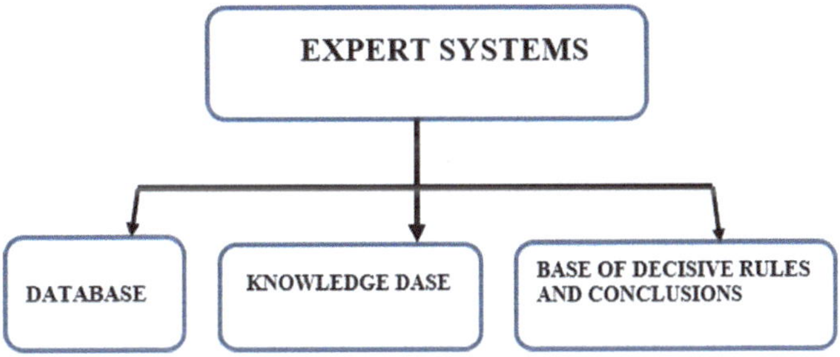

Fig. 10.1 Structure of the expert system (*Source* Compiled by the authors)

similar to the decisions experts would make in a given subject area. Expert systems have one big difference from other artificial intelligence systems: They are designed for high-quality problem solving in an area defined by developers (in rare cases, inpluralistic areas). Expert knowledge is a combination of theoretical understanding of the problem and practical skills of its solution, the effectiveness of which has been proven by the results of activities. The expert system is the fruit of the joint work of experts in this subject area, knowledge engineers, and programmers.

Knowledge management, including expert systems, is widely used by companies belonging to the fuel and energy complex. For example, *Lukoil PJSC* has been implementing a knowledge management system since 2013 and has received an economic effect of more than $200 million from the introduction of best practices, while *British Petroleum* has developed and implemented many expert systems. Perhaps the best-known system is an advisor system for the design of gas /oil distribution systems, using it provided cost savings, on average, $260 million per year.

10.2 Methodology

10.2.1 Research at the Level of the Government of the Russian Federation

The Ministry of Energy of the Russian Federation creates two concepts of digital transformation of the fuel and energy complex of Russia. As part of the digital transformation concept, a Council for the Digital Transformation of Fuel and Energy Industries has been established, the purpose of which is to create conditions for the development, implementation, and application of digital technologies in the field and the formation of a consolidated position and key decisions on the main issues of its digital transformation.

The Council, as of February 5, 2020, includes the following organizations (selectively): Ministry of Energy, Ministry of Communications, Ministry of Industry and Trade, Ministry of Economic Development, and the State Duma Committee on Energy; it also includes the group of companies, such as VEB.RF, Gazprom, Gazpromneft, EvrazHolding, Zarubezhneft, Inter RAO, Novamind, NOVATEK, Rossetti, RussNeft, SDS-Coal, Surgutneftegas, SUEK, Tatneft, Transneft, Russian Railways, the SKOLKOVO Foundation, MEI, Rusatom automated control systems (RACE), and Gubkin Russian State University of Oil and Gas.

The following tasks are key for the working group:

1. defining the target vision of the digital transformation of the oil and gas complex, critical technological solutions, and the possibility of replicating pilot projects;
2. assessing best practices in the development and implementation of digital technologies in the industry, the potential effects, and the costs of digitalization;
3. preparing proposals for the adjustment of the regulatory framework;
4. forming proposals for detailing the activities of the project of the Ministry of Energy—Digital Energy.

Thanks to the application of best practices in the development and implementation of digital technologies, by 2035, it is expected to reduce costs in exploration and production by 10–15 percent for comparable categories of reserves, as well as to reduce the time for the commissioning of new large projects by up to 40 percent. Particular attention is paid to improving the quality of public services (e.g., state expertise of stocks and design solutions), as well as the transition to electronic document management.

10.2.2 Research at the Corporate Level

Practice of knowledge management in oil companies has been neatly investigated by Bezginova et al. (2018). According to them, knowledge management is traditionally carried out using the following components:

information technology, organizational processes and structures, and corporate culture and personnel. In addition, there are complex knowledge management tools, which, as a rule, combine several components: a document library (knowledge base, corporate memory), communities of practitioners, lessons learned, etc. The purpose of this study is to illustrate the application of knowledge management practices in oil industry companies.

For *Eni*, the knowledge management system is a must have, as there is no other way to compete with Shell and BP—the most successful self-learning companies in the world. The creation of the system began in 2000 with the launch of Technical Network Groups—the prototype of communities of practitioners. For Eni, it is particularly valuable that the knowledge management system makes it possible to involve the maximum number of employees in solving strategic tasks and form their commitment to the company, to distribute innovative solutions very quickly, and to experience learned from mistakes. All this translates into a quite tangible reduction in costs and an increase in the company's profit. In the future, Eni plans to focus on the development of mobile applications, the development of a knowledge management culture, the quality of content in the system, and the use of augmented and virtual reality tools to train employees, as well as to facilitate the use of knowledge collected in the database in work processes.

British Petroleum is recognized as a world leader in knowledge management, having developed a solid and systematic approach to this process. British Petroleum has been the winner of Most Admired Knowledge Enterprises many times. The main factors, the overcoming of which stimulated the development of a knowledge management system, were:

- A variety of goal-achieving tools used in various departments of the company;
- A lot of new major projects in the company (in 2004, more than 70 projects, more than $100 million each), the experience of which should quickly spread throughout the company;
- Loss of valuable "know-how" due to a large number of experts retiring;
- The realization that many BP project teams prefer to solve problems in parallel, instead of cooperating with others.

10.3 Literature Review

Research related to all risk factors of the construction process can be divided into the following thematic areas:

- the study of the control system, as presented in the following works of Barkhi et al. (2018), Bezginova et al. (2018), Perevozova and Vozilova (2015), Osipov and Skryl (2016), Kotelnikov et al. (2012), and Borkovskaya (2014);
- knowledge management system, as touched upon by Loban (2011), Tsvetkova and Bachurin (2018), Bezginova et al. (2018), and Pershukova (2012);
- application of expert systems, as depicted in Kalimullina and Yartseva (2018);
- the influence of weather and climatic conditions, as has been studied by: Evseev et al. (2019), Sergeeva (2020), Sokolov (2015), Sall (2014), Makosko (2021), Komogorova (2014), and Barkhi et al. (2019);
- environmental problems which have been investigated by Polyakova et al. (2018), Kasyanov et al. (2018), Borkovskaya (2013), Borkovskaya and Passmore (2018), Piskulova and Pak (2017);
- research on risks and conflict situations, which has been undertaken by Bardenwerper et al. (2018), Serdtsev (2018), Bogachev et al. (2015), Nekrasova and Kaloshina (2016), Pleshivtsev et al. (2018), Evseev and Seleznev (2021);
- research on the problems of human factor, which has been done by Skryl et al. (2017), Passmore and Borkovskaya (2018), Passmore et al. (2018), Evseev (2017), Evseev and Skryl (2018);
- related analytical methods, as have been applied by Ivanova (2012), Evseev (2020), Evseev et al. (2020);
- study of damage problems, which can be derived from Boyarshinov et al. (2019), Savin (2013), Avdotyin et al. (2012);
- the study of territorial problems, which has been performed by Osipov et al. (2016).

In this turn, Piskulova and Pak (2017) exemine the problem of creating a scientific and technical knowledge management system (using the example of the fuel and energy complex), which are characterized by the presence of a full technological cycle of regulation of all processes and requirements, a large volume of control and accounting data (e.g., accounting documents) from various sources, and the territorial location of organizations, users, and processes. It is also noted that for industrial companies that are natural monopolies, the need for special attention to the efficiency of scientific and technical activities is the most important factor in their sustainable development. In the knowledge management system, the author included expert systems and knowledge bases about the experience of employees.

10.4 Building an Expert System: Research Methods

In a knowledge management system, expert systems are an instrumental part that is filled with data, knowledge, and rules according to a certain procedure. The structure of the expert system for the analysis and management of construction risks in the northern zone consists of three main blocks: (1) databases; (2) knowledge bases; (3) bases of decisive rules and conclusions; and (4) blocks of simulation modeling of events.

Database. In this case, the database is the source of information revealing the features of the construction of the fuel and energy complex in the northern territories and the Arctic Zone in particular. This block could be broken down to:

1. *Database—Legislative Information (basic)*:
    - "On the Strategy for the Development of the Arctic Zone of the Russian Federation and Ensuring National Security for the Period up to 2035," Decree of the President of the Russian Federation (2020);
    - "On the Strategy for the Development of the Information Society in the Russian Federation for 2017–2030," Decree of the President of the Russian Federation (2017);
    - The National Program "Digital Economy of the Russian Federation," approved by the minutes of the meeting of the Presidium of the Council under the President of the Russian Federation for Strategic Development and National Projects (2019);

- "On approval of the Energy Strategy of the Russian Federation for the period up to 2035" Order of the Government of the Russian Federation (2020);
- "On the Fundamentals of the State policy of the Russian Federation in the Arctic for the period up to 2035" (2020);
- Federal Law No. 256-FZ "On the Safety of Fuel and Energy Complex Facilities" (2011);
- Digital Transformation Program of the Ministry of Energy (2018).

2. *Database—Statistical Information*:

- Rosstat, section "Construction";
- Rostechnadzor (statistical information on the results of scheduled and unscheduled inspections; Analysis of the state of power, drilling, and heavy machinery equipment in fuel and energy sector organizations);
- Construction climatology, SNiP 23-01-99.

Database—The results of scientific research shows the share of risks in the construction process: violation of construction deadlines (25.8 percent); violations during the conclusion of contracts (15.2 percent); violation of technology (15.2 percent); default (21.7 percent); insolvency of the contractor (13.7 percent); change of circumstances (8.4 percent); increase in prices for construction resources (28 percent); high level of taxes (36 percent); competition from other construction firms (23 percent); insufficient qualification of workers (16 percent); weather conditions (11 percent); deterioration of machinery and mechanisms (6 percent); and lack of financing (21 percent).

Based on the analysis of all the information, a database model was formed (see Table 10.1), which includes the following knowledge related stages of the construction process with the corresponding risk factors: (1) Development of feasibility study; (2) Design; (3) Construction; (4) Start-up and adjustment works; (5) Commissioning of the facility; (6) Installation supervision of the construction process; (7) Construction project/process management system; (8) Specific Arctic risks; (9) Risks of fraud in construction; and (10) Political risks.

Knowledge base . Knowledge has a complex classification, according to which there are more than thirty types, and therefore, they have a complex interconnected structure of dependencies and levels of mutual

Table 10.1 Information model of the Expert system database

1. Development of feasibility study	**6. Stage-installation supervision of the construction process**
1.1. Risk of inaccurate calculation of project payback	6.1. Control over dismantling and assembly of equipment components
1.2. Risk of underestimation of-the construction budget	6.2. Control over the connection of communications (electricity, air. recycled water, etc.)
1.3. Risk of underestimation of construction time	6.3. Control over the configuration and adjustment of equipment
1.4. Errors in the development of the **organizational** scheme of project management	6.4. Checking the operability of the equipment at idle and in operating mode with the check of compliance with the performance characteristics
1.5. Risk of making an error when choosing the main technological scheme and the matin technological parameters of the object	6.5. Consultation and instruction on issues related to operation and maintenance
2. Stage-Design	**7. Construction project/process management system**
2.1. Errors in the development of project documentation	7.1. Project management (implementation strategy, administrative management)
2.2. Errors in the development of estimate documentation	7.2. Cost and financial flow management
2.3. Risk of affiliation of the general designer and general contractor	7.3. Supply Chain Management
2.4. Risk of biased design decisions	7.4. Risk management and implementation of control procedures
2.5. Risk of using materials that have no analogues	7.5. Time and Resource Management (Network Planning)
3. Stage-Construction	**8. Specific Arctic risks**
3.1. Opacity in the selection of contractors and suppliers	8.1. Technical lists (technical risk insurance)
3.2. Non-compliance with technological regulation in the production of construction and installation works	8.2. Transport risks (transport insurance: "casco": "cargo"; carrier's liability insurance)
3.3. Inflated cost of materials, works, equipment	8.3. Environmental risks (liability insurance for harm)
3.4. Manipulation of costs and manure of the project	8.4. Investment risks: payback period; the degree of adaptation of projects to risk: taking into account the risk: in individual elements and in the results of calculating the effectiveness of investments (investment insurance)

(continued)

Table 10.1 (continued)

3.5. Risks of increasing the construction time	8.5. Increased risks of morbidity and injury of personnel (li& insurance)
4. Stage-Start-up find adjustment works	**9. Risks of fraud in construction**
4.1. Installation and assembly of equipment components	9.1. Construction: (1) Change of project documentation. (2) Manipulation of reporting. (3) Overestimation of volumes and cost of works. (4) Fictitious services
4.2. Connection of communications (electricity, air, recycled water, etc.)	9.2. Purchases: (1) Purchases at inflated prices. (2) Fictitious purchases. (3) Presence of affiliates
4.3. Adjustment and adjustment of equipment	9.3. Realization: (1) Offsets with fictitious accounts payable. (2) Sales of products with low margin
4.4. Checking the operability of the equipment at idle and in operating mode with checking the compliance of performance characteristics	9.4. Finance: (1) Availability of bill schemes. (2) Issuance of non refundable advances, credits and loans. (3) Financial leasing through intermediary companies
4.5. Consulting and instructing on issues related to operation and maintenance	9.5. Personnel: (1) Overestimation of labor costs for the project. (2) Use of salary schemes for cashing out money. (3) The presence of "dead souls"
5. Stage—Commissioning of the facility	**10. Political risks**
5.1 Failure to receive a package of permits	10.1. Impact of economic sanctions on construction
5.2. High corruption component in the legalization of construction projects	10.2. Currency fluctuations
5.3. Risk of failure to achieving the technical indicators of the project	10.3. Change of the key rate of the Bank of Russia
5.4. Risk of untimely commissioning of the facility	10.4. Risk of changes in legislation
5.5. Risk of incompleteness of iter and workers	10.5. Impact of economic sanctions on product certification

Source Compiled by the authors

influence. The proposed expert system has the following knowledge (basic): subject, functional, evaluative, regulatory, normative, probabilistic, and dynamic. The knowledge base is formed accounting for: (1) the knowledge and experience of experts; (2) the results of scientific research; (3) rules and regulations on organization, technology, management, and planning of construction and installation works.

Expert knowledge. Based on the knowledge and evaluation characteristics of experts, the following input data is generated for the expert system (see Figs. 10.4 and 10.5):

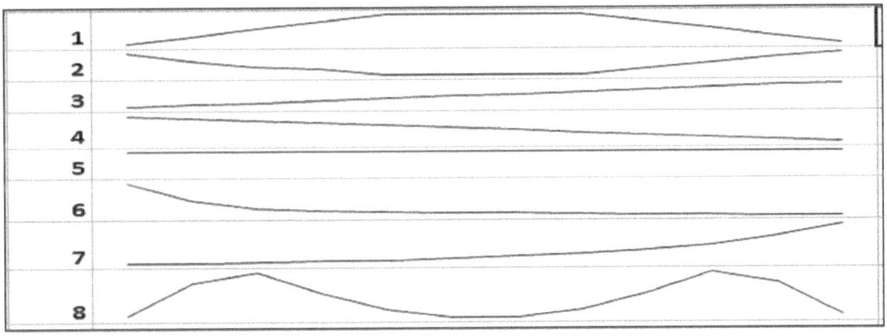

Fig. 10.2 Possible trajectories/trends in the development of risk factors by construction time (*Source* Compiled by the authors)

- Importance/Significance of the factor (in total = 10)
- Presence of a risk factor (1—yes; 0—no)
- Expected/Probability of event (1–5)
- Dynamics/Increase in the significance of the event, percent (±1–5 percent)
- Selection of the trend of change in significance (1–5) (see Fig. 10.2).

Figure 10.2 shows the possible trends of risk level changes in the considered time interval of the construction process: nonlinear, linear, and uniform. In principle, any trend/trajectory of the risk level is possible, from cyclical to impulse, but all these trends should have a scientific justification and mathematical expression.

10.5 Results

The results of the conducted studies include the identified dependencies linking the level of risk and its impact on the increase in construction time and the increase in the cost of construction. A general view of the dependencies is shown in Fig. 10.3.

Also in the expert system, based on the level of risk obtained, possible consequences associated with this level of risk are issued; for example, for a risk with a level of 70 points, the following possible consequences are issued (Figs. 10.4 and 10.5):

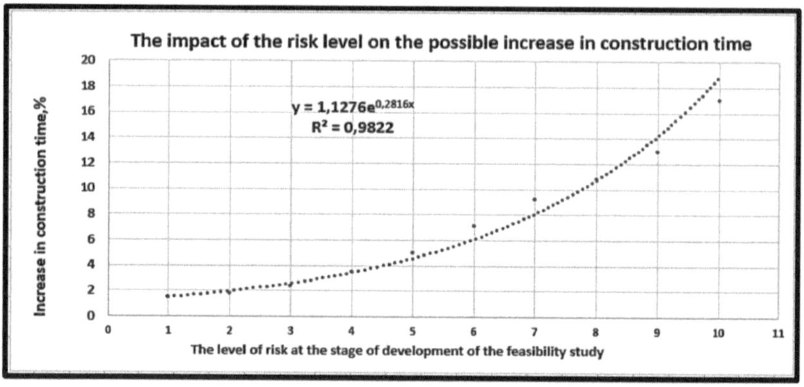

Fig. 10.3 Relationship between the level of risk and the increase in time construction, percent (*Source* Compiled by the authors)

1.Development of feasibility study	Importance/ Significance of the factor (in total = 10)	Presence of a risk factor (1-yes; 0-no)	Expected/ Probability of event (1-5)	Dynamics/ Increase in the significance of the event,% (+/-)(1-5%)	Selection of the trend of change in significance (1-5)	Total in statics
1.1.Risk of inaccurate calculation of project payback	3	1	3	3	1	3
1.2.Risk of underestimation of the construction budget	1	1	2	2	2	1
1.3.Risk of underestimation of construction time	1	1	3	3	4	1
1.4.Errors in the development of the organizational scheme of project management	2	1	4	-5	5	2
1.5. Risk of making an error when choosing the main technological scheme and the main technological parameters of the object	3	1	4	3	4	3

Fig. 10.4 Realization of knowledge about risk factors in the development of a feasibility study of a construction project (*Source* Compiled by the authors)

8. Specific Arctic risks	Importance/ Significance of the factor (in total = 10)	Presence of a risk factor (1-yes; 0-no)	Expected/ Probability of event (1-5)	Dynamics/ Increase in the significance of the event,% (+/-)(1-5%)	Selection of the trend of change in significance (1-5)	Total in statics
8.1. Technical risks (technical risk insurance)	2	1	3	3	1	2
8.2. Transport risks (transport insurance: "casco"; "cargo"; carrier's liability insurance)	2	1	2	2	2	2
8.3. Environmental risks (liability insurance for harm)	2	1	3	3	4	2
8.4. Investment risks : payback period; the degree of adaptation of projects to risk; taking into account the risk in individual elements and in the results of calculating the effectiveness of investments (investment insurance).	2	1	4	-5	5	2
8.5. Increased risks of morbidity and injury of personnel (life insurance)	2	1	4	3	4	2

Fig. 10.5 Realization of knowledge about Arctic risk factors (*Source* Compiled by the authors)

(1) Increase in cost, ranging from 20 to 40 percent;
(2) Increase in time, from 10 to 20 percent;
(3) Reducing the content, which is unacceptable to the customer;
(4) Quality reduction, which is unacceptable for the customer.

10.5.1 The Base of Decisive Rules and Conclusions

In this case, depending on the level of risk, the following recommendations are issued as examples:

Neutralization of financial risks provides for the use of the following main methods: risk avoidance; limiting the concentration of risk; hedging; diversification; transfer of risk; self-insurance; and other methods.

Organization of risk management and internal control is recommended proceed from the objectives of RM&VAC defined in the Corporate Governance Code 9: ensuring reasonable confidence in achieving the Company's goals; ensuring the efficiency of financial and economic activities and economical use resources; the identification of risks and management of such risks; ensuring the safety of the Company's assets; ensuring the completeness and reliability of accounting (financial), statistical, managerial, and other reporting; control over compliance with legislation, as well as internal policies, regulations, and procedures Companies.

Risk management is implemented within the framework of control systems: (1) Definition of the objectives of the regulatory system; (2) Asset management (traceability requirement); (3) Identification of risks; (4) Risk analysis and assessment: understanding the most important risks; (5) Choice of risk management strategy; (6) Application of risk management strategy; (7) Contingency planning and crisis management (including the development of a plan to eliminate the risk of destabilization); and (8) Monitoring, re-analysis, and improvement of the risk management process.

10.5.2 Determination of Climate Hard Conditions

As mentioned earlier, the expert system is a program shell for storing the knowledge of leading specialists, experts, and practitioners in the subject area under consideration. Knowledge in the expert system is stored in the form of tabular forms, graphic forms, and calculation formulas. Since the database of expert systems includes probabilistic characteristics (which describe risks), the expert system is based on a complex algorithm of the simulation method. Based on the probabilities of occurrence of the corresponding risks specified in the initial data, random numbers are generated; here, this happens according to a uniform distribution law, and as a result, for each stage of the construction process and for the entire situation as a whole, it is possible to obtain the mathematical expectation of the cumulative level of risk and its average deviation. The mathematical expectation is adjusted in accordance with the given trends of risk development (see Fig. 10.2). The possible change in the construction time and the cost of construction are calculated based on the calculation equations (see Fig. 10.3).

The expert system assumes the possibility of risk management; for this, it is necessary to conduct research to identify the dependencies between the resources affecting the change of risks and the share/percent of the change of the risk itself. An example of such a dependency is shown in Fig. 10.6.

The working expert system includes, as a rule, three thousand interdependencies or more, which are revealed between the factors of the construction process. In the proposed expert system for risk assessment during construction, five hundred dependencies are considered and implemented, which are presented as logical and mathematical expressions.

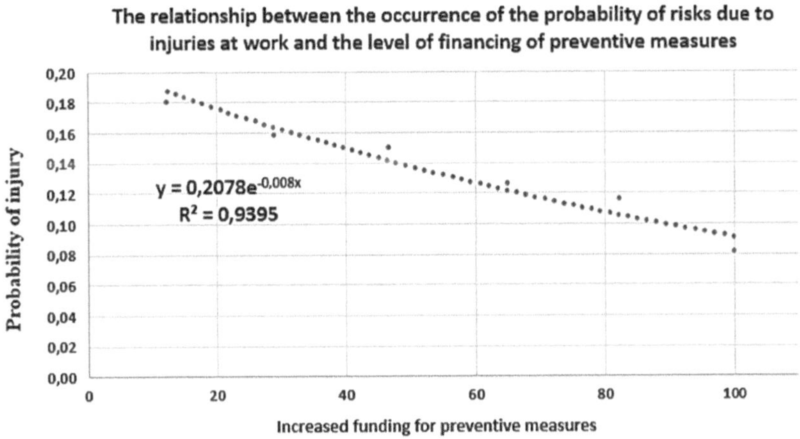

Fig. 10.6 The impact of the level of labor protection financing (preventive measures) on the likelihood of injury cases, percent (*Source* Compiled by the authors)

10.6 Discussion

An integral part of the expert system is the interactive information, showing the results of the impact of risks on the characteristics of the construction process. The composition of the interface information is formed according to the user's requests and may include text, or tabular and graphical information. In the developed expert system, two types of information are provided to the user: text and graphic. Text information produces two types of messages: (1) Depending on the level of risk, possible negative consequences that affect the estimated cost of construction, construction time, and personnel characteristics; (2) Depending on the level of risk and possible consequences, information is provided both on reducing the level of risk, and recommendations are given on possible ways to reduce the level of exposure to negative consequences.

The graphical information (see Figs. 10.7 and 10.8) gives an idea of the change in risks for the selected factors over the entire time interval, and Fig. 10.8 specifically shows the value of the total risk dynamics over the time intervals of construction.

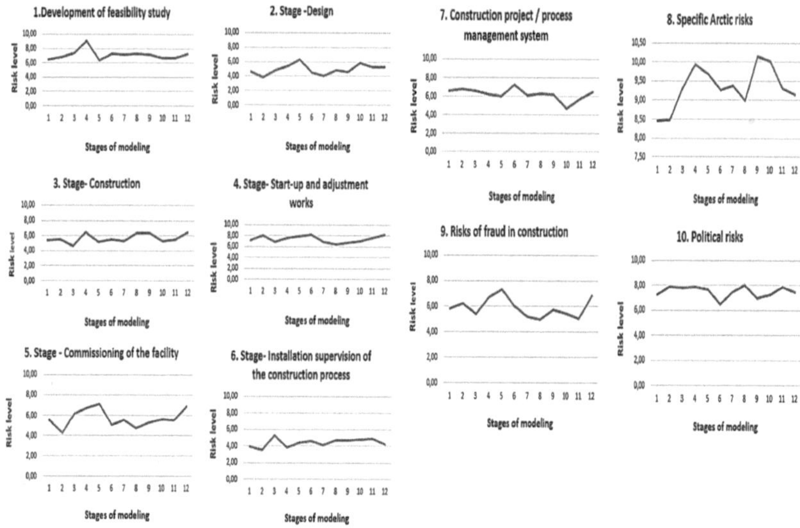

Fig. 10.7 Dynamics of changes in the level of risks for the considered components/factors of the construction process (*Source* Compiled by the authors)

10.7 Conclusions

Firstly, given the rising international competition in the fuel and energy sector for the development of the Arctic Zone, the level of implementation of the knowledge management system is of increasing importance. The instrumental part of the knowledge management system is expert systems, the software shell of which is designed to create a database, a knowledge base, a base of decisive rules, and conclusions.

Secondly, in the expert system proposed for consideration for the Arctic Zone, the following ten factors of the construction process were applied, for each of which the corresponding risks were selected:

1. *Development of feasibility study;*
2. *Design;*
3. *Construction;*
4. *Start-up and adjustment works;*
5. *Commissioning of the facility;*
6. *Installation supervision of the construction process;*

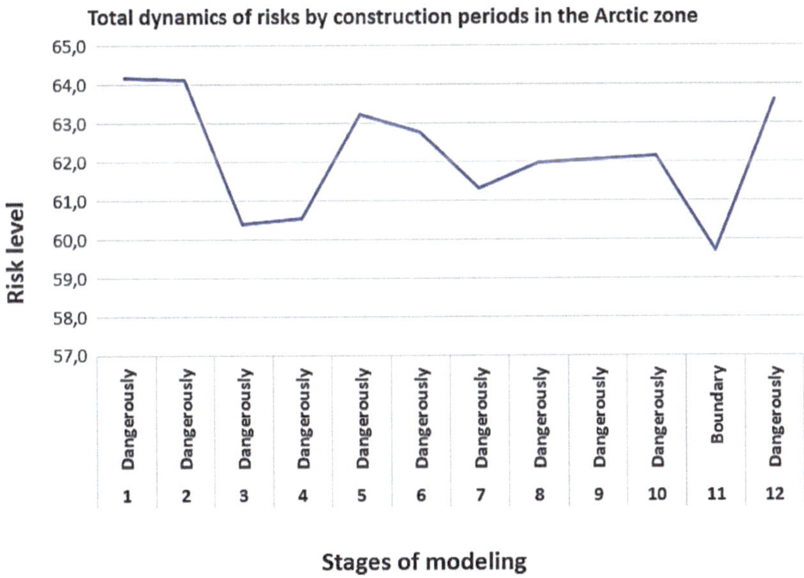

Fig. 10.8 Total risk dynamics in the Arctic Zone by construction periods (100—maximum risks) (*Source* Compiled by the authors)

7. *Construction project/process management system;*
8. *Specific Arctic risks;*
9. *Risks of fraud in construction;*
10. *Political risks. It should be also noted that it is possible to build/create an expert system with a much larger number of factors and with a large number of risks.*

Thirdly, the creation of expert systems is impossible without extensive research work, both on the creation of a database and on the creation of a knowledge base, where causal relationships are identified and recorded, both between the factors of the construction process and between the risks associated with these factors. The scope of the study also includes identifying the dependencies of the influence of relevant resources (finance, time, weather, qualifications, sanctions, etc.) on reducing or increasing the level of risk.

Fourthly, the simulation results showed an increasing level of risks in the region, which reach their maximum value at the beginning of spring and autumn. The graph of the total risk dynamics shows that a significant level of risk occurs at the stages of the beginning and end of the construction process.

Finally, the use of expert systems in construction in the Arctic Zone is fully consistent with the Decree of the President of the Russian Federation, No. 633 dated 08.11.2021, "On approval of the Fundamentals of State Policy in the field of strategic planning in the Russian Federation," in section IV, "Scientific-methodological, information-analytical and personnel support of strategic planning."

References

Avdotyin, V. P., Dzybov, M. M., & Samsonov, K. P. (2012). *Assessment of damage caused by natural and man-made emergencies*. FSBI VNII GOCHS (in Russian).

Bardenwerper, W., Borkovskaya, V. G., & Roe, R. (2018). Interactive teaching of risk management in the Russian construction industry. In *IOP Conference Series: Materials, Science and Engineering* (Vol. 365, p. 062030).

Barkhi, R., Burkov, V. N., Burkova, I. V., & Berlinov, M. (2018). Qualitative risk assessments in project management in construction industry. *Journal MATEC Web of Conferences, 251*, 06027. https://doi.org/10.1051/matecconf/201825106027

Barkhi, R., Evseev, V. O., & Borkovskaya, V. G. (2019). Simulation model of the influence of climatic conditions on the characteristics of the construction process. *CITISE, 5*(22), 516–525. http://ma123.ru/en/2019/12/5-barkhi-evseev/

Bezginova, Y. A., Garanina, T. A., et al. (2018). *Knowledge management practices in oil companies*. Saint-Petersburg: Economics and Business (in Russian).

Bogachev, S. N., Shkolnikov, A. A., Rosentul, R. E., & Klimova, N. A. (2015). Construction risks and opportunities to minimize them. Academia. Architecture and Construction (Vol. 1, pp. 88–92).

Borkovskaya, V. G. (2013). Environmental and economic model life cycle of buildings based on the concept of green building. *Applied Mechanics and Materials, 467*, 287–290. https://doi.org/10.4028/www.scientific.net/amm.467.287

Borkovskaya, V. G. (2014). Complex models of active control systems at the modern developing enterprises. *Advanced Materials Research,* 3012–3015.https://doi.org/10.4028/www.scientific.net/amr.945-949.3012

Borkovskaya, V. G., & Passmore, D. L. (2018). Application of failure mode and effects analysis in ecology in Russia. In *MATEC Web of Conferences* (Vol. 193, p. 05027). https://doi.org/10.1051/matecconf/201819305026

Boyarshinov, A. L., Noev, I. I., & Ivacheva, O. A. (2019). The influence of weather and climatic conditions of the republic of Sakha (Yakutia) on road accidents. In *Proceedings of the 106th International Scientific and Technical Conference* (pp. 109–115).

Evseev, V. O. (2017). *Human resources in the system of socio-economic studies*. INFRA-M (in Russian).

Evseev, V. O. (2020). *Computer modeling of socio-political processes*. Center Catalog (in Russian).

Evseev, V. O., Barkhi, R., Pleshivtsev, A., & Scrynnik, A. (2019). Modeling the influence of weather and climatic conditions on the safety characteristics of the construction process. In *E3S Web of Conferences* (Vol. 97, p. 03035). https://doi.org/10.1051/e3sconf/20199703035

Evseev, V. O., Borkovskaya, V. G., & Barkhi, R. (2020). Modelling the reliability of the implementation of the schedule building work. *Journal of Physics Conference Series, 1425*, 012175. https://doi.org/10.1088/1742-6596/1425/1/012175

Evseev, V. O., & Seleznev, I. A. (2021). Conflict management and methods of research of regional conflicts. *CITISE, 3*, 79–95. https://doi.org/10.15350/2409-7616.2021.3.07 (in Russian).

Evseev, V. O., & Skryl, T. V. (2018). Simulation of the voting process at polling stations. In *Proceedings of the 2018 3rd International Conference on Modern Management, Education Technology, and Social Science (MMETSS 2018)* (pp. 202–205). https://doi.org/10.2991/mmetss-18.2018.42

Guterres, A. (2021). The UN Secretary-General on the global digitalization and international cooperation. *Digital Law Journal, 2*(2), 10–13. https://doi.org/10.38044/2686-9136-2021-2-2-10-13

Ivanova, S. S. (2012). *Mathematical modeling in construction: Educational and methodical manual*. Publishing House of ISTU (in Russian).

Kalimullina, O. V., & Yartseva, K. A. (2018). Advantages and prospects of using expert systems in business and economy. *ASTERON* (pp. 98–106) (in Russian).

Kasatkin, P. I., Salnikova, L. S., & Fatykhova, V. M. (2021). Scientific, technical and educational cooperation in the EAEU. In N. A. Piskulova (Ed.), *The economic dimension of eurasian integration*. Palgrave Macmillan. https://doi.org/10.1007/978-3-030-59886-0_7

Kasyanov, V., Danilchenko, V., Amelin, V., & Tolmacheva, V. (2018). Environmental risk management: Forecasting and modeling of emergency risk management situations. In *MATEC Web of Conferences* (Vol. 251). https://doi.org/10.1051/matecconf/201825106030

Komogorova, A. D. (2014). The influence of weather and climatic conditions on the state of human health. In *Proceedings of All-Russian Scientific and Practical Full-Time and Part-Time Conference with International Participation* (pp. 133–134) (in Russian).

Kotelnikov, V. V., Chetverikov, N. P. Andrievsky, R. A., & Ananiev, A. A. (2012). *Safety of construction and implementation of the construction control method.* Moscow: JSC "Scientific and technical center "Industrial Safety" (in Russian).

Kravchenko, S. A., & Salygin, V. I. (2015). A new synthesis of scientific knowledge: The making of interdisciplinary science. *Sotsiologicheskie Issledovaniya, 10*(378), 22–30 (in Russian).

Loban, D. (2011). The Experience of Hewlett-Packard and British Petroleum in creating knowledge management systems in multinational corporations. *Audit and Financial Analysis, 5*, 10–17 (in Russian).

Makosko, A. A. (2021). Climate risks and the economic complex of Russia in the XXI century. *Designing the Future. Glimpses of Digital Reality, 1*(4), 64–72 (in Russian).

Nekrasova, M. V., & Kaloshina, S. V. (2016). Risks of construction participants in the conditions of the economic crisis. *Bulletin of the Perm National Research Polytechnic University. Construction and Architecture, 7*(1), 67–73 (in Russian).

Osipov, V. S., & Skryl, T. V. (2016). The strategic directions of the modern Russian economic development. *International Business Management, 10*(6), 710–717.

Osipov, V. S., Evseev, V. O., & Skryl, T. V. (2016). An analysis of economic issues of territories of priority development. *Research Journal of Applied Sciences, 11*(9), 833–842.

Osipov, V. S. (2020). Digital future: Economic growth, social adaptation, and technological perspectives. In T. Kolmykova & E. Kharchenko (Eds.), *Advances in intelligent systems and computing*. Springer.

Passmore, D., & Borkovskaya, V. G. (2018). Behavioral engineering model to identify risks of losses in the construction industry. In D. Solovev (Ed.), *Smart technologies and innovations in design for control of technological processes and objects.* Springer.

Passmore, D., Chae, C., Kustikova, Y., Baker, R., & Yim, J. (2018). An exploration of text mining of narrative reports of injury incidents to assess risk. In *MATEC Web of Conferences* (Vol. 251). https://doi.org/10.1051/matecconf/201825106020

Perevozova, O. V., & Vozilova, E. V. (2015). Risks of managerial incompetence: Sources and causes. *Innovative Science, 1*(3), 40–42.

Pershukova, V. (2012). *Rosatom shares knowledge*. State Corporation Rosatom.

Piskulova, N. A., & Pak, E. V. (2017). Prospects of Russian engagement in green economy. *International Trends, 15*(4), 40–58. https://doi.org/10.17994/IT.2017.15.4.51.3 (in Russian).

Pleshivtsev, A., Korol, O., & Barkhi, R. (2018). Risks on optimization of life cycle of technology of installation of transformed low-rise buildings from sandwich panels. In *MATEC Web of Conferences* (Vol. 251, p. 06024). https://doi.org/10.1051/matecconf/201825106024

Polyakova, V., Degaev, E. N., & El Haddad, P. (2018). Reduction of ecological and economic risks in utilization of solid domestic wastes and construction waste. In *MATEC Web of Conferences* (Vol. 251, p. 06017). https://doi.org/10.1051/matecconf/201825106017

Popova, N. V. (2020). Hightech as a platform for economic growth. *Digital Law Journal, 1*(3), pp. 8–13. https://doi.org/10.38044/2686-9136-2020-1-3-8-13

Sall, M. A. (2014). Weather and climate risks as an object of management. In *Proceedings of the Main Geophysical Observatory Named After A.I. Voeikov* (Vol. 575, pp. 183–203) (in Russian).

Savin, A. V. (2013). The influence of weather and climatic conditions in Russia on the economic security of industrial organizations. *Vestnik Universiteta, 12*, 129–132 (in Russian).

Serdtsev, P. O. (2018). Economic risks of cost estimation in the production of engineering and environmental surveys. In *Proceedings of the Fourteenth All-Russian Scientific and Practical Conference and Exhibition of Survey Organizations* (pp. 119–120) (in Russian).

Sergeeva, A. A. (2020). Statistical analysis of the influence of climatic conditions on the economic development of russian regions. In *Proceedings of the 11th International Scientific and Practical Conference of Students and Postgraduates* (pp. 133–134) (in Russian).

Skryl, T. V., Osipov, V. S., & Evseev, V. O. (2017). The interaction model of business and state as a factor in human resource management. In *Proceedings of the Second International Conference on Economic and Business Management* (FEBM2017) (pp. 862–868). https://doi.org/10.2991/febm-17.2017.116

Sokolov, Y. I. (2015). Russia's climate risks. *Problems of Risk Analysis, 12*(5), 66–85.

Stepanov, S. K. (2021). Deconstruction of the legal personhood of artificial intelligence. *Digital Law Journal, 2*(2), 14–30. https://doi.org/10.38044/2686-9136-2021-2-2-14-30

Tsvetkova, V. I., & Bachurin, A. I. (2018). Features of information support in knowledge management systems on the example of the fuel and energy complex. *Information Resources of Russia, 3*(163), 2–8 (in Russian).

Vasiliev, A. A., & Pechatnova, Yu. V. (2020). The position of the artificial intelligence among the elements of the legal relationship. *Digital Law Journal, 1*(4), 74–83. https://doi.org/10.38044/2686-9136-2020-1-4-74-83

Yankovskaya, V. V., Osipov, V. S., Zeldner, A. G., Panova, T. V., & Mishchenko, V. V. (2021). Institutional matrix of social management in region's economy: Stability and sustainability vs. innovations and digitalization. *International Journal of Sociology and Social Policy, 41*(1–2), pp. 178–191. https://doi.org/10.1108/ijssp-03-2020-0088

PART III

Infrastructure for Energy Projects in the Russian Arctic

CHAPTER 11

Overview of Practical Experience of Projects' Implementation in the Arctic

James C. Towers and Sergey V. Volchkov

11.1 Introduction

James C. Towers and Sergey V. Volchkov hold unique and profound field expertise gathered in multiple Arctic and Sub-Arctic projects executed in Russia, USA, Canada, and Kazakhstan. Both of these professionals are recognized experts in the design, construction, and operation of cold region oil and gas facilities, both onshore and offshore.

Firstly, it is necessary to understand exactly where the "Arctic" and Sub-Arctic regions are. That is best defined by use of the map depicting the Arctic Circle as shown on the projection of the Northern half of the world as viewed of above the North Pole, as shown in Fig. 11.1.

J. C. Towers
O&G Engineer, Port Alto, TX, USA
e-mail: jctowers@jcti.biz

S. V. Volchkov (✉)
Independent Consultant, Moscow, Russia
e-mail: sergey-volchkov@yandex.ru

V. I. Salygin (ed.), *Energy of the Russian Arctic*,
https://doi.org/10.1007/978-981-19-2817-8_11

Fig. 11.1 View of the Arctic Regions (*Source* Adapted by the authors from Pinterest [2021]. *Arctic Regions*. URL: https://www.pinterest.ru/pin/92886811042959122/)

It is worth mentioning here that Russia's Arctic region stretches about half way around the top of the world, more than any other country. The USA's Arctic region is limited to Alaska and is much smaller than Arctic lands in Russia, Canada, or Greenland.

It is important in Russia for definitions to be precise, as this is directly connected with rates, wages, and other details with direct impacts on cost estimations, projects planning, operations, and other sensitive features. Please see Sect. 10.4 for details.

The Arctic, and to some extent the Sub-Arctic, presents many special challenges to resource development and recovery due to the very challenging weather conditions that can have fatal consequences, environmental preservation issues that have little to no scientifically factual or proven technology impact mitigation measures, and significant psychological consequences for the people that think they can endure the isolation just for the money they can make there. There are also many other challenges such as permafrost[1] in both Arctic and Sub-Arctic regions, which is undergoing the constantly changing impacts of climate change as well as the inherent destabilizing consequences this has on the structures based upon it. Logistics is a great challenge since, as advances can only successfully be made as far as their supply chain can reach. This is true for any project in the Arctic regions where there are no population centers, road and/or rail networks, and ease of mobility. In the chapters that follow, we will give our views on each of these factors.

For any mega project, a Project Execution Plan (PEP) and the Construction Execution Plan (CEP) should be prepared and updated throughout the life of the project. The Construction Execution Plan (CEP) is a critical part of any permit filing and is a driver for many of the engineering aspect of the project. The Construction Execution Plan will need to address all aspects of the construction process in a practical and understandable document to support the project licensing and permitting applications. An early start on this work is needed as a part of the engineering contractor's work effort. A number of universal issues that should be considered part of the CEP include:

- Construction Contract Strategy
- Project Cost Reduction and Co-Use of Existing Assets
- Central Project Field Office
- Camps and Associated Infrastructure
- Project Organization
- Road and Bridge Improvements and or New Infrastructure Requirements
- The Pipeline Construction Plan and Pace of Construction.

[1] Permafrost is soil that stays frozen for 18 months or more.

11.2 Construction Execution Plan: Technical Aspects

11.2.1 Weather Hazards

Temperature hazards are related to severe injuries. The challenge of working outdoors in Arctic temperatures is real, and severe injuries do occur. The authors recommend the chart in Fig. 11.2 be used to decide what reasonably safe limits should be used when establishing safe exposure times for workers.

The NWS Wind Chill Temperature (WCT) index uses advances in science, technology, and computer modeling to provide an accurate, understandable, and useful formula for calculating the dangers from winter winds and freezing temperatures. The index does the following:

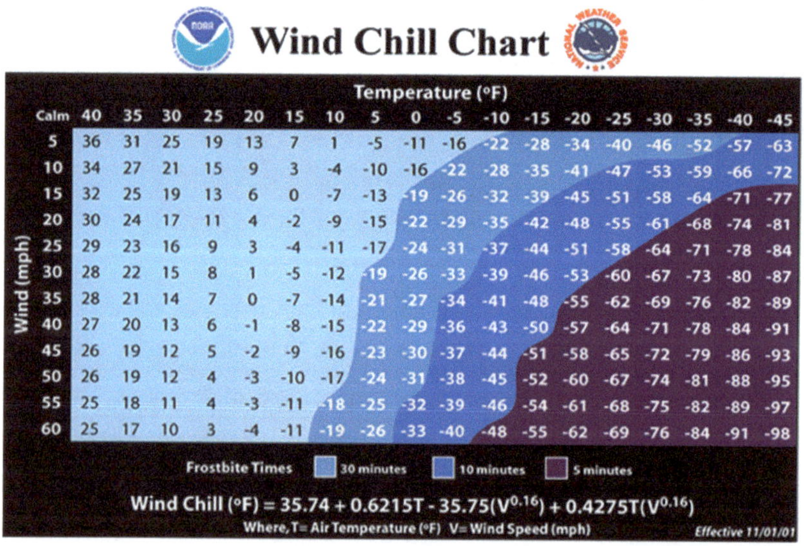

Calm	40	35	30	25	20	15	10	5	0	-5	-10	-15	-20	-25	-30	-35	-40	-45
5	36	31	25	19	13	7	1	-5	-11	-16	-22	-28	-34	-40	-46	-52	-57	-63
10	34	27	21	15	9	3	-4	-10	-16	-22	-28	-35	-41	-47	-53	-59	-66	-72
15	32	25	19	13	6	0	-7	-13	-19	-26	-32	-39	-45	-51	-58	-64	-71	-77
20	30	24	17	11	4	-2	-9	-15	-22	-29	-35	-42	-48	-55	-61	-68	-74	-81
25	29	23	16	9	3	-4	-11	-17	-24	-31	-37	-44	-51	-58	-64	-71	-78	-84
30	28	22	15	8	1	-5	-12	-19	-26	-33	-39	-46	-53	-60	-67	-73	-80	-87
35	28	21	14	7	0	-7	-14	-21	-27	-34	-41	-48	-55	-62	-69	-76	-82	-89
40	27	20	13	6	-1	-8	-15	-22	-29	-36	-43	-50	-57	-64	-71	-78	-84	-91
45	26	19	12	5	-2	-9	-16	-23	-30	-37	-44	-51	-58	-65	-72	-79	-86	-93
50	26	19	12	4	-3	-10	-17	-24	-31	-38	-45	-52	-60	-67	-74	-81	-88	-95
55	25	18	11	4	-3	-11	-18	-25	-32	-39	-46	-54	-61	-68	-75	-82	-89	-97
60	25	17	10	3	-4	-11	-19	-26	-33	-40	-48	-55	-62	-69	-76	-84	-91	-98

Fig. 11.2 Wind-chill temperatures based on ambient temperature and wind speed, along with the corresponding risk of frostbite (*Source* Adapted by the authors from National Weather Service [2021]. *Wind Chill Chart*. URL: https://www.weather.gov/safety/cold-wind-chill-chart)

- Calculates wind speed at an average height of 5 feet (the typical height of an adult human face), based on readings from the national standard height of 33 feet (which is the typical height of an anemometer)
- Uses a human face model
- Incorporates heat transfer theory based on heat loss from the body to its surroundings, during cold and breezy/windy days
- Lowers the calm wind threshold to 3 mph
- Uses a consistent standard for skin tissue resistance
- Assumes no impact from the sun, i.e., clear night sky.

Valuable data and practices are included in Gudmestad et al. (1999).

11.2.2 Human Factor

It is cold, wet, and slippery on the ground, structures, or generally in the Arctic; slip, trip, and fall injuries are very common. A study by Towers on one Arctic project indicated that the minor injury rate due to harsh working and living conditions was 16 times more frequent than in temperate climates. During the cold months, light summer clothing is replaced with dreaded “winter weight.” For crews, this means long underwear, heavy sweaters, winter jackets, double gloves, wool hats, heavy wool socks, and more. This can represent an additional 13.6 kilograms (30 pounds) of clothing, significantly slowing movement. At times when the combined wind speed and low temperatures fall in the range where flesh can freeze in 5 minutes or less, worker exposure time must be limited, and “warm up” shacks with temperatures must be held between 4 degrees Celsius (~40 degrees Fahrenheit) and 7 degrees Celsius (~45 degrees Fahrenheit) where workers can recover core temperature for 15 or so minutes.

Regardless of how well work crews are dressed, the cold plays a significant role toward the risk of frostbite to fingers, toes, noses, and earlobes, as well as hypothermia, when working in extreme temperatures. According to the Canadian Center for Occupational Health and Safety (CCOHS), the risk of moderate frostbite can occur between a wind chill of –10 and –27 degrees Celsius (14 and –16.6 degrees Fahrenheit). Figure 11.3 illustrates the exposure risk as it relates to ambient temperature and wind.

WINDCHILL (°C)	EXPOSURE RISK	HEALTH CONCERNS	WHAT TO DO
0 to -9	Low risk	• Slight discomfort.	Dress warmly. Stay dry.
-10 to -27	Moderate risk	• Uncomfortable. • Risk of hypothermia and frostbite if outside for long periods without adequate protection.	• Dress in layers of warm clothing, with an outer layer that is wind-resistant. • Wear a hat, mittens or insulated gloves, a scarf, and insulated, waterproof footwear. • Stay dry. • Keep active.
-28 to -39	High risk. Exposed skin can freeze in 10-30 minutes.	• High risk of frostnip (milder frostbite) or frostbite: Check face, fingers, and toes for numbness or whiteness. • High risk of hypothermia if outside for long periods without adequate clothing or shelter from wind and cold.	• Dress in layers of warm clothing, with an outer layer that is wind-resistant. • Cover exposed skin. • Wear a hat, mittens or insulated gloves, a scarf, neck tube or face mask, and insulated, waterproof footwear. • Stay dry. • Keep active.
-40 to -47	Very high risk. Exposed skin can freeze in 5-10 minutes. *(In sustained winds over 50 km/h, frostbite can occur faster than indicated)*	• Very high risk of frostbite: Check face, fingers, and toes for numbness or whiteness. • Very high risk of hypothermia if outside for long periods without adequate clothing or shelter from wind and cold.	• Dress in layers of warm clothing, with an outer layer that is wind-resistant. • Cover exposed skin. • Wear a hat, mittens or insulated gloves, a scarf, neck tube or face mask, and insulated, waterproof footwear. • Stay dry. • Keep active.
-48 to -54	Severe risk. Exposed skin can freeze in 2-5 minutes. *(In sustained winds over 50 km/h, frostbite can occur faster than indicated)*	• Severe risk of frostbite: Check face, fingers, and toes for numbness or whiteness. • Severe risk of hypothermia if outside for long periods without adequate clothing or shelter from wind and cold.	• Be careful. Dress very warmly in layers of warm clothing, with an outer layer that is wind-resistant. • **Cover exposed skin.** • Wear a hat, mittens or insulated gloves, a scarf, neck tube or face mask, and insulated, waterproof footwear. • **Be ready to cut short or cancel outdoor activities.** • Stay dry. • Keep active.
- 55 & colder	Extreme risk. Exposed skin can freeze in less than 2 minutes. *(In sustained winds over 50 km/h, frostbite can occur faster than indicated)*	DANGER! Outdoor conditions are **hazardous.**	• **Stay indoors.**

Fig. 11.3 Human Exposure Risk in Ambient Temperatures and Wind Conditions (*Source* Adapted by the authors from Environment Canada [2018]. *Wind Chill*, p. 5. URL: https://ec.gc.ca/meteo-weather/80b0f2af-9697-4bee-ab17-d401ebba5b4b/windchill_factsheet_en.pdf)

In the contract documents, the question of where, when, and under specifically what conditions the crews will be required to use the owner's facility for warm up breaks is asked. Contractor storage needs and whether the contractor has to provide onsite heated storage containers must also be defined in contracts. The logistics issues associated with working in the Arctic require a lot more forward planning than necessary for projects in temperate climates. Temperate climates are generally defined as environments with moderate rainfall spread across the year or portion of the year with sporadic drought, mild to warm summers, and cool to cold winters (Simmons & Poli, 2015). Thus, there will be no quick, easy, or low-cost fixes for any shortcoming in preplanning and provision.

Working over the winter months will also have an impact on crews' overall productivity and on the project schedule. From the authors' experience, depending on the type equipment and task, production can decrease by 20 percent or, in some instances, more. This can be attributed to the need for additional site preparation, temporary heating, storage, and additional material handling. Before the winter snow falls, photographic records should be made of all equipment and material laydown areas where these are stored outdoors so that they can be found and retrieved during the dark snow-covered periods. Additionally, there is only so much daylight available in the winter. During the summer, crews can have additional working hours, but, in the authors' experience, workers' productivity and morale suffer whenever they work more than 10 hours per day and more than 6 days per week. Morale has been widely recognized as having a direct influence on overall on the job safety statistics.

11.2.3 Money, the Irresistible Lure

The amount of potential for Arctic Development in Russia is obvious. Russia has been taping some of the reachable resources for over a century. In the future, step-out project will have to be economically justifiable. Therefore, the lure of huge financial gains to the Russian Federation (and, perhaps, joint ventures with major Western corporation) and will no doubt be the driver for "step-out" projects far away for the infrastructure already developed in locations like parts of the Yamal Peninsula. Sub-Arctic projects—such as the developments on Sakhalin Island by the Russian Federation in partners with such Western corporations as Exxon-Mobil and Shell in the 1990s—were just such economically attractive projects.

As if the weather, logistics, and communications challenges are not enough, finding the workers that are willing to endure the inconvenience and dangers associated with Arctic development is a significant challenge. Even more of a challenge is keeping workers on the job. From experience, the promise of significant gains in earnings will make many think about it, and some will even agree to try it out. Sadly, of 100 workers arriving at a very remote camp site in the Arctic, only about 20 will remain after 2 months, and some of those will not come back to the camp after they leave on their first Rest and Realization (R&R) period back at their point of origin. Ultimately, the authors' experience was that about one out of eight workers will stay to see the construction and startup/checkout era complete, and very few of those will transfer to long-term operations. Unfortunately, their experience and local knowledge experience leaves with them. Some information might be found in Roscow (1977).

11.2.4 Enabling Infrastructure, Supply Chain, and Communication Consideration

"The Obskaya – Bovanenkovo railroad (572 kilometers in length) has proven vital to Gazprom's Yamal megaproject. The railroad ensures year-round, fast, cost-effective, and all-weather delivery of technical equipment, construction materials, and personnel to Yamal fields in the harsh Arctic climate. Cargo used to be delivered to the Yamal Peninsula by sea through the Kharasavey port during the summer navigation period. However, it would have been impossible to create large-scale infrastructure in Yamal without expanding railroad communication" (Gazprom, 2021a).

This railroad is a remarkable achievement and has the potential to facilitate further and faster resource development, but the Yamal Peninsula only represents a very small part of Russia Arctic lands that also may hold great potential. Its construction experience as documented by Gazprom. Gazprom's experience appears to parallel the authors' experience on the Trans-Alaska Pipeline where, in the 1970s, a 358 mile (576 kilometers) long haul road was constructed from the Yukon River to Prudhoe Bay. It was started on April 29, 1974, and completed on September 29, 1974 (154 days), requiring 3 million labor hours and costing $125 million. The road passes through the very challenging Brooks Mountain Range at the Atigun Pass. Thirty-two million cubic yards of gravel were utilized to build the roadbed which has a thickness of 3–8 feet (Alyeska Pipeline, 2007) (Fig. 11.4).

Gazprom has predicted on its website that the development and operation in the Yamal Project will last 100 years (Gazprom, 2021b).

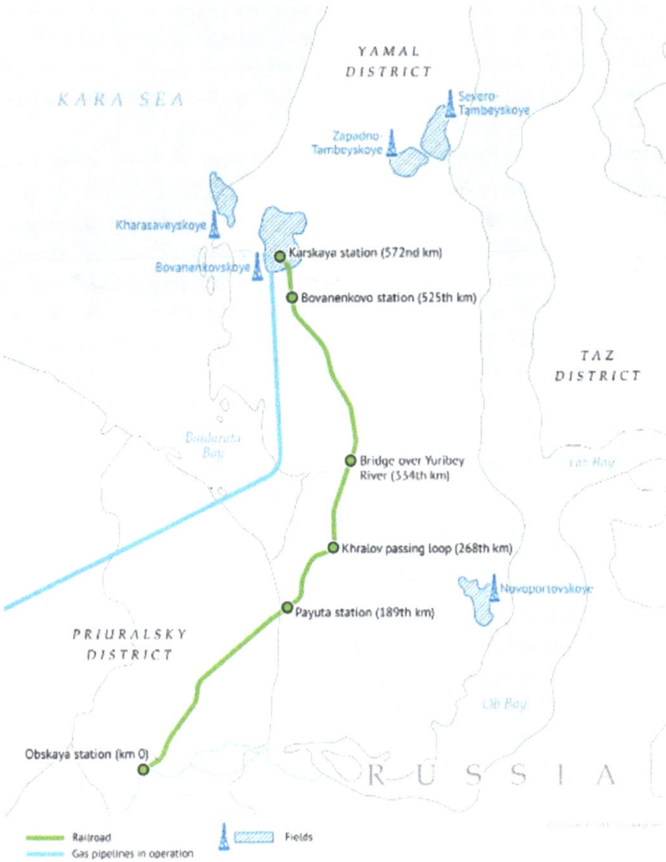

Fig. 11.4 Map of the Obsjaya-Bovanekovo Railway (572 km, all within the Arctic Circle) (*Source* Adapted by the authors based from Gazprom [2021a]. *Obskaya-Bovanenkovo Railroad*. URL: https://www.gazprom.com/projects/obskaya-bovanenkovo/)

Future step-out projects in the Russian Arctic will no doubt benefit from the lessons learned by Gazprom in the Yamal. The most anticipated improvements in the future are better communication facilities and Internet availability and usage from the days of earliest exploration. The authors have witnessed a remarkable improvement in the availability of

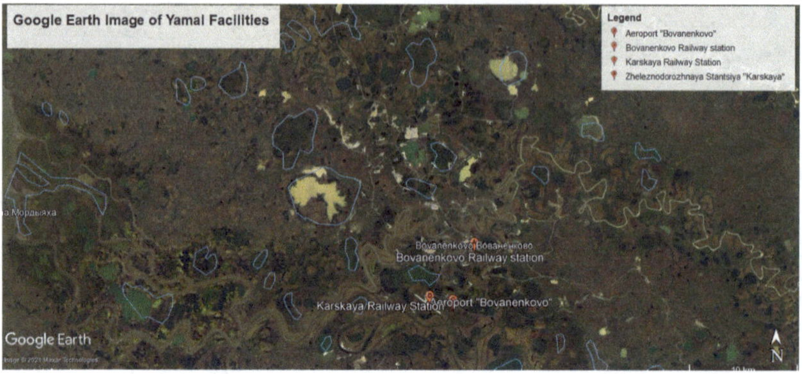

Fig. 11.5 Locating the Karskaya Railway Station, the Bovanenkovo Railway Station, and the Bovanenkovo Airport in the vicinity of major Gazprom Yamal Gas Processing Facilities (*Source* Adapted by the authors from Google Earth [2021]. *Yamal Facilities*. URL: https://www.google.ru/intl/ru/earth/)

portable relay equipment that can be easily move to the field in a short time and provide remarkable performance. Such excellent communication facilities will no doubt have a beneficial cost and schedule impact on long-range future projects and make even more remote resources economically recoverable.

Figure 11.5 shows the location of the Bovanenkovo Airport (with a 2730-meter (8956 foot) landing strip and several large plants within a few kilometers). There are some other projects being implemented in Yamal Peninsula, such as Novatek and partners Yamal-LNG project (already operating), Gazprom Neft's project Novyi Port, and others, including a number of infrastructure projects. Again, there are some other projects in the Russian Arctic region at various stages of development.

11.2.5 Integrity Management of Structures—The Permafrost Hazard

Extensive planning on how to maintain the integrity of the operating systems and its efficiency in an Arctic Environment pays off in the long run. Redundant equipment can sometimes be economically justified. Larger spare part inventories than normal can be considered. Wider spacing between components (pumps, valve, compressors, and vessels) can speed recovery from shutdowns if complete component replacement can reduce down time and keep production flowing. A lot of valuable information is contained in (Mahlbauer, 2003). Another source of valuable information is Oswell (2021).

11.2.6 Winter Breakup Hazard

When the snow and ice deposits begin to melt, all kinds of new hazards appear. For example, the Yuribey River, which is normally 200–300 meters in width, overflows its banks and becomes 3–4 kilometers wide. It is possible that the original designers or those responsible for assuring the civil integrity of oil and gas pipelines river approaches may have to install River Training Structures to prevent delay river coarse changes that would damage the pipelines.

11.3 Russian Regulations, Approvals, and Standardization: Brief Overview

Firstly, Russian Federal authorities pay a lot of attention to the Arctic region; the success of any significant project depends on them, as well as regional authorities.

In this regard, the first document to mention is a strategic publication "Decree of the President of the Russian Federation dated November 12, 2021 No. 651 "On Amending the Strategy for the Development of the Arctic Zone of the Russian Federation and Ensuring National Security for the Period until 2035, approved by the Decree of the President of the Russian Federation No. 645 dated October 26, 2020, and invalidating some acts of the President of the Russian Federation approved by President Vladimir Putin on November 12, 2021. The Strategy is a strategic planning document on ensuring the national security of the Russian Federation. It determines the measures aimed at fulfilling the main tasks of developing the Arctic Zone and ensuring national security, as well as the stages and expected results of carrying out these measures. The document establishes strategic directions and identifies threats associated with the successive warming of the Arctic, aiming to advance development of the region's abundant resources (first and foremost oil and gas) and to improve living conditions for the population. In the longer term, it also aims to establish the Northern Sea Route as a new global shipping artery.

Rebuilding Russian military presence there is not excluded. One of the objectives is to preserve the region's ecological balance.

Another important document is Russian Government Resolution of November 16, 2021 No. 1946 "On approval of the list of regions of the Far North and localities equated to the regions of the Far North, in order to provide state guarantees and compensations for persons working and living in these regions and localities, invalidation of some acts of the Government of the Russian Federation and recognition as invalid on the territory of the Russian Federation some acts of the Council of Ministers of the USSR." It has direct effect on Arctic and Sub-Arctic projects implementation as it re-specifies the list of regions of the Far North and localities equated to the regions of the Far North, in order to provide state guarantees and compensations for persons working and living in these regions and localities.

One needs to know there is a state approved system of schedule of values. The system shall be used in case of working with state-owned companies. In many cases, private companies are using this system as well. For areas in the Arctic and in the Sub-Arctic, considerable multiplication factors are being used, depending on the location.

The Russian Federation is in the process of adopting international standards, though this is not a very fast activity. As to design and construction standards, Russia went through a difficult journey of reforming a whole set. Some have been updated and transformed from mandatory to recommended level; some have been transformed to the so-called GOSTs (Russian state standards). In relatively frequent cases when new technology or new materials are being implemented, there might well be a situation not supported by these standards.

The authors have experienced a good practice recommended for further implementation. The first step is to compare the most relevant standards (Russian and International). Afterward, the most stringent requirements are selected. Then, the appropriate RDI (Russian Design Institute), with support from the customer, develops the document or even a few of them under the title Project Specific Technical Specification (PSTS). When ready, the document goes into approval process normally via *Rostehnadzor* (in case of the so-called dangerous facilities) and some other agencies depending on the scope and finally via *Glavgosexpertiza* (GGE—Federal autonomous institution "Main Department of State Expertize"). Their scope includes State Expert Appraisal of Design Documentation and/or Engineering Survey Results, and Cost Estimate

Validation Pricing in Construction Validation. After all this, the document can be used, along with standards and other norms.

Some of the most relevant standards are listed in the references, as for Site Investigation on the Continental Shelf for Offshore Oil and Gas Facilities Construction see SP 11-114-2004 (2004), for engineering hydrometeorological survey for construction see SP 482.1325800.2020 (2020), for main and field pipelines construction in permafrost see VSN 013-88 (1990), and for geotechnical surveys for construction in permafrost see RSN 31-83 (2004).

11.4 Environmental Preservation

Based on the authors' experience in both countries and the Canadian Arctic, similar regulations with very similar goals and objectives exist. These studies take a lot of hours in the field by skilled archeologists (identifying antiquities and archeological sites that should not be disturbed or require mitigation and avoidance along a pipeline, power line, highway, or railway route). Likewise, biologists have to preform field studies to identify critic habitats, threatened or endangered species along a route, and all the engineering disciplines must collaborate on the route challenges. The authors have included a route "Ground Truthing"[2] guide in Appendix 1 that has been used by firms in the USA, Canada, and Russia to evaluate possible route alternatives that have been mapped in advance using topographic maps, images, and every technical resource available before the very expensive fieldwork begins (Kreig & Metz, 2005). This includes *Temporary Land Allocation* or permission the survey from land owners or other stake holders such as indigenous tribes and original rights holders.

Environmental studies, surveys, scoping meetings, and permit review and approval processes range from a few months to as long as several years depending on challenges from non-governmental organizations, and various regulatory or governmental authorities.

[2] "Ground Truthing" is the process where a team of experts from multiple disciplines in pipeline design construction archeology and environmental sensitivities fly, drive, or walk a pipeline or other linear project route to verify the accuracy of maps and other references used to choose a pipeline router at the desktop level.

11.4.1 The Critters of the Arctic

A complete environmental assessment of the site for new facilities is required in the Arctic, regardless of the country where facilities are to be constructed. Gazprom mentions that such studies were done for the railway to the Yamal and identified precious fish, birds, and terrestrial habitats that require appropriate consideration.

11.4.2 Oil Spill Prevention and Recovery

Any new development in the Arctic is faced with the risk of an oil spill. Their prevention or early detection is essential to the success of the overall economics of the project, as well as being required by regulatory authorities. The terrain on land and the trajectory of a spill in streams or at sea require extensive computer modeling, followed by planning and implementation of clean-up and repair capabilities.

11.4.3 Removal and Restoration Plans

As the initial designs are being prepared, so must the plan for how the facilities will eventually be removed and the environment returned to a natural state. Recognition of this requirement will likely influence how facilities are designed also.

11.4.4 Waste Management

Waste management in the Arctic is an astoundingly difficult issue. It has to be considered from the very outset of the project. Due to the low temperatures, biological process that normally consumes human and eatable waste simply does not occur unless they are managed in a temperature-controlled structure that is designed for Arctic conditions.

Solid waste generated by construction is a problem that simply does not solve itself in the Arctic either. Most has to be hauled out to designated waste facilities hundreds of kilometers south of the Arctic.

11.4.5 Operations Integrity Management System (OIMS)

OIMS stands for Operations Integrity Management System.[3] It is a system or collection of procedures, training, reporting, and documentation to manage systems transporting hydrocarbons and maintaining facilities with a very high level of integrity; therefore, accidents are rare, there are no losses of containment, and everyone and every piece of equipment is reliable and cost-effective. In short, OIMS sets high standards of achievement and makes sure each and every employee, contractor, and vendor does their job correctly the first time. This is essential in the Arctic where corrective action costs are astronomical. The concepts that apply in temperate climates apply in the Arctic, but are much more important in this region (Scrivner & Alexander, 2008). Equipment maintenance needs to be scheduled by computer databases. Several such platforms are available, such as Maximo (one of the authors has experience with). Any such system should generate Work Orders that:

- Optimize and automate asset management and maintenance to improve operational performance;
- Use best-practice industry data models and workflows to accelerate industry transformation;
- Customize workspaces, work plans, and add-ons to guide teams to critical tasks and data;
- Build preventive and prescriptive approaches to maintenance;
- Enhance investor return on assets with financial and performance analytics.

To ensure operations, integrity workers must be provided with:

- A written job description clearly communicating management expectations;
- Thorough training on their specific responsibilities following well-established principals and procedures;
- Information of the laws and regulations that impact their work and the environment mitigation requirements required under the permits governing operations;

[3] Document authored by Mr. J. B. Holt Jan 28, 1993, in the archives of the author.

- The appropriate resources to accomplish the job without failure;
- The correct tools for every job they are to perform.

They must also constantly seek and document ways to improve performance, reduce cost, and avoid waste.

11.5 Conclusions

Heavily rooted in the industry field, the contributors have arrived at the following findings:

- Nature preservation measures are of the most significant conditions need to be met for successful Arctic projects.
- The majority of the Russian Arctic is a remote area, with hostile conditions and very few exceptions. Therefore, any project which could be considered a medium or a large project automatically becomes a megaproject in the Arctic. Capital investments are rocketing up if compared with the central part of Russia. The same is applicable to Sub-Arctic regions in Russia. Those who come to the Arctic shall know the fact that the original cost estimate for the Trans-Alaska pipeline system—prepared by experts that had never worked in the Arctic—turned out to be more than ten times less than the final cost.
- With respect of the aforementioned, a special attention needs to be paid to hydrometeorological data and appropriate surveys. These surveys are one the most important underlying components for project implementation in the Arctic.
- A meticulous planning is needed for these conditions. As mentioned above, many activities are limited by weather windows. A typical situation might be when someone is late with their activity by 3–4 days; under different conditions, this would be not a huge disaster, but the weather window closes in the Arctic, and therefore, this activity is moved to the next year with all the corresponding consequences. One of the items that need to be planned is lighting during the polar night. Nothing could happen without lighting during almost half of the year. Another aspect is the rotation plan, which needs to be carefully though through.

- Modern best practices should be widely implemented—e.g., Advanced Work Packaging, among others—helping to catch the schedule and improve efficiency.
- Design and engineering solutions are greatly affected by the following requirements multiplied by hostile Arctic conditions:
  - The vulnerability of the Arctic nature;
  - The need to keep the permafrost frozen and minimize the impact on it;
  - The lack of sunlight during polar nights;
  - Excessive sunlight during polar days;
  - Extremely low temperatures in the winter period;
  - A large temperature range;
  - Excessive wind loads;
  - Icing of the structures;
  - Excessive snow load;
  - Heavy snowfall and snow transfer;
  - Corrosion;
  - Reliability requirements;
  - Oil spills preventive planning;
  - Delivery of materials;
  - Polar day—polar night considerations;
  - Maintenance considerations;
  - Fire safety, including secondary containment.

Appendix 1: Pipeline Route Recon and Ground Truthing Guideline Used on Russian and USA Mega Projects

Guidelines for Field Personnel

Collecting Data for Construction Feasibility Studies or Pipeline Construction Estimates and Ground Truthing

Planning and Preparation

Field trips are expensive, difficult to arrange, and time consuming. Every effort should be made to plan for the most efficient and effective execution of field data gathering far in advance of the occurrence of any actual site visits. Prior to departure, the field representatives should become completely familiar with any previous screening estimates of the pipeline construction effort. Thorough study of maps and previously published material about the terrain along the anticipated route should be reviewed and understood. Using such resources, the field representative should prepare a prioritized list of sites of highest interest (such as difficult swamps, mountain passes, congested areas, and natural hazards) that are to be documented.

Where possible, field notes should be supplemented with latest technology innovations such as GPS coordinates, photos, a well-organized and annotated photo diary, videotapes, satellite photography, and locally acquired maps.

Field Notes

Field notes should be taken and recorded on standard, mutually accepted forms that might be similar to the examples attached, entitled "Tabulation of Pipeline Field Observations." The words "mutually acceptable" mean that the field representative has met with the estimators before leaving on the field trip, and they had reached mutual understandings on the level and kind of data to be gathered, as well as the format the data is to be recorded in.

Appendix 1 outlines the various data required for a construction estimate. Items I, II, III, & IV of the Appendix 2 define terrain classifications, ground cover, soil conditions, and special conditions, as they are to be interpreted when taking and recording field notes.

Item V of the Appendix 2 outlines additional general information that must be collected in order to establish the best construction approach to be employed.

Item VI of the Appendix 2 outlines other information required to prepare an accurate estimate of installation costs.

In cases where a field office has been established for the project, every effort should be made to enlist as much assistance as possible from said office in obtaining relevant data for *Items V & VI of the Appendix*2.

Spread Design

"Gang Sheets" for construction operations should be evaluated in the field, during the site visitation process, and marked with the estimator's impression of how they might be adjusted for observed field conditions. The field evaluation of the adjusted gang sheets should include estimates of the anticipated production of each crew. Operations included in the attached gang sheets are:

R/W Grade—Regular
R/W Grade—Rock
Stringing
Above Ground over Permafrost
Trench—Regular
Trench—Rock
Bending
Line Up
Weld
Treat
Lower & Tie In
Backfill & Clean Up
Casing
River Crossings

A set of blank gang sheet forms is included for documentation/estimation of crew needs that can be foreseen by visiting the site. If an estimate has been prepared before the field reconnaissance document, any changes felt appropriate due to obvious but previously unanticipated site conditions can be made.

Logistics Support Plan Data Review

While in the field, the observer should visit all the proposed beach landing points, rail yards, receiving ports, pipe staging yard sites, pipe storage yard sites, and the proposed sites for double jointing yards.

Proposed construction camp sites should be visited and then evaluated, photographed, and commented on in field notes with respect to any unfit-for-purpose issues spotted in the field. If floating camps are contemplated, mooring locations and nearest infrastructure should be visited and documented.

Spread Breaks Field Confirmation

The field observer should visit the vicinity of the proposed spread breaks to verify that they are appropriate and document the field conditions.

Appendix 2: Outline for Notes on Field Observation for Pipeline Construction Estimates

Title: Recon standards	
Field reconnaissance and ground truthing of Onshore and offshore pipeline transportation alternatives	Rev. Page: 18 Date: 12/7/21

All field notes should be taken and recorded by segments along the mainline and reference by Mileposts shown on maps to accompany notes. Segments should not be longer than about 5 miles in bad terrain or 10 miles in good terrain.

Route Surface Condition Classifications

Terrain conditions	*See page pictorial representation of terrain classifications*
Classifications	
1. Flat	Generally flat ground, most bends at roads or creeks. 0–10 bends/mi
2. Slightly rolling	Few bends at crests and hollows of slopes. 10–20 bends/mi
3. Rolling	Rolling hills with several small bends or some large ones to get over tops or through bottoms. Usually, most bends at top or bottom of hills, with relatively straight slopes. 10–30 bends/mi
4. Washboard	Lots of light bends, very few large ones. Choppy ground but slope is flat enough to cause no problems. 20–80 bends/mi
5. Sharp choppy	Short steep slopes, lots of bends with many of them large. Some winch work and some shooflies. 30–80 bends/mi
6. Rough	Considerable winch work and need for shooflies
7. Mountainous	Long steep slopes, many too steep to work on
8. Swamp (wetlands)	Wet ground that will slow up progress. Amount should equal total of soil condition (trench). Classifications No. 4, 5, and 6

Ground Cover Classifications

1. Open country.
2. Cultivated farm land.
3. Predominately brush and trees to 4-inch diameter (note orchards).
4. Light timber (predominately trees 4–12-inch diameter).
5. Medium timber (predominately trees 12–24-inch diameter).
6. Large timber (predominately trees over 24-inch diameter).

NOTE: For classifications 3, 4, 5, and 6, density should also be noted:

(a) Scattered
(b) Medium
(c) Dense

Soil Condition Classifications (Trench)

1. Good	Dug with ditching machine
2. Boulders and loose rock	Dug with hoe without shooting. Indicate availability of padding material
3. Rock	Requires drilling and blasting to excavate. Indicate amount top soil over rock, and availability of padding material
4. Soft ground	Soft or wet ground but which can be crossed without building corduroy work pad
5. Crossable swamp	Swamp where timber available and which can be crossed by building corduroy work pad
6. Non-crossable swamp	Swamp where timber not available for corduroy or which cannot be crossed even with corduroy. Will be necessary to shove (push) pipeline through these areas. Indicate if hoe can work on mats. Also indicate how mainline spread can get around it

Special Classification Conditions

1. Creek crossings	Can be forded by equipment on R/W. Indicate number
*2. Small river crossings	Not fordable on RIW; up to 300 feet wide. Indicate number and names
*3. Large river crossings	Over 300 feet wide. Indicate number and names *NOTE: On small and large rivers indicate the following: • Width of water • Width to high ground, both side • Nature and height of banks • Nature of bottom (materials and unevenness) • Depth of water (both high and low water) • Probable current (at both high and low water) • Possibility of scour and bad ice conditions • Access to river and means of getting mainline spread around or across
4. RR crossing	Number and casing length
5. Cased highway crossings	Number and lengths
6. Uncased road crossings	Average number/mi (indicate farmers' driveways)
7. Fences	Average number/mi
8. Restricted or congested construction R/W	Amount
9. Abnormal amount of side-hill on construction R/W	Amount
10. Special backfill and clean-up requirements, such as disposal of timber, stumps, rock, or backfill of R/W cuts	Amount
11. Nature and extent of rock which will require blasting in order to build adequate construction RIW or access roads	Amount
12. Spans	Describe with dimensions and foundation conditions, together with suggested crossing design

General Information

1. Nature and availability of access to pipeline right-of-way:

   (a) Where railroads exist, determine spacing, capacity, and distance from pipeline right-of-way, of sidings suitable for unloading pipe. Also indicate spacing of sidings with land suitable for stockpiling pipe.
   (b) Describe nature and availability of existing access roads.
   (c) Where navigable rivers or lakes exist, document data (similar to (a) above), regarding unloading docks or wharfs, or sites suitable for their construction.
   (d) Where no access other than the pipeline route is available, describe the road construction work necessary on and off the right-of-way, in sufficient detail to estimate cost of construction. Particularly note non-fordable water crossings that must be bridged.
   (e) Expand on items (a), (b), (c), and (d) above as regards the question of movement of human resources, equipment, and materials and supplies.

2. Nature and availability of camp sites.
3. Describe communities where adequate housing is available, in relation to pipeline location.
4. Obtain copies of all available maps (preferably topographic), which would be useful in subsequent work on the construction estimate.
5. Climatological data—precipitation, temperatures, wind, etc. If seasonal, investigate local experience concerning the best period for construction and its duration. If period of construction has been predetermined, document details of past weather conditions for the period to facilitate anticipation of construction conditions.

Other Estimating Data to Be Documented on Field Trips

1. Determine labor rates. If union agreements exist, obtain copies.
2. Determine applicable payroll taxes, insurance, and other fringes.
3. Determine availability and productivity of skilled and unskilled labor.
4. Determine availability of construction equipment and rental rates.

5. Determine other taxes, duties, and insurance rates applicable to the project.
6. Determine freight rates.
7. Determine availability of contractors and subcontractors capable of performing the work.
8. Where subcontractors are proposed (for such items as clearing, stringing, river crossings, etc.), attempt to get prices.
9. Determine availability of camps, caterers, and prices per day with minimum and maximum manpower guarantees.
10. Determine sources for food, supplies, fuel, grease, tools, parts and materials, and costs relative to USA.
11. Obtain copies of or information relating to local government regulations or practices affecting design and construction.

E.g.

Timber disposal requirements
Brush & stump disposal requirements
Casing requirements
Road use and loading restrictions (check for special seasonal limitations)
Blasting regulations
Rights of eminent domain
Other—document any other information which would assist in the evaluation of the project.

References

Alyeska Pipeline. (2007). The facts. Trans Alaska Pipeline System. http://large.stanford.edu/courses/2011/ph240/mina1/docs/finalfacts-2007.pdf

Decree of the President of the Russian Federation dated November 12, 2021, No. 651 "On Amending the Strategy for the Development of the Arctic Zone of the Russian Federation and Ensuring National Security for the Period until 2035, approved by the Decree of the President of the Russian Federation No. 645 dated October 26, 2020, and invalidating some acts of the President of the Russian Federation". http://publication.pravo.gov.ru/Document/View/0001202111120001 (in Russian).

Environment Canada. (2018). *Wind chill*. https://ec.gc.ca/meteo-weather/80b0f2af-9697-4bee-ab17-d401ebba5b4b/windchill_factsheet_en.pdf

Gazprom. (2021a). *Obskaya-Bovanenkovo Railroad*. https://www.gazprom.com/projects/obskaya-bovanenkovo/

Gazprom. (2021b). *Yamal*. https://www.gazprom.com/projects/yamal/

Google Earth. (2021). *Yamal facilities*.https://www.google.ru/intl/ru/earth/

Gudmestad, O. T., Zolotukhin, A. B., Ermakov, A. I., et al. (1999). *Basics of offshore petroleum engineering and development of marine facilities: With emphasis on the Arctic Offshore: Norwegian petroleum directorate*. Statoil.

Kreig, R., & Metz, M. (2005). *Lessons learned from fifty years of terrain evaluation in northern regions, especially as applied to major pipelines*. Institution of Civil Engineers. http://www.kreig.net/archive/misc/kreigmetz2005.pdf

Mahlbauer, W. K. (2003). *Pipeline risk management manual*. Gulf Professional Publishing.

National Weather Service. (2021). *Wind chill chart*. https://www.weather.gov/safety/cold-wind-chill-chart

Oswell, J. M. (2021). *Soil mechanics for pipeline stress analysis: 2nd edition*. Naviq Consulting Inc.

Pinterest. (2021). *Arctic regions*. https://www.pinterest.ru/pin/92886811042959122/

Roscow, J. P. (1977). *800 Miles to Valdez*. Prentice Hall.

RSN 31-83. (2004). *Code of practice for geotechnical surveys for construction in permafrost*. PNIIS (in Russian).

Russian Government Resolution dated November 16, 2021, No. 1946 "On Approval of the List of Regions of the Far North and Localities Equated to the Regions of the Far North, in order to Provide State Guarantees and Compensations for Persons Working and Living in these Regions and Localities, Invalidation of some acts of the Government of the Russian Federation and Recognition as Invalid on the Territory of the Russian Federation some acts of the Council of Ministers of the USSR". http://government.ru/docs/43820/ (in Russian).

Scrivner, R., & Alexander, C. (2008). Elements of an engineering-based integrity management program. In *Proceedings of IPC2008 7th International Pipeline Conference*. http://compositerepairstudy.com/downloads/ipc_2008-64492.pdf

Simmons, A. J., & Poli, P. (2015). *Arctic warming in ERA-Interim and other analyses*. https://rmets.onlinelibrary.wiley.com/doi/abs/10.1002/qj.2422

SP 11-114-2004. (2004). *Site investigation on the continental shelf for offshore oil and gas facilities construction*. PNIIS (in Russian).

SP 482.1325800.2020. (2020). *Engineering hydrometeorological survey for construction: General regulations for execution of work*. Standardinform (in Russian).
VSN 013–88. (1990). *Main and field pipelines construction in permafrost*. VNIIST (in Russian).

CHAPTER 12

The Northern Sea Route as a New Global Route for Energy Resource Transportation: Problems and Prospects

Alexey G. Kazanin, Marina A. Kazanina, Alexey I. Andreev, and Vladimir S. Osipov

12.1 Introduction

Modernization and development of the infrastructure of the Arctic transport system are identified as strategic priorities of the state policy of the Russian Federation in the Arctic.

A. G. Kazanin · M. A. Kazanina
OJSC "Marine Arctic Geological Exploration Expedition", Moscow, Russia
e-mail: a.kazanin@mage.ru

M. A. Kazanina
e-mail: marina.kazanina@mage.ru

A. I. Andreev · V. S. Osipov
Lomonosov Moscow State University, Moscow, Russia
e-mail: andreev@fgp.msu.ru

V. I. Salygin (ed.), *Energy of the Russian Arctic*,
https://doi.org/10.1007/978-981-19-2817-8_12

The key element of the Arctic transport system is the Northern Sea Route (NSR). In addition to the water area directly related to the NSR, it is adjacent to the vast territory of Russia, with such large rivers as the Northern Dvina, Pechora, Ob, Yenisei, Lena, Yana, Indigirka, and Kolyma. These river highways form a united system of waterways, connected in the north by the sea route.

The Northern Sea Route is one of the main transport corridors—in many aspects the only one—for the movement of goods in the Arctic regions of Russia. Developed logistics is utterly necessary for the large-scale development of the Arctic resources and the socio-economic growth of the Arctic macroregion.

12.2 The Northern Sea Route as the Key Element of the Development of Energy Resources in the Russian Arctic

For a long time, logistics professionals and shipowners did not view the Arctic as a viable option to reduce traffic from Asia to Europe or from Asia to North America. However, in the twenty-first century, the situation has changed dramatically. Today, the Arctic region is warming up twice as fast as the rest of the planet, with sea ice, snow cover, glaciers, and permafrost declining sharply in recent decades. In the past, only powerful nuclear-powered icebreakers were able to navigate the Arctic seas; today, even commercial ships are able to navigate the region from about July to October, although sometimes only with the help of experienced pilots and icebreaker escorts (Zelenkov, 2019). Research shows that by 2060, ships with reinforced hulls will be able to operate in the Arctic for nine months a year. According to the National Snow and Ice Data Center in Colorado (NSIDC), in July 2018, the Arctic was covered with 5.7 million kilometers squared of ice, and in the first two weeks of August, the area of ice cover decreased by 65 thousand km^2 every day. Scientists at the University of Newcastle in northern England estimate that ice has receded by about 10 percent every 10 years for the past 30 years.

V. S. Osipov (✉)
MGIMO University, Moscow, Russia
e-mail: vs.ossipov@gmail.com

Of the northern polar routes between North America, Europe, and Asia, two can be considered commercial: the Northwest Passage (NWP) through the Canadian Arctic and the Northern Sea Route (NSR) which runs along the northern shores of Russia.

The NSR crosses the seas of the Pacific and Arctic oceans (Kara, Beringovo, Barents, Chukotskoe, Laptev, East Siberian) and runs along the Siberian coast, connecting the Russian Far East ports and the mouths of northern rivers into a single shipping transport system. According to experts, by 2030 the waterways of the NSR will be ice-free for six months a year. In the more distant future (between 2035 and 2050), the NSR will be fully navigable for most of the year—or even all year round—without the obligatory escort of icebreakers. The route is expected to be ice-free in summer by the middle of the twenty-first century. Thus, global warming makes the NSR the most promising commercial route. This conclusion is also confirmed by the statistics of cargo transportation along the NSR for 2018: 8399 thousand tons of liquefied gas (LNG), 7810.5 thousand tons of oil, 805.4 thousand tons of condensate, 290.8 thousand tons of coal, 43 thousand tons of ore, and 2340.1 thousand tons of general cargo (Morskiye Porti, 2019). According to the Russian Federal Agency for Maritime and River Transport, the volume of cargo transportation along the NSR tends to increase. The growth in the volume of cargo transportation is provided by real economic needs—the transportation of LNG, coal, the increase in gas and oil production, and northern delivery through the Arkhangelsk transport hub.

In 2019, the tonnage of shipments along the Northern Sea Route amounted to 31 million tons, and in 2020, it was 32 million tons. Before that, starting in 2014 (when their volume was only 4 million tons), they grew annually. The main growth since 2018 has been observed in liquefied natural gas. While in 2018 its volume in transportation along the NSR amounted to just over 8 million tons (before which its share in traffic was very small), in 2019–2020 it approached 20 million tons. From 2014 to 2017, growth was mainly driven by oil. In the structure of transportation through NSR in 2020, the share of LNG was 59 percent, oil 24 percent, general cargo 11 percent, gas condensate 3 percent, coal 1 percent, oil products 2 percent, and ore concentrate (gold) 0.2 percent. Since 2017, the main share of transportation has been carried out by ice-class vessels Arc7; in 2020, their share in transportation exceeded 80 percent. Since 2019, ships with a deadweight of 90–100 thousand tons have been passing along the NSR.

The lion's share of LNG supplies in 2020 was carried out to Europe; only a small part of liquefied gas went to Asia (only in September and October 2020 were supplies to Asia equal to supplies to Europe). LNG goes to Europe mainly to ports in France, Belgium, the Netherlands, Spain, and Great Britain. LNG supplies to Asia were carried out mainly to China and to a much lesser extent to Japan and Taiwan. A very small amount of LNG—less than 0.05 million tons—was supplied to Asia via LNG from Europe. Oil and condensate are transported mainly from the Novoportovskoye field. Iron ore accounted for 78 percent of the volume of transit traffic along the NSR. In 2020, the volume of transit traffic on the NSR exceeded 1 million tons (before that, it reached such values in 2011–2012), while more than 90 percent of them were carried out from west to east, and not from east to west (it should be noted that this excess has taken place in the last decade; China and other Asian countries are still weakly involved in the development of the NSR). Iron ore received its overwhelming share in transit traffic along the NSR only in 2020; before that, in 2011–2012, the largest share of transit cargo was oil, oil products, and gas condensate. All iron ore and oil products over the past decade were transported from the east, while coal and equipment were transported from partly east to west. Transit transportation over the past decade has been carried out mainly by vessels of the Arc4 class. In 2020, the tonnages of 40–80 thousand tons and 80–120 thousand tons prevailed in the deadweight of ships. The icebreakers Arktika, Yamal, 50 Let Pobedy, Taimyr, Vaygach (Atomflot), Krasin, and Admiral Makarov (Rosmorport) operate in the water area of the NSR, navigating in the ports of the NSR Sabetta, Dudinka, Dixon, Khatanga, Tiksi, and Pevek.

In accordance with the general plan for the implementation of the "Strategy for the development of the Arctic Zone of Russia and ensuring national security until 2035," mechanisms for accelerated economic and social development of the Arctic territories should be developed in 2020–2024, including the creation of a regulatory framework for the functioning of the special economic regime of the Arctic Zone. Already at this first stage, it is planned to use a certain new model for the implementation of economic projects on the continental shelf, to actively develop the Northern Sea Route, to construct a number of ships, to modernize the energy infrastructure of the Russian Arctic (including the transition from diesel fuel to LNG), to stimulate research within the framework of creating a state system for monitoring and preventing the negative consequences of the degradation of permafrost, and to intensify international cooperation—including in science and technology—on the development of the Arctic.

The second stage of the Strategy implementation covers 2025–2030 and is intended to ensure a significant increase in the competitiveness of the economy of the Russian Arctic through the introduction of a new special economic regime and additional investments. Year-round navigation in the Arctic should be ensured, appropriate ships built, a communications infrastructure for the Northern Sea Route and the Arctic in general should be created within the framework of a highly elliptical space system for "obtaining hydrometeorological data of high temporal resolution for the polar region of the Earth," and finally, new technology (including robotic, unmanned, and portable equipment) should be created and put into operation providing for both an increase in labor productivity and an increase in the safety of industrial activities.

At the third stage of the implementation of the Strategy (2031–2035), it is actually planned to complete the transformation of the regional economy with the formation of a highly efficient, environmentally friendly, transport-secured NSR industrial complex that would ensure profitable oil and gas production (including on the shelf), as well as the processing (LNG, petrochemicals) and transportation (NSR, pipelines) of products to foreign markets and domestic consumers with minimal negative impact on the environment.

Among the most important national goals for 2024, identified in 2018 by the President of Russia, there is a goal of reaching the volume of 80 million tons in the transportation of goods along the Northern Sea Route (Krutikov et al., 2020). In 2035, however, only 10 million tons are "strategically reserved" for transit cargo. This is somewhat at odds with the notion of the Russian Arctic as the future transport corridor of global importance (Andreeva et al., 2018). The transport development of the Arctic is significant for the transportation of oil and gas products (Pospelova et al., 2020). In addition, transport development is highly important for ensuring food security of the Arctic macroregion, including in terms of the sustainable development of modern agriculture and the formation of sustainable food systems (Shchevev & Bykov, 2020). The fishery industry in the Arctic also needs scientifically substantiated support and coordination with other industries that ensure the development of the region's economy and its food security (Beljaev et al., 2020).

With regard to the oil and gas industry, strategic planning based on creating a digital image of the Russian fuel and energy complex within the global world would also be advisable. Work is underway to create strategies for digital transformation of individual companies in the fuel and energy complex.

An example of the current approach to the digitalization of fuel and energy companies is the decision reached after a joint meeting of the sections of the Scientific and Technical Council of PJSC Gazprom and the Scientific Council of Gazprom VNIIGAZ: they agreed on the need for the early deployment of a complex of scientific and technical work to develop promising technologies for the creation of domestic sensor and control equipment for the development of offshore oil and gas fields to overcome the existing lag behind the leading foreign countries in this area. At the same time, plans for a general digital modernization (transformation) of companies and the industry as a whole are still lagging behind, especially in comparison with foreign experience. However, it is evident that such processes are expedient as soon as possible, because digital modernization of oil and gas production processes in the Arctic Zone will ensure environmental friendliness, the safety and high efficiency of commercial gas and oil production, and a significant reduction in the impact of the human factor on the nature and reliability of technology (Eremin & Dmitrievskij, 2018). This approach—and, indeed, the entire process of digitalization in Russia—can be called "digitalization from below," a process that is institutionalized with the need to address newly arising problems, or when the lag behind advanced countries becomes all too evident (not so much in terms of the use of foreign developments, but rather in terms of a simple understanding of how to correctly apply foreign systems in Russian practice). The difficulties of digitalization are also due to the relatively weak readiness of business entities (more precisely, local staff) for this process. Thus, the obligatory transition of fishing vessels operating north of the 75th latitude to the use of the domestic Gonets satellite communication system has been postponed to 2020, since workers in the fishing industry simply did not have time to master and debug this system (Chereneva, 2019).

At the same time, the creation and development of a distributed system of digital modeling of the Russian Arctic as a unified ecological and economic system, taking into account the factors of strategic national security in Russia and the global world, should not be hindered by the obvious distribution—and one might even say "polycentricity"—of the institutions and instruments of public administration in the Russian Arctic. Thus, the development of the Northern Sea Route is entrusted to the State Atomic Energy Corporation Rosatom. In accordance with the Order of the Government of Russia dated December 21, 2019, No. 3120-r, it will become the parent organization implementing a list of measures for the development of the NSR infrastructure until 2035 (together with the Ministry for the Development of the Russian Far East, providing an annual report to the government on the progress of these events). The main part of the specific activities on this list is scheduled for the next 5 years. In particular, the list provides for the introduction of digital technologies for the development of multimodal transportation in the water area of the NSR (until 2023, Rosatom State Corporation alongside the Ministry of Transport) and the preparation of proposals for the creation of a platform for digital services serving the NSR (until the end of 2021, Rosatom State Corporation alongside the Ministry of Telecom and Mass Communications of the Russian Federation and the Ministry of Industry and Trade of the Russian Federation). The participation of oil and gas companies alongside the Ministry of Energy of the Russian Federation in the implementation of the plan is not envisaged, which does not allow judging the integrity of the intentions for the unified digital management of the region in cooperation with the industries.

The modern concept of digital public administration includes not only and not so much the state provision of services to the population and organizations, but rather the "intellectualization" of the governance process itself through the automation of the data collection, data analysis, and the development of specific proposals for policies and governance. This is best achieved through the formation of a "digital image" or "digital twin" of a controlled system, which is a mathematical model or a set of models that describe the properties and dynamics of the system depending on internal processes and external influences. At the same time, such a model can also be associated with the system of rendering public services, as well as other horizontal and vertical interactions from a managerial point of view within the framework of the formation of a single state ecosystem (possibly in the next decade). This path is, apparently, invariant, since attempts to form separate GIS (state information systems) for individual industries and regions (not counting IS' own state organizations) have so far led to success only in the case of maximum centralization and generalization of the tasks (e.g., "Gosuslugi" state services). In the case of the Arctic, such a system that summarizes huge amounts of data and proposes (or recommends) governance decisions for discussion by stakeholders seems to be the only promising one. It could form the basis of the Arctic direction of the Ministry for the Development of the Russian Far East, a ground for testing this approach throughout the country. This approach, in turn, perfectly reflects the consideration of the Arctic as a key megaproject for the country's development.

12.3 Soviet Policy in the Arctic: The Early Stages of Energy Resource Development in the Russian Arctic

Soviet policy toward the Arctic is a separate complex and multidimensional area of research, which was actively developed by Soviet and Russian scientists. Timoshenko, of the specialists in the field of USSR policy in the exploration and development of the Arctic, pays attention to the issue of continuity in the Arctic policy of the USSR and modern Russia, and concludes that a number of strategic directions of this policy persist from Soviet times into modern Russia's policies. Such directions are connected, first of all, with ensuring state interests in the Arctic territories and the desire of the state (regardless of the political system) to preserve its influence in the Arctic (Timoshenko, 2011).

In this matter, we agree in many respects with Timoshenko. Indeed, an analysis of the Arctic policy of modern Russia gives grounds to assert that a number of its directions were first identified in the so-called Murmansk initiatives—this name was assigned to a series of foreign policy proposals put forward by the General Secretary of the CPSU Central Committee Mikhail Gorbachev in his speech delivered on October 1, 1987, in the city of Murmansk (Gorbachev, 1987).

The first initiative was related to the creation of a nuclear-free zone in Northern Europe and was aimed at reducing the level of militarization of the Arctic region, the scale of military activity, and the risks of possible international military conflicts in the Arctic. The second initiative concerned the issue of limiting naval activity and the scope of the Navy and Air Force in the seas adjacent to Northern Europe (Baltic, North, Norwegian, and Greenland Seas). In the waters of international straits and on the intensively used shipping routes, it was proposed to completely prohibit such activities.

Of greatest interest for our study is the third initiative related to the development of peaceful international cooperation in the development of Arctic resources. Gorbachev proposed developing a common international concept for the rational development of the northern regions (a proposal that has not yet been fully implemented, despite some individual proposals of the Arctic Council in this area). Moreover, Gorbachev put forward an even bolder proposal (especially considering the context of his time) to develop a unified energy program for the countries of Northern Europe, the "material" basis of which was to be the exploration and production of "boundless" Arctic hydrocarbon reserves, and the conceptual basis of which was the cooperation of these countries in the development of equipment and technologies necessary for effective exploration and development of Arctic deposits, taking into account their complex (and often extreme) natural and climatic conditions. As concrete examples showing the USSR's readiness for such cooperation, Gorbachev cited proposals to attract Canada and Norway (as well as, possibly, other states) to the creation of mixed enterprises for the extraction of oil and gas from fields located on the North Sea shelf of the Soviet Union (in particular, the resources of the Kola Peninsula).

At the moment, this initiative also cannot be considered fully implemented, since a unified energy program for the Nordic countries has not yet been formed. However, to a large extent, Gorbachev was perhaps the first to identify the key priorities for the development of Arctic offshore fields—more than 30 years later, the main Norwegian oil company Equinor (Statoil) declared that the key innovation in the exploration and development of Arctic oil and gas reserves should be international cooperation, which will allow standardizing the developed technological solutions and will reduce the cost of the production process. Certain steps toward the implementation of this initiative did take place: for example, in the form of cooperation between Russia and other Arctic states in the energy field.

The fourth initiative concerned the scientific study of the Arctic; in this area, the leadership of the USSR proposed intensifying the international exchange of accumulated scientific experience in the study of the Arctic in the framework of various disciplines. It is noteworthy that a proposal was put forward to create an Arctic Scientific Council—this proposal was not implemented, but, in all likelihood, the idea itself contributed to the formation of the international discourse that led to the creation of the Arctic Council in 1996.

A certain role in the deployment of international environmental discourse in relation to the Arctic was played by the fifth initiative, which called for the intensification of cooperation between the Nordic countries in matters of environmental protection in this region. Almost a decade later, this initiative has found practical reflection in the priority areas of the Arctic Council's activities.

Finally, the sixth initiative, concerning the possibility of opening the Northern Sea Route, is gaining more and more relevance today due to the melting of the polar ice, as a result of which this route remains accessible for navigation for more days a year than before.

Russia's Arctic policy in the 1990s can be characterized in different ways. Some researchers are convinced that after the collapse of the USSR, Russia "disappeared from the landscape of Arctic security" and was revived as an active participant in the international Arctic discourse only in the 2000s (Laruelle, 2014: 16). However, this statement should not be understood as a complete absence of Arctic policy in Russia during this period—it was promoted, albeit not very actively, in this difficult time both in institutional and in legal terms (as will be seen in more detail below); however, during this period, attention was focused not on the Arctic as an international region, but on the problems of the Russian Arctic territories proper. Moreover, both the list of problems identified as the most important and the measures for their solution often present a striking contrast to the current priorities of Russia's Arctic policy.

Indeed, during the 1990s, characterized by a sharp drop in the standard of living of the Russian population as a whole, there was simply no talk of a purposeful Arctic policy aimed at the strategic development of the northern territories—the Arctic territories also experienced a drop in the standard of living of the population (in particular, northern benefits were greatly reduced, important sectors of the Arctic economy experienced a dramatic drop) (Ponomarev, 2009). Leonov notes that "the crisis of the planned economy and its dismantling in the 1990s hit the industrial potential of the Far North more than in other regions of the country" (Leonov, 2013: 62). During this period, the state policy was based on the idea that the North does not need a permanent population, but rather temporary residents; that for the non-indigenous population, it is necessary to move "from a policy of residence to a policy of stay"; and, accordingly, a rotational expeditionary method should be introduced as widely as possible. This was expected to make it possible to minimize the necessary investments in the development of the infrastructure of the northern regions. For all of the above reasons, the country's Arctic Zone during this period increasingly lost contact with the rest of the territories of Russia.

12.4 Russia's Arctic Policy: The Development of Energy Resources in the Russian Arctic

The institutionalization of the Arctic policy of modern Russia dates back to the creation in 1991 of a special authorized body—the Interdepartmental Commission for Arctic and Antarctic Affairs (named "the State Commission of the RSFSR for Arctic and Antarctic Affairs" between 1991 and 1992). This body functioned until 2004. However, the activities of this commission can hardly be considered identical to the pursuit of a proactive policy in the Arctic region—its tasks did not imply promotion of their own projects and initiatives (or, moreover, holistic strategies for the development of the Russian Arctic), but were reduced to the consideration of various kinds of scientific, technical, environmental, and economic programs and projects, as well as proposals from other government agencies related to the Arctic. In the first half of the 1990s, a number of structures were formed that were connected with the Arctic in one way or another (in 1992, the All-Russian Scientific Coordination Center for Complex Problems of the North, the Arctic and the Life of Indigenous Peoples of the North; in 1994, Committee for the Affairs of the North and Indigenous Peoples in the Federation Council). However, the focus of their attention was one specific aspect of the development of the Arctic: namely the situation and problems of the indigenous peoples of the North. They did not make any attempts at planning and implementing Russia's comprehensive Arctic policy.

In general, according to the calculations of the researcher of the history of Russian Arctic policy Tamitsky, the number of regulatory documents on the social, economic, and environmental aspects of Russia's Arctic policy, adopted in the period from the beginning of the independent history of modern Russia until 2012, amounted to about a hundred (Tamitsky, 2012). Among these documents, Tamitsky considers the Federal Law of June 19, 1996 No. 78-FZ "On the Foundations of State Regulation of the Socio-Economic Development of the North of the Russian Federation" a prominent document, which, in his opinion, was a "prototype" of the strategies of Russian Arctic policy adopted in subsequent years. The key point uniting this document with subsequent strategies, according to Tamitsky, was the priority of ensuring the development of the Russian Arctic and protecting Russia's long-term national interests in the Arctic region. Indeed, this law was essentially the first attempt to systematically reflect the main priorities of Russian policy in the Arctic region. Some of these priorities will now be outlined closely.

In the Decree of December 31, 1997 No. 1664 of the Government of the Russian Federation, the priority of the state policy with regard to the labor resources of the North was still proclaimed "optimization of the population size"—as was discussed above, in No. 78-FZ, this wording hid actual support for population outflow from the northern territories, to which the state policy did not oppose any powerful counter-incentives. This document also dealt mainly with measures to support the population leaving the northern territories. It was proposed to attract only specialists to temporary residence in the North through labor contracts (a significant part of the work involved a rotational method). The resolution also included a number of measures aimed at improving the socio-economic situation of the indigenous peoples of the North (e.g., it was proposed to lower the retirement age for representatives of these peoples). It should be noted that the resolution instructs the relevant state bodies to prepare proposals for ensuring the functioning of the Northern Sea Route.

The most complete strategic vision of the new Arctic policy of Russia was reflected in the "Fundamentals of the state policy of the Russian Federation in the Arctic for the period up to 2020." The draft of this document was approved by the Government of the Russian Federation back in 2001; however, another 7 years passed before the document entered into force—in 2008, the Fundamentals were approved by Presidential Decree, where the list of subjects of the Russian Federation belonging to the Russian Arctic was defined. The following points were indicated as the main interests of Russia in the implementation of its national Arctic policy:

- Using the Arctic Zone as a strategic resource base, ensuring the solution of the tasks of the country's socio-economic development,
- Preserving of the Arctic as a zone of peace and cooperation,
- Saving the unique ecological systems of the Arctic,
- Using the Northern Sea Route as a national unified transport communication of the Russian Federation in the Arctic.

The most elaborated and comprehensive document of strategic planning in relation to the Arctic is the Fundamentals of State Policy of the Russian Federation in the Arctic until 2035. This document refers to

Russian national security and is globally significance, declaring interests and defining strategic priorities and guidelines for the activities of the Russian state in the Arctic. Issues of ensuring military security receive a lot of attention in this document.

Among the primary national interests of Russia in the Arctic, this document lists the development of the Arctic Zone of the Russian Federation as a strategic resource base and its rational use to accelerate the economic growth of the Russian Federation. In terms of threats, low rates of geological exploration of promising mineral resource centers in the Russian Arctic are mentioned. The goals include acceleration of economic development. Particular tasks include the expansion of private investors' participation in offshore projects; infrastructure development of mineral resource centers associated with the Northern Sea Route; increasing the volume of geological exploration work through public and private investments; bringing the system of secondary vocational and higher education in the Arctic Zone of the Russian Federation in line with the forecast of the need for qualified human resources; and state support for the economically active population ready to move to the Arctic. An important place is given to the greening of the development of oil and gas resources (the creation of a system for the prevention and elimination of the consequences of emergency oil spills).

In general, the document focuses on the issues of ensuring a steady improvement in the quality of life of the population, which is the main condition for the sustainable development of the Russian Arctic. Considerable attention is paid to the development of research. In general, despite some novelties (such as accelerated development of the Northern Sea Route), this document is a continuation of the evolution of strategic documents in the sphere of Russia's Arctic policies.

In November 2009, the new Energy Strategy of Russia for the period up to 2030 came into force. First of all, this featured an extremely important novelty (from an administrative and managerial point of view): a direct indication that the provisions of this Strategy were to be used in the development and further refinement of socio-economic development programs and energy strategies of the constituent entities of the Russian Federation. This instruction was intended to ensure hierarchical coherence and consistency between the federal and regional levels of energy policy, conceptual and logical connection of documents adopted at these levels, and subordination of the goals and objectives of the energy policy of the regions to the goals and objectives of such a policy set at the federal level. The provisions of the Strategy were also to be taken into account at the level that can be considered "intermediate" between the federal and regional, and designated as the "macro-regional" level—these included, for example, comprehensive programs for the energy development of the regions of Eastern Siberia and the Far East, the North-West region Russia, the Yamal Peninsula, and the continental shelf of the Russian Federation.

As in the previous Strategy, a change in the geographical distribution of hydrocarbon production in Russia is noted as one of the leading trends in subsoil use. This change was mainly due to the fact that the resources of the fields in Eastern Siberia and the Far East, the Yamal Peninsula, and the continental shelf of the Arctic seas began to be developed and exploited. To a large extent, this is connected with the second leading trend—a change in the structure of proven hydrocarbon reserves toward more and more difficult-to-recover resources. Arctic oil and gas reserves are increasingly shifting into the focus of energy policy: among the strategic initiatives for the development of the Russian fuel and energy complex, the current Strategy includes, among other things, the initiative to develop the hydrocarbon potential of the continental shelf of the Arctic seas and northern territories of Russia.

At the second (2015–2022) and third (2022–2030) stages of the Strategy implementation, the importance of the Arctic hydrocarbon reserves in the Russian fuel and energy complex becomes even more obvious: during these periods, the Arctic shelf oil and gas fields are assigned the role of a "stabilizer" of the dynamics of hydrocarbon production in Russia, a kind of strategic "reserve," which will be put into motion if necessary. They are viewed as a compensation for a possible decline in the level of production in traditional oil and gas producing regions (mainly located in Western Siberia). In addition to the stabilizing role, it is assumed that the development of these territories would contribute to the development of various types of transport infrastructure, various industries that are in some way associated with various operations of the oil and gas sector, and, finally, strengthening the role of the Northern Sea Route.

In 2017, a number of significant amendments were made to one of the fundamental documents governing the implementation of Russia's Arctic policy—namely, to the State Program of the Russian Federation "Social and Economic Development of the Arctic Zone of the Russian Federation" adopted in 2014. To achieve the main goal of the state program—an increase in the level of socio-economic development of the Arctic Zone of the Russian Federation—the document set a number of updated tasks (note that it was very laconic):

- Improving the quality of life and security of the population on the territory of the Arctic Zone of the Russian Federation;
- Creating conditions for the development of the Northern Sea Route as a national transport artery of the Russian Federation in the Arctic and the development of a system of hydrometeorological support of navigation in its water area;
- Developing science, technology, and increasing the efficiency of using the resource base of the Arctic Zone of the Russian Federation and the continental shelf of the Russian Federation in the Arctic;
- Increasing the efficiency of state governance of the socio-economic development of the Arctic Zone of the Russian Federation.

12.5 Conclusions

The Strategy for the development of the Arctic Zone of the Russian Federation and ensuring national security until 2035 sets out the task of organizing the training and retraining of human resources, but only from the angle of the human resource needs of the Northern Sea Route and not the Arctic oil and gas sector. At the same time, the emergence of new technical and technological complexes in this sector will obviously require retraining human resources and the search for new qualified human resources. A separate acute problem is the training of highly qualified researchers and scientists to create the actual fundamental and practical knowledge base of these technologies. The Strategy provides for the creation of a system of research and educational centers, the formation of a register of critical technologies, and, most importantly, their coordination; already at the level of the Strategy, a number of priority research areas have been identified. However, none of these activities is specifically tied to the needs of the Arctic oil and gas complex.

References

Andreeva, T. A., Khmara, D. S., & Firsova, A. V. (2018). Information and analytical support for the implementation of a long-term strategy for the development of the transport system in the Arctic. In *Scientific Dialogue: Young Scientist. Collection of Scientific Papers Based on the Materials of the XIII International Scientific Conference* (pp. 16–19) (in Russian).

Beljaev, V. A., Ridiger, A. V., & Sogrina, A. V. (2020). Science—For the fishery industry in the Arctic Zone: Cooperation, the main tasks of the industry and scientific ways to solve them. *Arktika 2035: Aktual'nye Voprosy, Problemy, Reshenija, 2*(2), 23–29 (in Russian).

Chereneva, V. (2019, April 8). Communication without borders. *Rossijskaja Gazeta*. https://rg.ru/2019/04/08/chto-pomozhet-arktike-stat-silnym-zvenom-v-cifrovoj-ekonomike.html (in Russian).

Eremin, N. A., & Dmitrievskij, A. N. (2018). Digital development of the Arctic zone of Russia: State and best practices. *Rossijskaja Arktika*, 3. https://energy.s-kon.ru/n-a-eremin-a-n-dmitrievskii-cifrovoe-razvitie-arkticheskoi-zoni-rossii-sostoyanie-i-luchshie-praktiki/ (in Russian).

Gorbachev, M. S. (1987). The unfading feat of the heroes of the Arctic. *Pravda*. 10/02/1987. No. 275 (25262) (in Russian).

Krutikov, A. V., Smirnova, O. O., & Bocharova, L. K. (2020). Development strategy for the Russian Arctic: Results and perspectives. *Arktika i Sever, 40*(40), 254–269 (in Russian).

Laruelle, M. (2014). *Russia's arctic strategies and the future of the Far North*. ME Sharpe.

Leonov, S. N. (2013). Development of the Arctic Zone of the Far East: State and problems. *Upravlencheskie Nauki, 4*(9), 57–70 (in Russian).

Morskiye Porti. (2019). Transportation of goods by the Northern Sea Route for 2018. *Mor. porty, 1*, 78 (in Russian).

Ponomarev, I. A. (2009). North of Russia: Ways of its socio-economic development. *Nacional'nye Interesy: Prioritety i Zezopasnost', 5*, 78–81 (in Russian).

Pospelova, I. V., Pomogaeva, L. N., & Mostovoj, A. D. (2020). Prospects for the development of supplies of liquefied natural gas along the Northern Sea Route. In *Synthesis of science and education in solving global problems of our time. Collection of articles on the results of the international scientific and practical conference* (pp. 66–68) (in Russian).

Shchevev, A. N., & Bykov, A. A. (2020). Strategy and management mechanisms for the development of food systems in the regions of development, the North and the Arctic of Siberia. *Jekonomicheskij Obzor, 4*(5), 27–31 (in Russian).

Tamitsky, A. M. (2012). State policy of modern Russia in the Arctic: Stages, priorities and some results. *Arktika i Sever, 6*, 1–8 (in Russian).

Timoshenko, A. I. (2011). Russian regional policy in the Arctic in the 20–21 centuries: Problems of strategic continuity. *Arktika i Sever*, *4*, 1–13 (in Russian).

Zelenkov, M. Y. (2019). Transport and logistics system of the Northern Sea Route: Prospects, problems and solutions. *Arctic: Ecology and Economy*, *4*(36), 131–140. https://doi.org/10.25283/2223-4594-2019-4-131-140 (in Russian).

CHAPTER 13

The Advantages of the Northern Sea Route as an Energy Transport Artery and Its Environment

Simona Chuguryan, Zuzana Čičková, and Tetyana Zubro

13.1 Introduction

Global warming and the interrelated issue of climate change are causing the Arctic glacier to melt and subsequently its ice sheet to decline. The Arctic region is changing and acquiring a completely different shape. These processes are so significant that they cannot be ignored; their effects on climate change are obvious all around the world. According

S. Chuguryan (✉) · Z. Čičková · T. Zubro
University of Economics, Bratislava, Slovakia
e-mail: simona.chuguryan@euba.sk

Z. Čičková
e-mail: zuzana.cickova@euba.sk

T. Zubro
e-mail: tetyana.zubro@euba.sk

V. I. Salygin (ed.), *Energy of the Russian Arctic*,
https://doi.org/10.1007/978-981-19-2817-8_13

to Gail Whiteman, Director of the Pentland Centre for Sustainability at Lancaster University, the Arctic is an incredibly important system in the global climate complex, so just as the Amazon is the lungs of the world, the Arctic resembles our circulation system and feeds into global climate change everywhere. We now know, based on observational data, that we have lost 50 percent of Arctic ice over about 50 years (Pomeroy, 2020).

With the retreat of the ice sheet, several issues are coming to the fore, such as whether the Arctic region can be developed while maintaining the sustainable development of its natural resources, the use of energy potential will damage the nature of the Arctic, whether its security will not be jeopardized, and whether the Arctic will become a new arena of power struggles or not. Can the symbiosis and the mutual integrity of the indigenous peoples and the Arctic itself be preserved, and will the uniqueness of the Arctic nature remain intact? Moreover, will the amount of drinking water be endangered for future generations? Many more such existential questions arise.

These problems are of interest to many experts internationally, especially scientists and science workers, as well as researchers and environmentalists working to eliminate the negative effects of climate change. There are also, conversely, many commercial and transport companies, entrepreneurs, businesspeople, and experts in the exploration and extraction of minerals, fishing, telecommunications networks, shipping, and other infrastructure; these see new—currently inaccessible—opportunities for their activities in the region, with the opposite interest.

The Arctic can be defined in a few different ways. In strictly scientific terms, the definition has often been limited to the area north of the Arctic Circle, or from the borderline that marks the beginning of permafrost, or the area in which the average daily temperature in the summer does not exceed 10 °C. In the context of international politics, however, the most accepted definition of the Arctic is characterized by political cooperation between the states and nations whose people live in the Circumpolar North (Thisted & Gremaud, 2020).

In a joint communication of the European Commission, the Arctic is defined as the "Arctic region", i.e. the area around the North Pole, north of the Arctic Circle (66°, 32 ′N). This area includes the Arctic Ocean and the territories of eight Arctic states: Canada, the Kingdom of Denmark (including Greenland and the Faroe Islands), Finland, Iceland, Norway, the Russian Federation, Sweden, and the USA (European Commission, 2016).

The melting of the Arctic glacier has been extremely severe in recent times, leading the Arctic states to rethink their policies and strategies, in the field of energy as well as in environmental protection. At the same time, they realized to implement changes in their interests and commitments, and to adapt them to the new opportunities offered by the Arctic region in its current state.

The old power relations between centre and the periphery are being disestablished, new centres are constituted in the Arctic, and perspective is shifting. Hence, the vantage point is moving to the north, looking south towards Europe, rather than the other way around. Consequently, identities are being reconsidered based on demands for self-determination and new awareness of the region's geopolitical significance. In the wake of all this, new alliances are being formed, and old relationships and interdependencies terminated or redefined (Thisted & Gremaud, 2020).

The effects of climate change have been very significant, especially in recent decades, and it is in the Arctic region where its effects are most visible and striking. Nevertheless, scientists have seen some cooling, although this is only a temporary phenomenon without long-term effects on the climate.

For the interior of the Antarctic continent—specifically the region near the South Pole—the winter of 2021 was among the coldest on record. At the National Science Foundation's Amundsen-Scott South Pole Station, temperatures for June, July, and August were 3.4 degrees Celsius (6.1 degrees Fahrenheit) lower than the 1981 to 2010 average at −62.9 degrees Celsius (−81.2 degrees Fahrenheit). This is the second coldest winter (June–July–August months) on record, behind only 2004, in the 60-year weather record at the South Pole Station. For the polar darkness period (from April through September), the average temperature was −60.9 degrees Celsius (−77.6 degrees Fahrenheit), a record for those months. The unusual cold was attributed to two extended periods of stronger-than-average encircling winds around the continent, which tend to isolate the ice sheet from warmer conditions. A strong upper-atmosphere polar vortex was observed as well, leading to a significant ozone hole. The ozone hole appears to have peaked as of this post, with initial measurements reporting that it is in the upper quartile (top 25 percent) of ozone reduction events since 1979 (NSIDC, 2021).

On the one hand, the warming of the region opens new opportunities for scientific research, economic development, energy policy, alternative transport, and mineral mining, which makes the Arctic one of the most interesting and attractive areas in the world. On the other hand, it is necessary to pause on the fact that increasing human intervention and activity is likely to lead to the risk of greater pollution and the devastation of nature in the Arctic, which in turn will lead to further climate change in the Arctic and the world.

In August 2021, the IceBird Summer campaign continued its observations of sea ice thickness north of Greenland and within northern Frame Strait by means of airborne electromagnetic sounding. Belter, Krumpen, Hendricks, and Haas at the Alfred Wegener Institute presented the observed average sea ice thickness of 1.7 metres (5.6 feet), among the lowest since 2001 (NSIDC, 2021).

It is already clear that climate change, technological progress, and growing global demand for resources are important accelerators for the growing interest in the Arctic region and its economic and energy potential. New opportunities are opening in the areas of international cooperation, global security, and economic development. This development potential in the Arctic makes it a new space for great power competition between Russia, the USA, and, last but not least, China. Despite their efforts to gain power, there is a degree of cooperation between them.

More recently, in 2018, Russia and the USA cooperated in creating a scheme for managing a two-way shipping traffic through the Bering Strait and Bering Sea, and in February 2021, the US Coast Guard and Russia's Marine Rescue Service signed an agreement updating a 1989 bilateral joint contingency plan for responding to transboundary maritime pollution incidents (O'Rourke et al., 2021).

Despite this, there is still a significant degree of cooperation at various levels. For example, the National Science Foundation (NSF) is a federal agency responsible for implementing Arctic research policy; the Arctic Council, created in 1996, is an international forum dedicated to addressing issues relating to the Arctic; and the United Nations Convention on the Law of the Sea (UNCLOS) sets forth a comprehensive regime of law and order in the world's oceans. There is also coordination through the Arctic Coast Guard Forum (ACGF), whose goal is safe and environmental maritime activity in the Arctic region (Brocková et al., 2020).

The aim of this article is to point out the possibilities and limitations of the Northern Sea Route (NSR) as an international arterial road in the future. In addition to assessing the current situation, we offer our views on the preferences and motivations of several actors in this conflict who are pursuing their own economic and energy interests in the Arctic region. In contrast to the economic and energy benefits resulting from the use of the NSR, there is a threat of ecological consequences, which can be considered a negative externality threatening the whole of society. Many conflicts in international relations can be explained using game theory tools. Game theory is a scientific field of applied mathematics dealing with decision-making situations in which several entities (players) participate, where the term *game* represents any conflict situation between these entities (individuals, states, political parties, and the like).

In the field of international relations, game theory tools have been used for more than 50 years (Zagare & Slantchev, 2018). However, the first important applications of game theory in international relations were developed during the most intense period of the Cold War and are associated with the application of non-constant sum games (bimatrix games). Because two global actors (China and Russia) play an important role in the Northern Sea Route conflict, we will conclude with an overview of relevant bimatrix games that can be applied to predict the possible outcomes of this conflict. We will use the analysis of Nash equilibrium as the basic concept for solving bimatrix games.

13.2 Russia's and China's Energy Interests in the Arctic as a New Geopolitical Area

The Arctic is gaining the position of an important strategic and geopolitical area, becoming the focus of attention of many global actors, as well as regional and international groups. Apart from the Arctic states, other actors outside the region are also interested in gaining influence in the region. At the same time, the Arctic is once again becoming a space for geopolitical competition, the main actors of which are the aforementioned USA, Russia, and China.

At a meeting with US Secretary of State Mike Pompeo in Copenhagen on July 22, 2020, Jenis av Rana, the current Minister of Foreign Affairs and Culture of the Faroe Islands, expressed worries that the Arctic could become a battleground for the USA and other major global powers, including Russia and China (Associated Press, 2020). Both Russia and China are working to increase their influence in the Arctic and to enforce their personal interests and preferences, although each comes from a different position, and with different goals. Their aim is to strengthen their position and actively participate in activities and development, in the economic, energy, and research spheres in the Arctic.

Stephanie Pezard, Senior Political Scientist at the RAND[1] Corporation, introduces two changes that have altered the geopolitical environment in the Arctic over the past 5–10 years. One change is the increased assertiveness of one Arctic nation, Russia, in the region. The second change is the rising presence of non-Arctic states—including, but not limited to, China—in a part of the world that used to be almost exclusively of interest to Arctic states (Pezard, 2018). Russia belongs to a group of eight countries that have territory north of the Arctic Circle, so it is an Arctic state that is a member of the Arctic Council. It also belongs to the group of coastal states of the Arctic, because it has mainland coasts that face onto waters north of the Arctic Circle. This position gives it the opportunity to meet economic, political, economic, energy, and environmental interests in the region.

The Arctic region is important to Russia for several reasons:

1. Firstly, it contains major population centres; Arkhangelsk with 350,000 inhabitants and Murmansk with 300,000 inhabitants. In comparison, the largest town in the North American Arctic is Nuuk (Greenland), which has a little over 17,000 inhabitants.
2. Secondly, Russia's economy relies heavily on its oil and gas industry, and such resources are heavily present in the Arctic; for example, in the Yamal region, where Russia has recently developed a massive liquified natural gas (LNG) plant and terminal.
3. Thirdly, the NSR is important for destination shipping. A record 9.74 million tons of goods—particularly gas, oil, grain, and coal—transited through the NSR in 2017. The NSR is a major economic artery of Russia, which it intends to protect and keep under its control.
4. Fourthly, Russia—the only non-NATO Arctic coastal state—sees the Arctic as central to the protection of its territory. Seasonal sea ice makes Russia's northern coastline largely inaccessible, although

[1] The RAND Corporation is a research organization that develops solutions to public policy challenges to help make communities throughout the world safer, more secure, healthier and prosperous.

it is becoming more accessible every day as ice recedes. Russia's Northern Fleet is based in the Kola Peninsula, near Murmansk, and stores two-thirds of Russia's nuclear submarine fleet (Pezard, 2018).

In May 2021, Russia took over the presidency of the Arctic Council, listing economic growth, sustainable development, and, last but not least, national security among its top priorities during the two-year presidency. Recently, there has been a clearly visible growing interest in the Arctic from countries outside the region. One such country is China, whose involvement is increasing. China has been an observer state to the Arctic Council since 2013. It is one of the most important and powerful states to attempt to assert itself in this area. China has joint projects with several Arctic nations like Russia, Canada, Greenland, Norway, and Iceland.

China's interest in the Arctic has been brewing for a long time, but it has always had a different area of interest. In the 1980s, China participated in scientific and research expeditions organized in the Arctic region. In the 1990s, its interests gradually changed, and trade, transportation, and mining became a priority. China sees the Arctic as a global issue that cannot be solved without cooperation with countries outside of the Arctic. China describes itself as a "Near Arctic State" (January 2018, "China's Arctic Policy" White Paper), which has significant economic, energy, and investment interests in the Arctic and would like to execute the "Polar Silk Road" project:

> *While Russia and China are formally allies through the Shanghai Cooperation Organization, Russia remains wary of China's military power on its southern border and, as an Arctic nation, is irritated by the intrusion in Arctic affairs of non-Arctic states, as evidenced by its long-standing reluctance to grant observer status to these countries in the Arctic Council.* (Pezard, 2018)

China is trying to establish itself in the Arctic region in several possible ways. For example, in 2017, China's warships participated in a military exercise called "Common Sea" in the Baltic Sea along with the Russian Navy (Martin, 2017). This form of cooperation between Russia and China is perceived very cautiously by the other Arctic states, especially the USA.

The interests of Russia and China in the Arctic can be described to some extent as identical: both players try to gain more influence in the area and the most economic profit from it. However, both actors carry out their activities from different positions. Russia has an advantage in this respect, as it is included in the group of Arctic states, providing it with more advantageous position as well as more bargaining options compared to China. Geographically, Russia is the most prominent of the eight Arctic states. According to one assessment, Russia "has at least half of the Arctic in terms of area, coastline, population and probably mineral wealth. About 20 percent of Russia's land mass is north of the Arctic Circle" (O'Rourke et al., 2021).

On the other hand, China's possibilities as a non-Arctic state predispose it to considerably weaker negotiating opportunities and less influence in the region. It is true that it has recently been working with several Arctic countries to strengthen its position in the region. The main priority for both actors is the economic profit they see in the development of trade, energy policy, and transport in this area. The receding ice is uncovering new mineral deposits, which is an important and positive indicator for both countries. New mining opportunities are opening for Russia, as there are large reserves of minerals indicated. This will have a significant impact on Russia's energy policy.

Large Arctic oil and natural gas fields (those that exceed 500 million barrels of oil equivalent of recoverable oil and natural gas) are particularly crucial with respect to future oil and natural gas development, as the cost of developing oil and natural gas fields in the Arctic is so high that large fields are initially necessary to pay for the infrastructure required to later develop the smaller oil and natural gas deposits (Arctic Oil, 2009).

One Geological Survey estimates that the Arctic contains approximately 13 percent of the world's undiscovered conventional oil resources and about 30 percent of its undiscovered conventional natural gas resources. Out of the Earth's total resources of oil and natural gas, this represents approximately 22 percent (King, 2016).

A substantial fraction of Russia's oil and gas production and reserves are in the Arctic. This ultimately makes Russia a major geopolitical player. China, on the other hand, as one of the largest producers in recent times, needs a lot of minerals in order to develop its production capacity and business potential on an international scale. The Arctic is thus a "new" economic, energy, and investment potential for both Russia and China.

As for the relations of both actors with other Arctic states, they externally show stability and have the character of positive relations. We can talk about a high level of working cooperation, especially Russia and the Arctic states, as evidenced by the existence of a number of bilateral and multilateral treaties, agreements, and conventions. Their cooperation within the region is reaching a good level. However, it must be noted that Russia's increasing efforts to govern the region raise some concerns among the Arctic states. Russia carries out certain military activities in the Arctic region, modernizes radio stations, deploys ground-to-air missile systems, and the like. They declare the development of these military capabilities as a necessity for their defence purposes. Nevertheless, these activities lead to fears that Russia would be able to restrict or block access to certain parts of the region, leaving them under their sole control. These concerns create tensions between Russia and the other Arctic states. For instance, Russia's other activities, such as Crimea's case (in 2014), contribute to Russia's being perceived as an unpredictable player. This step affected the loss of traditional Russian investors. Russia's assertive behaviour leads to greater cooperation and coherence among the other Arctic states without the participation of Russia. They conclude agreements that are bilateral or multilateral in nature, but without Russian participation.

Relations between China and the Arctic states are at a good level as well and can be described as pragmatic (Torkunov, 2019). China recognizes their need to penetrate the region and carry out its activities. More active cooperation between the Arctic states is a certain advantage for China, as it can also promote its interests through these states as opposed to only through its cooperation with Russia. On the other hand, it presents a partial disadvantage for Russia, as it weakens mutual relations and the possibilities of different mutual trade agreements between Russia and China. An important project in Russian-Chinese relations is the Yamal LNG project. China's strategy is easier to read and is set within the limits of existing treaties and rules, which in the case of Russia is not always clear, and many Arctic rules and regulations are subject to their own interpretation. However, there is some concern among Arctic states that China will be given a greater role in the Arctic than is appropriate, and that it would interpret the international maritime law as they find convenient.

Another problem is that the existing legal regimes governing activities in the Arctic do not lay down precise rules for the extraction of cross-border oil and gas resources, leading to confrontations between the Arctic states and beyond. Following the announcement of the US Geological Survey on the hydrocarbon potentials of the Arctic, tensions have increased. It is true that Chinese investment is a good opportunity for Arctic communities to develop their infrastructure as well as to improve the conditions of their daily lives. Russia is also showing efforts to establish itself in this area, as the majority of Arctic's population is under Russian administration: "4.2 million people reside in the Arctic region, 2.1 million in Russia" (Borgerson et al., 2021). However, with new investment into communications and new technologies in the field of energy in the Arctic, China may ultimately indirectly gain stronger political influence, which raises some concerns on the part of Arctic states. For this reason, each of the Arctic states sets certain limits in relation to China and its presence in the Arctic:

> *In 2011, Iceland blocked the sale of a large plot of land to a Chinese investor; in 2016, Denmark declined to sell a vacant naval base in Greenland to a Chinese mining company; and in that same year, a projected Chinese resort in Svalbard, under Norwegian sovereignty, was cancelled.* (Pezard, 2018)

China would like to be actively involved in the development of maritime and submarine routes in the Arctic, such as the Northern Sea Route, as does Russia, which also wants to gain significantly greater control over their management and administration. In terms of the position of the Arctic state, Russia has an advantage over China. With regard to this, an important role is played by the USA, whose growing influence in the Arctic region has resulted in a slowdown or even suspension of China's attempts to infiltrate the region, alongside a partial elimination of Russian influence in the region. The growing domination of the USA is supported, for example, by the reopening of its consulate in Greenland's capital Nuuk after ten years.

The Arctic is evidently a top strategic priority for Russia as well, as it adopted strategic documents dedicated to the Arctic Region in 2008, 2013, 2014, 2017, and 2020.

13.3 Northern Sea Route and Its Potential as a Transport Energy Artery in the Future

The advent of climate change has brought about a number of different changes in the Arctic, including increased accessibility to Arctic ports as well as the opening of new Arctic shipping lanes. With new trans-Arctic

> *routes, including the Northern Sea Route (NSR) and the Northwest Passage (NWP), as well as newly built and refurbished ports from Russia, political and military interests are re-evaluating the region as one of geopolitical competition.* (Gricius, 2021)

The ongoing natural changes in the Arctic and its parts offer new opportunities for lucrative yields in the areas of mineral resources, energy, oil, gas, and shipping. The decline of the ice cap in the Arctic may lead to an increase in commercial shipping in the future, although the rate of commercial utilization is still questionable.

One of the alternatives is a sea trade route called the Northern Sea Route. This Arctic route runs along the Arctic shores of Russia and connects Europe with Asia. The possibility of its use divides experts into two clusters. On the one hand, there are those who defend its high potential and the possibility of its year-round navigability, especially more recently after significant global warming. What helped this discourse was the blockage of the Suez Canal in March 2021, when transport was suspended for several days. Attention is also focused on the use of minerals, the deposits of which are gradually being uncovered as the ice sheet recedes; these are important mineral sources that can be used in the future, and their transport can be conducted through the NSR. These mineral resources will have a significant impact on energy security policy.

On the other hand, there is another group who are sceptical, claiming that the Northern Sea Route will never be able to match the routes already used between Europe and Asia, and its use will be marginal, even limited to a certain type of ship. They also argue that this route has never been used for commercial purposes; this means that no shipping company has operated regular transport services there. At the same time, they point out that more intensive use of the Arctic for mining will cause high environmental damage, which will be irreversible. The unique natural ecosystems in this part of the world will be disrupted. This may be due to oil pollution as well as various accidents that may occur during transport. It also will not be possible to ensure the sustainable development of natural resources in the region.

Despite the above, maritime transport in the Arctic is increasing, which is contributed to by change in the existing and applied environmental, social, and economic patterns and also by the survey on the occurrence and mining of mineral resources in this region.

> *Changes to the marine environment itself including sea ice thinning and reduction has significant implications for longer seasons of navigation and new access to previously difficult to reach coastal regions.* (Arctic Council, 2020)

The trend of the Northern Sea Route being increasingly used in the future is also indicated in the results of the first Arctic Shipping Status Report released by the Arctic Council's Protection of the Arctic Marine Environment Working Group. Their results show by how much Arctic ship traffic has increased between 2013 and 2019: notably, during the six-year period between 2013 and 2019, the number of ships entering the Arctic Polar Code area grew by 25 percent. The total distance sailed by ships in this area grew by 75 percent, from 6.51 million nautical miles in 2013 to 9.5 million nautical miles in 2019 (Arctic Council, 2021).

According to Russian Deputy Prime Minister Yuri Trutnev, Russia has invested in the development of the Northern Sea Route infrastructure, and their intention is for it to become the main sea route throughout the year. This plan should take place in 2022–2023. At present, the NSR, which runs along Russia's northern bank, is currently used to ship hydrocarbons and other resources for up to nine months a year.

Carriers, on the other hand, are sceptical and consider the likelihood of the Arctic opening up to new international competition unlikely. Many carriers have expressed concern that making full use of the Northern Sea Route would have a negative impact on Arctic nature and its ecosystem. France's CMA CGM, the world's fourth-largest container carrier, has pledged not to use the Northern Sea Route for fear that increased shipping would have a negative impact on the environment in such a fragile ecosystem as the Arctic.

> *With this decision, CMA CGM makes the resolute choice to protect the environment and the planet's biodiversity despite the major competitive advantage this route represents for shipping companies," Rodolphe Saadé, Chairman and Chief Executive Officer of the CMA CGM Group, said.* (World Maritime News, 2019a)

Two other major international shipping companies have taken the same position. One of them is German company Hapag-Lloyd, which does not plan to use the Northern Sea Route for transport, as this poses an environmental threat to the Arctic. "*'Hapag-Lloyd does not use the Northwest*

Passage or the Northeast Passage as shipping routes right now, nor are there any plans to do so in the future,' Jörg Erdmann, Senior Director Sustainability at Hapag-Lloyd, pointed out" (World Martime News, 2019b). The second one is MSC in Switzerland, which has raised similar environmental concerns as other previously mentioned companies.

Russia's stance on the NSR is perceived as quite controversial by other Arctic states. Russia considers certain parts of the NSR to be internal Russian waters and has asserted a right to regulate commercial shipping passing through these waters.

The Arctic states also disagree with the condition of having a Russian maritime pilot on board, what they consider absurd, and which discourages them from using this route.

> *The Russian government has elaborated a set of rules for foreign naval vessels sailing on the Northern Sea Route, Izvestia informs. The newspaper has obtained a copy of the document that states that all vessels are obliged to comply. The foreign state must send a notification about the voyage at least 45 days ahead of its start. Included will have to be the name of the ship, its objective, route and period of sailing, as well as ship characteristics such as length, width, deadweight, draft and type of engine power. Also, the name of the ship captain must be listed. The ships must also have a Russian maritime pilot on board. In case the voyage is not conducted in line with the regulations, Russia will have the right to take extraordinary measures including a forced halt, arrest and in extreme cases, elimination.* (Stalesen, 2019)

Russia and China are most interested in the development of the Northern Sea Route. In March 2021, China presented its 14th Five-Year Plan (FYP), which will guide the country's economic policies until 2025. The plan included Beijing's interest in the Polar Silk Road, along with China's increased Arctic engagement, which will be one of China's foreign policy priorities in the future. Economic and energy activities will continue to take precedence over diplomatic or scientific goals, as has been the case in the past.

The FYP statement is the most coherent signal of Beijing's aspirations to be accepted by the global community as a stakeholder at both poles, including being accepted as a de facto "near-Arctic state," in relation to China's evolving international power (Lanteigne, 2019). China will seek to strengthen its global influence as well as to develop trade routes in the region.

Russia is trying to contribute to the attractiveness of the NSR and to increase traffic on this route. Some planners hope that, because of Russia investing tens of billions of dollars in their northern infrastructure, including the construction of new ports of call and nuclear-powered icebreakers, the region will emerge as a "Suez of the North" (Borgerson et al., 2021).

In order to determine whether the Northern Sea Route will be suitable for international transport, it is necessary to list particular advantages and disadvantages of both the NSR and the Suez Canal and to point out their limits.

The advantages of the Northern Sea Route compared to the Suez Canal include the length of the route, which is shorter by about 10–15 days, resulting in shorter delivery times. Ultimately, fuel consumption will be lower, which will be reflected in lower fuel costs. For instance, a voyage from Shanghai to Hamburg via the NSR shaves roughly 30 percent of the distance off a similar trip via the Suez Canal, while also avoiding the heavily pirated Strait of Malacca and waters off the Horn of Africa. Operators can either arrive at their destinations earlier or use the extra time for super-slow sailing, reducing fuel costs and emissions (Borgerson et al., 2021).

At the same time, the operator of the Russian nuclear icebreaker fleet Atomflot plans to speed up transit traffic on the Northern Sea Route from 14 days to 10–11 days. A new generation of nuclear icebreakers will accompany merchant ships and break their way. Shortening the route will help to protect the environment as it will reduce the emissions of greenhouse gases.

The absence of any restrictions on vessel size and tonnage is also a positive feature of the NSR. There are also non-existent transit fees in support of the NSR, which eliminates any further increase in the price of transported goods. Last but not least, the great advantage is the elimination of political instability in the form of pirate raids.

In addition to the benefits, the use of the NSR also has certain pitfalls. One is its navigability, which lasts only a few months a year, making transportation significantly more expensive. It is true that the transport period has been prolonged due to climate change, but it still cannot be described as a year-round navigable route. Another negative is the unstable weather typical of the whole area of the Arctic; during the period when the route is open, vessels can expect to encounter moderate to strong winds, low air temperatures, a high number of days with fog, and, late in the season, long polar nights, snowstorms, and possible blizzards (ABS, 1996).

Another disadvantage is that it is not a common commercial route which has precisely defined conditions of transport. Certain transport limits can also be caused by possible route management disputes or territorial disputes in the Arctic, which will disrupt transport safety. This is not to forget the need for icebreakers to clear the route, which can increase costs. Russian ice-breaking fees (US $500,000 per transit) would have to decrease by 25 percent in order for the Arctic Sea route to be competitive with the current costs of utilizing the Suez Canal (Brady, 2020). The carrier will have to use more expensive ships capable of crossing and overcoming more demanding sailing conditions. This will require higher investment costs into shipbuilding and construction. Finally, due to environmental reasons, the lack of interest major transport companies have in transport via the NSR deprives it of its attractiveness. A disadvantage is also the underdeveloped infrastructure and the lack of international ports, which limits the commercial potential of the route.

As for the Suez Canal, its main advantage is that it is a well-known commercial transport route, having been used for a long time. Transport conditions are relatively stable, a considerable advantage for carriers. It is a route that is known and used by contemporary, world-famous transport companies, which gives it credit for reliability and transparency. Another advantage is the year-round navigability and less demanding conditions for the execution of the cruise, which does not incur additional cost increases. Compared to the NSR, it is more suitable for container transport, which is carried out exactly on schedule, and unpredictable conditions are unacceptable. Its disadvantages are its longer route, which prolongs the delivery time and leads to higher fuel consumption and transport costs. More disadvantages are high transit fees due to the proximity to the zone of military conflict in the Middle East, as well as the threat of pirate assault, forcing ships to get insurance which is significantly high given the problem.

Russia and China are among the most important international global players seeking active participation in the management and use of the Northern Sea Route. The extent to which the NSR will be used depends on its interconnection with the world economy and international markets, as well as the security situation, which will build upon the activities of both powers. The development and streamlining of transit to the NSR should contribute to the simplification of oil and gas supplies to Southeast

Asia. The NSR is an artery that connects the Pacific and Atlantic Oceans in a much shorter time than before. Russia is an important strategic partner of China, although their interests are not always aligned. Both partners are working to increase their economic and energetic potential through modernization and better use of transit to the NSR.

China's involvement in the polar region also has growing potential. It is interested in the extraction of oil, gas, mineral resources, and other non-fossil energy sources. One of the reasons for the growing interest in this region is China's major share in the Russian liquefied gas project Yamal, which should supply liquefied natural gas to China. Russia has been the biggest beneficiary of Beijing's Arctic investment by far, as illustrated by the degree of financial support in oil and LNG enterprises in the Russian Arctic , including the Yamal LNG project in Siberia (Lanteigne, 2019).

Russia hopes to improve and increase the volume of international transit to the NSR. However, this depends to a large extent on the international shipping industry to determine whether the NSR is sufficiently safe, efficient, economically viable, and environmentally friendly. More efficient use of the NSR can contribute to intensified economic and energetic cooperation between Europe and Asia.

Oil and gas are produced in four of the states that possess land or sea territories belonging to the Arctic: Russia, Alaska, the Northwest Territories of Canada, and Norway. About 61 large oil and gas fields have been discovered inside the Arctic Circle territories of these four countries. Of the 61 large fields, 42 are located in Russia. 35 of these large Russian fields (33 natural gas and 2 oil) are in the West Siberian Basin. Of the remaining 8 large fields, 5 are in the Timan-Pechora Basin, 2 in the South Barents Basin, and 1 in the Ludlov Saddle. Of the 18 large fields outside Russia, 6 are in Alaska, 11 are in Canada's Northwest Territories, and 1 is in Norway (EIA, 2009).

The probability of the existence of large deposits of oil and gas predetermines Russia to be a major energy player in the Arctic, and thus, it claims its rights. The Arctic is estimated to hold nearly a quarter of untapped global natural gas resources. Oil production is gradually shifting from land to sea, where significant deposits of these resources are located; however, their mining is more demanding than onshore. Increasing oil prices in the global markets plays into Russia's policy to uncover as many of these resource sites as possible. China is also showing interest in doing so. It is already clear that the extraction and transportation of these commodities will be fully dependent on their international market prices (in 2014, the attractiveness of the NSR decreased due to lower fuel prices on world markets), as well as on demand in the European and Asian markets. This will ultimately affect the interests of Russia and China. The security situation in the region will also play a role; its disruption could jeopardize transit to the NSR. The extraction, processing, and transportation of oil as well as natural gas are very demanding due to the inhospitable conditions of the Arctic. It requires better, more modern technical equipment, transport infrastructure, and higher wage costs for staff working in such remote areas. The NSR is a kind of corridor for bringing energy resources east to Asian markets and west to Europe.

The extent to which the NSR will be used for transportation also depends on the availability of oil and gas in the Arctic.[2]

China sees several advantages in using the NSR: the shortening of commercial traffic between Europe and China, as well as the reduction of China's dependence on the southern sea routes. The country is in favour of making the NSR commercially available not only for economic reasons, but also as it sees an opportunity to gain legitimate claims to its influence in the Arctic region. In the last 20 years, China has become a global economic power dependent on the sea for much of its commerce: China now has the world's third-largest merchant marine; since 2010, it has been the world's largest shipbuilding nation; and as of 2014, China has been the world's largest goods exporter and oil importer (Brady, 2020).

Russia sees high potential in the development of the NSR. Vladimir Panov, Deputy Chairman of the State Commission for the Development of the Arctic, commented:

> *We believe that a new stage in the development of the NSR and the Great Northern Sea Route as a transport artery connecting Europe and Asia will begin after 2025. According to our conservative estimates, cargo turnover will reach 150 million tons in 2030.* (Neftegazru, 2021)

Interest in the NSR is gradually growing with the improving climatic conditions in the Arctic. Between 2016 and 2019, 1232 destination voyages took place in the NSR, of which 1108 were between the NSR and European ports and 124 were between the NSR and Asian ports

[2] Large Arctic oil and natural gas discoveries began in Russia with the discovery of the Tazovskoye field in 1962 and in the United States with the Alaskan Prudhoe Bay field in 1967.

(Gunnarsson & Moe, 2021). The development of shipping in the NSR is highly dependent on the development of the global market, whether the commodities transported will be in demand on Asian markets, and whether their prices will rise and international transit traffic in the NSR will develop.

Increasing activities in the NSR can lead to several geopolitical problems. It will therefore be necessary to lay down precise rules and management of the transit artery in order to avoid unnecessary disputes. Monitoring and control will also need to be improved to prevent illegal vessels from engaging in illicit fishing or human trafficking. In this respect, Russia has an important role to play as an Arctic state and superpower.

13.4 Application of Bimatrix Games in the Northern Sea Route Energy Conflict

The application of game theory in international politics is based on the fact that many conflicts of an international nature can be viewed from the perspective of two main actors (superpowers), where one party cannot take any decision unilaterally because the outcome of the game depends on the other actor's strategy. As the two superpowers—Russia and China—can be considered the main actors in the Northern Sea Route conflict, we will list several well-known bimatrix games that can be used to analyse and predict the outcome of this conflict. The conclusions will be based on Nash's equilibrium, which can be characterized as a state that is self-enforcing in which a player who attempts to change this state unilaterally harms themself, provided that the opponent behaves rationally.

One of the most famous and applied games is the *Hawk-Dove* game. It represents two players fighting for a resource. Players can choose one of two strategies, one more escalated than the other—Hawk and Dove. Value > 0 will be the value of the disputed resource and Cost > 0 will be the price of the escalated fight. In the basic version of the game, it is assumed that the value of the resource is lower than the cost of the fight, i.e. Cost > Value. If both players choose the Hawk strategy, a fight ensues, resulting in (Value − Cost) / 2 < 0. If one player retreats and chooses Dove's strategy while the other chooses Hawk's strategy, Hawk wins the resource prize. In the case of Dove and Dove strategies, players split the resource in half. A *Hawk-Dove* game is any game where the fight for a resource means the worst possible outcome for both entities, and the outcome for a player who "retreats" is worse than the outcome in the event of a resource share. The best possible outcome is tied to the strategy when the subject tries to get the whole resource and push the opponent out of the game. If players decide independently and do not cooperate, the game has two Nash equilibria in pure strategies. Nash's equilibria represent an outcome where one player adopts the Dove strategy and the other the Hawk. This game can illustrate a situation where the main players in the decision-making process try to push the opponent to Dove's strategy, take the Hawk's strategy, and win the whole prize.

Russia and China pursue several common economic and energy interests in the NSR project, which represent benefits for both players. However, the profit may be lower as they will have to split it. At the same time, however, both will have lower transport costs, which will allow them to diversify transport and reduce the transport time allowed by the NSR. Cooperation between Russia and China strengthens the NSR in terms of its use as an international transport artery as well as in terms of their position in the Arctic region. This may ultimately weaken other players (Arctic states) in the area. At the same time, increasing their influence in the Arctic is an advantage for both players, enabling them to have a lower degree of dependence on other international transport routes and giving them the opportunity to influence the world market.

However, the game also points to the fact that if both players go into conflict and play the Hawk strategy, the result of the game is the worst possible outcome for both. In the present case, this result can be described as a situation in which the negative environmental consequences of excessive intervention in the Arctic can ultimately affect all stakeholders in a negative way.

The *Hawk-Dove* game led to the possibility of modelling the negative externalities of the conflict in the case of the fight between the subjects, which resulted from the fact that the price of the resource was below the price of the escalated fight.

Another game serves to model the impact of negative externalities on player behaviour directly. This game is known as Chicken. It is assumed that each player has two strategies—Swerve and Straight—and the results of the game can be expressed verbally as Tie, Win, Lose, and Crash. The combination of two Swerve strategies leads to the result of Tie. If one player adopts the Straight strategy and the other player chooses the Swerve strategy, the game results in Win for the player with the Straight strategy and Lose for the other player. In the case of both Straight strategies, the result is a Crash of both players. Each player prefers Win over

Tie, Tie over Lose, and Lose over Crash: Win > Tie > Lose > Crash. The game has two Nash equilibria in pure strategy, which is a combination of strategies (Win and Lose), so each player would like to go with the Win strategy and use all possible means of power to push the opponent to the Swerve strategy. However, if both players follow the Win strategy, the game ends with a Crash result for both players. This result can be considered an environmental disaster that could occur if players implement an aggressive strategy without taking into account all environmental impacts.

Both *Chicken* and *Hawk-Dove* are anti-coordination games in which it is mutually beneficial for players to play different strategies. The given types of game lead a stronger player to use various means of power to force their opponent to play a less favourable strategy. From its position, Russia can use all possible means of power to push China, both as a teammate and as an opponent, to make concessions that will be acceptable and advantageous for Russia; these will also result in the highest possible economic profit for Russia and an increase in its sphere of influence in the Arctic, which will help it make the NSR a route with international operations.

The basic concept of the game is a rival resource, the sharing of which costs something. These costs of the Northern Sea Route can also be seen as a negative externality measured through environmental impact. The existence of two Nash equilibria in pure strategies points to the instability of the situation, and if neither player gives up, the game ends in the worst possible outcome for each of them. In the case of the struggle for the Northern Sea Route and the Arctic, it must be pointed out that inadequate pursuit of the main players' own interests can lead to global damage, which will ultimately affect both main players (China and Russia) in a catastrophic way. Economic profit versus environmental protection contrasts here. These games are closely related to the term Tragedy of the Commons (Hardin, 1968), which refers to poor results for the whole system when individual interests are at odds with the common good. This shows that if actors act independently in their own interests and against the common good, they cause resource depletion through their uncoordinated actions.

If, in the basic version of Hawk-Dove, the assumption of a great loss in the event of an escalated fight is abandoned, instead assuming Cost $\leq$ Value, it moves on to a Prisoner's dilemma. In this case, the Hawk strategy will be replaced by the defect strategy and the Dove strategy will be replaced by the cooperate strategy. This type of game can be any game where the worst possible result is the cooperate strategy, if the opponent adopts the defect strategy. The second worst result is the combination of defect-defect strategies, followed by the combination of cooperate-cooperate strategies. The best possible outcome is associated with the defect strategy, if the opponent adopts a cooperate strategy. The prison dilemma game can be used as a model for many real-world situations involving cooperative behaviour. It is used when two entities could gain important benefits from cooperation or suffer because they failed to do so, i.e. when it is considered difficult, but not necessarily impossible, to coordinate their activities. Nash's equilibrium results in a defect-defect strategy, which, however, leads to the worst common outcome of the game. This game usually points out the fact that two individuals who act in their own self-interest do not produce an optimal result, as opposed to a combination of cooperate-cooperate strategies. The game confirms that in international relations, all states (regardless of their internal policies or ideologies) will act in their own rational self-interest. Although the best possible outcome would be cooperation, it is rational to adopt the defect strategy. The game can also point to the fact that agreements between powers tend to be broken, because players tend to "trip" the opponent in order to increase their profit. The solution is to enforce mutual cooperation by binding agreements, the observance of which is enforceable. The Prisoner's dilemma is also one of the most applied games in the event of repeated conflicts. It proves that players tend to cooperate until the game is almost over, and then, they disrupt cooperation.

Entities are inclined to prioritize their own interests and attempt to achieve them in any way possible. The situation is similar in the case of the NSR, where both Russia and China are trying to manoeuvre in mutual relations in order to obtain the highest possible benefit. China is also interested in using other transport routes, which may be one of its strategies that will lead to the elimination of Russian influence on the transport of Chinese products to Asian and European markets. They are able to do so if it suits them. Russia is often unpredictable and unclear in its actions, which is in its favour, although it presents a major disadvantage for China (from the point of view of game theory; these are games with incomplete information). The reason may be the fear of a possible weakening of its activities in the NSR project as well as in the Arctic in general.

Recently, tensions in the Arctic region have been escalating. The reason is China's growing ambition to take advantage of the NSR, as well as the existing mineral resources that are being uncovered by the decline of the ice sheet. Russia's revitalized militarization along its Arctic coast is also contributing to the pressure. From a global perspective, however, it is necessary to eliminate sources of tension so that the Arctic can properly develop.

13.5 Conclusions

Climate change has opened up various Arctic regions and enabled the discovery of new extraction sites of mineral resources, oil, and natural gas. At the same time, it has brought more attention to commercial maritime routes such as the NSR. These facts make the Arctic a new phenomenon that is becoming of interest to many powers and superpowers. Their aim is to intensify activities in the fields of science, research, economic, and energy development, economic growth, considering natural conditions, and maintaining sustainable development. The desire is to eliminate confrontations and disputes, as well as to maintain the security of the Arctic region. The goal of geopolitical players is to enforce their interests and priorities and thus increase their sphere of influence to the highest possible level.

Many factors will affect these goals, such as natural conditions, the state of the sea ice, demand, and the level of prices of mineral resources, oil, and natural gas on world markets—not to forget the scientific and technological development opportunities in the Arctic region, political support of governments, and support of global actors. The protection of the Arctic ecosystem in order to prevent its disruption and possible destruction will also be a determining factor, taking into account the views of experts. It will be necessary to respect the diversity of cultures, historical traditions, and the character of the indigenous population, which is an integral part of the Arctic region.

An important role will be played by Russia, being one of the most important states in the Arctic and an Arctic economic and energy superpower, which is managing the largest part of this territory with the largest indigenous community. To pursue its economic, energy, political, and security ambitions, it will need a high level of cooperation within the Arctic states and beyond. Only a safe and stable environment will allow it to develop and make the Northern Sea Route a modern international artery that will be able to compete with existing routes.

China, as one of the global players that is gaining momentum, is likely to play an important role in developing and making the Arctic accessible. It is already clear that China's foreign policy towards the Arctic region shows that it is in its interest to use more than one route across the Arctic Ocean. This will strengthen its influence in the Arctic and increase the acceleration of economic and energy development.

Relations between China and Russia are contradictory. These are two strong geopolitical players. In order to fulfil their intentions, they will have to alleviate emerging tensions in the region, promote international cooperation, and respect international laws and agreements. Finally, they will need to step back from their positions and competition, and establish a new correlation of relations that takes into account existing conditions. Despite all this, they are strong players who will play a key role in promoting the Northern Sea Route as a relevant transport route that connects the Russian oil and gas industry with Asia. In the future, it will also lead to an intensification of the transport of Chinese goods to European markets.

The Arctic is changing from the periphery to an economic centre of investment, an important transport hub, and a transit point that connects the strategic interests of Europe, Asia, and America. The Arctic states themselves will have to take new positions on the changing situation in the region.

Acknowledgements This chapter was prepared within the VEGA project No. 1/0427/20 "Multi-criteria models of Game Theory in Economics and Politology", responsible researcher Zuzana Čičková.

References

ABS. (1996). *Navigating the Northern Sea Route status and guidance: Navigating the Northern Sea Route Advisory*. https://ww2.eagle.org/content/dam/eagle/advisories-and-debriefs/ABS_NSR_Advisory.pdf

Arctic Council. (2020, April 14). *First Arctic shipping status report from Pame Working Group highlights increase in Arctic shipping traffic*. https://arctic-council.org/news/first-arctic-shipping-status-report-increase-shipping-traffic/

Arctic Council. (2021). *Shipping status report*. https://www.pame.is/projects/arctic-marine-shipping/arctic-shipping-status-reports

Associated Press. (2020, July 22). *Pompeo says US to expand Arctic role to Deter Russia, China*. https://abcnews.go.com/International/wireStory/pompeo-meeting-danish-leaders-discuss-pipeline-71914448

Borgerson, S., et al. (2021). *The emerging Arctic*. Council Foreign Relations. https://www.cfr.org/emerging-arctic/#!/

Brady, M. A. (2020). China's interest in Arctic shipping. *Insight Magazine*. https://www.amcham-shanghai.org/en/article/insight-magazine-chinas-interest-arctic-shipping

Brocková, K., Kucharčík, R., & Mráz, S. (2020). Application of international law in international relations. *Pravo: Naukovij visnik Užgorods'kogo nacional'nogo universitetu, 60*(2), 206–209.

EIA. (2009, October 19). *Arctic oil and natural gas potential*. Independent Statistics & Analysis, U.S. Energy Information Administration. https://www.eia.gov/analysis/studies/archive/2009/arctic/index.html

European Commission. (2016, April 27). *An integrated EU policy for the Arctic—Frequently asked questions*. https://ec.europa.eu/commission/presscorner/detail/sk/MEMO_16_1540

Gricius, G. (2021). *Geopolitical implications of new Arctic shipping lanes*. The Arctic Institute, Centre for Circumpolar Security Studies. https://www.thearcticinstitute.org/geopolitical-implications-arctic-shipping-lanes/?cn-reloaded=1

Gunnarsson, B., & Moe, A. (2021). Ten years of international shipping on the Northern Sea Route: Trends and challenges. *Arctic Review on Law and Politics, 12*, 4–30. https://arcticreview.no/index.php/arctic/article/view/2614/5113

Hardin, G. (1968). The tragedy of the commons. *Science. New Series, 162*(3859). http://www.jstor.org/stable/1724745

King, M. H. (2016). *Oil and natural gas resources of the Arctic*. Geoscience News and Information. https://geology.com/articles/arctic-oil-and-gas/

Lanteigne, M. (2019, April 19). Snow fort or ice path? China's emerging strategies in the Arctic. *High North News*. https://www.highnorthnews.com/en/chinas-emerging-strategies-arctic

Martin, D. (2017, July 18). *Chinese warship in route to Baltic Sea*. DW Academie. https://www.dw.com/en/chinese-warships-en-route-to-baltic-sea/a-39735964

Neftegazru. (2021, September 6). *The Northern Sea Route discussed as a new global trade route on the Eastern Economic Forum*. https://neftegazru.com/news/exhibitions/695880-the-northern-sea-route-discussed-as-a-new-global-trade-route-on-the-eastern-economic-forum/

NSIDC. (2021). *Arctic sea ice news & analysis*. https://nsidc.org/arcticseaicenews/

O'Rourke, R., et al. (2021). *Changes in the Arctic: Background and issues for Congress*. Congressional Research Service. https://sgp.fas.org/crs/misc/R41153.pdf

Pezard, S. (2018). *The new geopolitics of the Arctic: Russia's and China's evolving role in the region* (pp. 1–2). RAND Corporation. https://www.rand.org/content/dam/rand/pubs/testimonies/CT500/CT500/RAND_CT500.pdf

Pomeroy, R. (2020). *A climate scientist explains what the melting Arctic means for the world*. World Economic Forum. https://www.weforum.org/agenda/2020/01/arctic-ice-melting-climate-change-global-impact/

Stalesen, A. (2019, March 7). What shall foreign military navy face in the NSR. *Barents Observer*. https://thebarentsobserver.com/ru/bezopasnost/2019/03/izvestiya-vot-chto-zhdet-inostrannye-voennye-korabli-na-sevmorputi (in Russian).

Thisted, K., & Gremaud, A. S. N. (Eds.). (2020). *Denmark and the New North Atlantic: Narratives and memories in a former empire* (Vol. 1). Aarhus University Press.

Torkunov, A. V. (2019). International studies: Chaos or pluralism? *Polis. Political Studies, 5*, 7–18 (in Russian). https://doi.org/10.17976/jpps/2019.05.02

World Maritime News. (2019a, August 26). *CMA CGM pledges not to use the Northern Sea Route*. https://www.offshore-energy.biz/cma-cgm-pledges-not-to-use-the-northern-sea-route/

World Martine News. (2019b, October 4). *Hapag-Lloyd ships will not use Arctic Sea Routes*. https://www.offshore-energy.biz/hapag-lloyd-ships-will-not-use-arctic-sea-routes/

Zagare, F., & Slantchev, B. (2018). *Game theory and other modelling approaches*. Oxford Research Encyclopaedia of International Studies. https://oxfordre.com/internationalstudies/view/10.1093/acrefore/9780190846626.001.0001/acrefore-9780190846626-e-401

CHAPTER 14

Global Climate Changes and Their Importance for the Building of Northern Sea Route and the Future of Energy Sector

Karol Janas and Barbora Jánošková

14.1 Introduction

As the consequences of climate change are already being felt, and it is assumed that the changes will influence the whole world, global climate changes have been the subject of enlarged attention in recent years. The polar ice is melting, so the levels of the oceans and seas are rising. Currently, there are extreme weather divergences in certain regions, some of which are fighting with the extreme dry seasons and heat, some of which experiencing strong rainfalls.

K. Janas · B. Jánošková (✉)
Department of Political Science, Alexander Dubček University of Trenčín, Trenčín, Slovakia
e-mail: barbora.janoskova@student.tnuni.sk

K. Janas
e-mail: karol.janas@tnuni.sk

V. I. Salygin (ed.), *Energy of the Russian Arctic*,
https://doi.org/10.1007/978-981-19-2817-8_14

Before proceeding with the context, it is necessary to define the most relevant terms used in the text. According to the United Nations Framework Convention on Climate Change, global climate change is defined as "the change of climate which is attributed directly or indirectly to human activity that alters the composition of the global atmosphere and which is in addition to natural climate variability observed over comparable time periods" (United Nations Framework Convention on Climate Change, 1992). In general, global climate changes cover a wide range of areas—climate alterations, land productivity, water resources including oceans, atmospheric chemistry, and ecological systems too.

Global warming directly affects the Arctic area on the northern part of the planet. The rise of the temperature causes the ice to melt, so annually the ice in the Arctic area is on the decrease. The melted ice opens new maritime routes important for international trade and enables the extraction of natural resources. The Arctic area is rich in mineral raw resources of different types. It is assumed that probably 22 percent of the undiscovered oil and gas world reserves are in the Arctic, as well as supplies of gold and rare metals (Didenko & Cherenkov, 2018). That is why many states are interested in the aforementioned area. After the melting of the ice in the Arctic area, an enormous amount of oil and gas will be accessible. It predicts that the states with access to these new mineral sources will have several advantages. Firstly, there will be a new production of oil and gas. Secondly, the interest in these sources will probably mean the opening of new markets and also new customers. For this purpose, new investment in the development of the required infrastructure will be necessary. This will have a huge impact on the energy sector and the future of the sources.

With this unexplored region, rich in raw minerals, the Arctic and non-Arctic States connect their various plans. One of the mentioned states is the Russian Federation, with its Northern Sea Route. The route is defined in Russian laws as the shipping route with the one point in Novaya Zemlya, along the Russian seaside to the other point in the Bering Strait. Besides this route in the Arctic, other sea routes have also been identified. One of them is the Northwest Passage, leading between the northern archipelagos of North America. The second route, Northeast Passage, was identified before the introduction of the Northern Sea Route in the northern coastline of Eurasia. The Arctic Ocean is associated with the Trans-Polar Route also going across the ocean. In recent years, more is being said about the Polar Silk Route, at the initiative of the People's Republic of China (Bekkers et al., 2015). China, as Russia, is looking for the most economically viable (i.e., cheaper) option of transporting goods, cargo, and energy commodities. They want to take advantage of the melting ice in the Arctic to build a suitable alternative to the Southern Sea Route leading through the Malacca strait and the Suez Canal.

14.2 Literature Review

The theme of climate change has been the subject of many works in recent years. Also, there are plenty of works about and research into the geopolitical context and global and strategic importance of the Northern Sea Route. There is no doubt that the meaning of this route will be strengthened in the future. The route shortens the distance between South-eastern Asia and North-western Europe, which is a valuable advantage. As an example is the longitude of the route between Yokohama and Rotterdam through the southern and northern route; through the Suez Canal (southern shipping artery), the distance between the mentioned economic centers is 20,900 km, while through the Northern Sea Route, it is "only" 13,700 km (Bekkers et al., 2015). It can reduce transport costs, which states may aim for in the time of the global COVID-19 pandemic and in the post-COVID time. The shortening is about 40 percent of transport time compared to the southern route leading through the Suez Canal. On the one hand, there is a considerable reduction in transport costs. On the other hand, the regular use of the route will introduce the real possibility for increasing the risk of pollution and accidents. We think that it is necessary to describe the broader context of building the Northern Sea Route. In this context, we would like to briefly describe the global importance of the Northern Sea Route in this modern world. As well as the global meaning of the route, we can underline the enormous internal meaning of the route for one of the players directly connected with the area—Russia. Russia is aware of the possibilities and importance of the stated area. The Arctic is mentioned in the strategic state documents. The Northern Sea Route, as mentioned above, is defined in Russian laws.

The topic is a subject of many domestic and foreign works. Besides this, in regular times, states accept the strategic documents about the policy in the Arctic. Martin Gress, in his conference paper, underlines the increased interest in the Northern Sea Route in recent years and suggests further research on this issue. He concludes that this area will have implications for the world economy and international policy. His conference paper was devoted to the analysis of the scientific and journalist literature of domestic and foreign origin (Gress, 2021).

In the discussion paper, Bekkers et al. (2015) devote attention to the melting ice caps and the economic impact of opening of the Northern Sea Route. In the context of the commercial use of the Northern Sea Route, they estimate that two-thirds of the world trade volume will be re-routed through this Arctic corridor, which is shorter than the traditional southern route. They assess that the cost reductions using the NSR will lead to the increase in the trade flows between two major economic regions—North-eastern Asia and North-western Europe—by, on average, 10 percent. In the relation to the environment, Bekkers et al. assume the slight increase of CO_2 emissions, not emissions associated with water transportation, but as a combination of higher trade volumes and emission-intensive production in countries of Northeast Asia (Bekkers et al., 2015).

The importance of the Northern Sea Route in the broader context is discussed by David et al., Kluge and Paul, Wan, Blunden, and many others. David and his colleagues in their studies from 2019 and 2021 thought about the Northern Sea Route as a new route for maritime transport between the Far East and Europe for two reasons. The first is the fact that global warming enables the extension of the duration of navigation times for several months of the year. The second reason is a big advantage of this route, which is the reduction of sailing times between Asian and European maritime ports (David et al., 2021; David & Hansut, 2019). Moe, in his research, suggests that the Northern Sea Route is a key component of Russia's broader ambitions in the Arctic. According to him, all expressions of national identity converge in a stated area. He examines the Russian policy for the Northern Sea Route through the state interests, economic opportunities, and the role of key stakeholders, especially Rosatom, Novatek, and other companies (Moe, 2020). Solski et al. in their study explain the shipping in the Arctic Ocean waters from the viewpoint of international law, national interests, and geopolitics (Solski et al., 2020).

From the environmental point of view, many articles about climate changes and the consequences of warming for the world were published and still are published to draw attention to this sensitive theme for mankind. One of the latest pieces of research analyzes the peer-reviewed scientific literature published between the years 2012 and 2020, which spotlighted climate changes and global warming. Among scientist, there is 99 percent consensus, as proven by the research of Lynas et al. (2021) that the originator of global climate change is human beings. The latest research of scientific works follows the study published in 2013, which researched the works on the theme of climate change published between 1991 and 2012. According to this study, 97 percent of all included works consider the human being as the main starter and initiator of climate change (Lynas et al., 2021).

14.3 Global Climate Changes: Consequences on Region's Energy Complex

The impact of the climate crisis is nowhere so perceptible than at the earth's poles. The most visible result of global warming is the melting of the Arctic ice caps (Bekkers et al., 2015). The ice melting in the Arctic will have several consequences: one of them is the possibility of opening up the Northern Sea Route for commercial traffic. From the environmental point of view, the ice melting induced by global warming causes the rising of the ocean and sea level. This directly impacts people all over the world. At first, it will threaten people living on small islands and in deltas of big rivers—the Ganges and Mekong River in Asia, or the River Nile in Africa. The further lifting of the water level could affect far more people and cause losses on the settlements, stocks, industries, and infrastructures. Climatologists, in their predictions as to when the ice in the Arctic will thaw, diverge somewhat. The worst scenario predicts that sea icebergs in the Arctic could be melted by summer 2030. This will be a catastrophe. The increase of the temperature by about one degree would mean that the level of snow will drop to 120 m in Chile and 150 m in the Alps. If the summer air temperature increases about three degrees, then the Alps will lose 80 percent of their glaciers. The increased temperatures will have an effect on people and their health too. The changes in the quality of the ecosystem will have a direct effect on the decline of species diversity. By the gradual increase in the average global air temperature, there is a danger of the imminent extinction of 20–30 percent of plant and animal mates (Vojtilla & Široký, 2009).

Mankind, with its wide range of activities, affects the global climate. Javaid (2014) underlines that humans affect climate change through industrial emissions, deforestation, destroying biodiversity, and increase in greenhouse gases. All these activities contribute to future instability over the world and in the Arctic too (Javaid, 2014). Mankind's activities will warm the atmosphere, but not the same everywhere. Above the oceans and seas, the air is heating slower than on continents, because water absorbs more energy. The continents are heating twice as fast as oceans. Even faster than the oceans, seas, and continents, the North Pole is also warming.

Javaid (2014) in his study points to the concrete example in the context of the extent of sea ice. This was reduced between the years 1979 and 2012 by about 13 percent per decade. According to Javaid's prediction, the majority of sea ice in the Arctic area will vanish in fifty years. It is assumed that one of the effects of global warming in the Arctic will be the increase in extreme waves deriving from the different parts of the Arctic Basin. On the one hand, the ice thawing is projected in the rising sea levels. On the other hand, together with the Arctic wind-wave activity, it creates unnatural and unpredictable climatic events (Javaid, 2014).

From the environmental point of view, commercial shipping through the Arctic area will have aftermaths. According to Osho (2021), allowing commercial ships to transit through the Arctic will lead to a climate disaster. Osho emphasizes that "increased shipping means an increased risk of oil spills, air pollution from fuel combustion, and accidents". Besides this fact, the most common shipping fuel in the Arctic area is heavy fuel oil, the use of which produces black carbon (known as soot). According to scientists, its use can accelerate the melting rate of Arctic ice. As the number of ships transporting through the Northern Sea Route will increase, black carbon emissions will grow. In the period from 2015 to 2019, emissions grew by 85 percent. Based on this fact, the UN Maritime Organization in November 2020 announced the paper ban on the use of heavy fuel oil, which will come into effect in July 2024, with some exemptions for certain types of ships until 2029. Talking about this ban, the use of heavy fuel oil for shipping cargo in the Arctic will be allowed indefinitely (Osho, 2021).

One of the main causes of the faster Arctic warming is the loss of sea floating ice. The explanation of this loss is simple and lies in the reflection and absorption of the sun's radiation. While the white ice or snow reflects most of the sun radiation, the water (especially dark water) absorbs it. In the absence of the white shield from snow and ice, Earth warms up faster, especially the Arctic, which is getting warmer three times more quickly than the rest of our planet, some parts even faster. This introduces the threat not only for the survival of the white polar bears; it fundamentally changes the entire Arctic ecosystem and has a serious impact on the planetary climate (Ač, 2021).

The situation with the reflection of the incoming solar radiation into space is similar in the Arctic. In the context of the black carbon emissions, it is necessary to underline that the black carbon is suspended in the lower atmosphere, absorbing the heat. However, when the black carbon floats down onto the snow, the reflectivity of solar radiation is reduced. The aftermaths are clear; the snow absorbs the heat and melts faster. According to the study published in the journal titled Geophysical Research Letters in 2019, "the black carbon motivated loss of reflectivity could advance global warming by 25 years, causing catastrophic and irreversible damage" (Osho, 2021).

Alongside the snow and ice in the Arctic, the permafrost is also gradually thawing. By this process, the trapped carbon dioxide and methane are released. It is also important to mention that both greenhouse gases cause more warming; for example, methane is more than eighty times more potent. All permafrost stores around 1700 billion tons of carbon and, based on this observation, it is possible to see the gradual thawing of permafrost—3.4 million square kilometers out of 15 million square kilometers have already thawed (Osho, 2021).

In the context of global warming and climate change in the Arctic, it is necessary to consider the thawing of the permafrost that already has, and in the close future, will have devastating effects. It is expected that process of thawing the permafrost will affect 70 percent of the infrastructure that will require the new construction features and techniques (Kluge & Paul, 2020).

According to State of the Global Climate 2020, the Arctic has been undergoing drastic changes. The temperatures have increased, the Arctic surface air temperature has warmed at least twice as fast as the global average, the ice has been melting, its extent has reduced, and the permafrost temperatures have increased too. In the year 2020, the Arctic belonged to the region with the largest temperature deviations from the long-term average. In individual parts of the Arctic, the contrasting conditions of ice, wildfire, and heat were seen. In the Siberian Arctic region, there were temperature anomalies too: in the periphery more than 3 degrees above average, and in the central coastal parts more than 5 degrees. The preliminary record temperature in 2020 was 38 degrees in the northern part of the Arctic Circle. It was measured in Verkhoyansk during the prolonged heatwave on June 20, 2020. In 2020, the Northern Sea Route was almost without ice for four months (from July to October), and the sea ice extent across the Russian Siberian coastline was particularly low. The high spring temperatures in Russia affected the June snow cover, which was the lowest for the Eurasian Arctic in the 54-year satellite record. It is also important to mention that although the Arctic was a warmer region than average for 2020, some parts of Alaska and Greenland were close to average or below-average temperatures. The State of the Global Climate 2020 pays attention to the wildfire season too. The most active wildfire season was in the region north of the Arctic Circle, breaking an 18-year data record in terms of fire radiative power and also CO_2 emissions released from the fire. Eastern Siberia was categorized as a drier region than average in 2020. The forest fire season started in this area earlier than average, and in some parts had a longer duration than usual, which had long-term damage to local ecosystems (State of the Global Climate 2020, 2021).

The summer in 2021 was record-breaking in temperatures in the western European part of Russia, where the temperatures were about 2–3 degrees higher than in the last summer. In August, which was the warmest month of the summer, it was about 5 degrees. However, the situation was different in the Arctic. The summer in the Arctic did not have a record level, but as a whole it was cold and windy. In the half of September 2021 (September is the peak of the navigation season in the Arctic), the ice was situated in the Novosibirsk area and the archipelago Northern Land. In comparison with the last years, the sea route along the seaside of the Russian Arctic was in this time without ice (Staalesen, 2021) (Fig. 14.1).

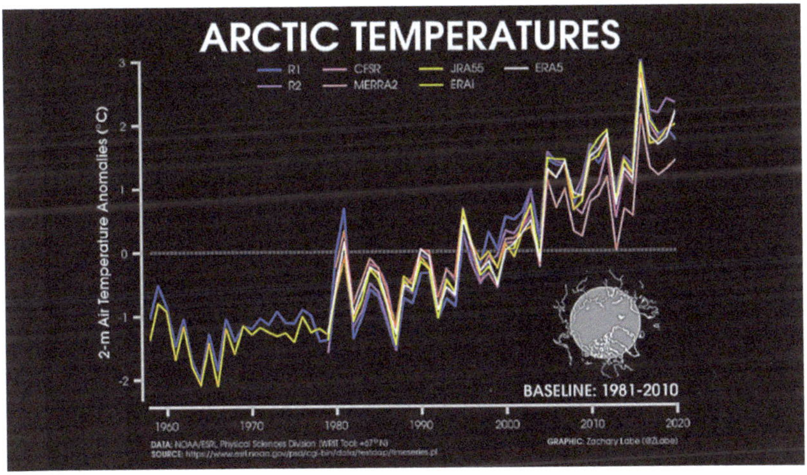

Fig. 14.1 Development of the Arctic temperatures (Labe, 2021) (*Source* Adapted by the authors from Labe, Z. [2021]. *Arctic Temperatures*. https://sites.uci.edu/zlabe/arctic-temperatures/)

14.4 Arctic Northern Sea Route in the Context of the Global Climate Changes and Future of Energy Resources

As stated above, the rise of the temperature causes the ice to melt, so annually the ice in the Arctic area is on the decrease. The melted ice opens new maritime routes important for international trade, enables the extraction of natural resources, and makes tourism possible. It is important to stress that the Arctic is also the place of the meeting of military interests.

Interest in finding a way to the Far East through the frozen sea began in the sixteenth century. It was much later, in the 1930s, that the Northern Sea Route was founded by Russians. However, it has not been used for many years because there is ice coverage throughout the whole year. The route across the Russian seaside was usually used for the transport of raw minerals. The democratic reforms promoted by Mikhail Gorbachev had an enormous effect on the Northern Sea Route; the transportation system in this area of the country was broken. It automatically meant the traffic decline (Javaid, 2014). At the turn of 1991 and 1992, the Union of Soviet Socialistic Republics was dissolved. In this period, the interest in the Northern Sea Route and its building was domestically and internationally low too. The political importance of the Arctic has been growing since 2000. It became part of the strategy papers and appeared in the political expression of Russia's representatives (Kluge & Paul, 2020).

When it was evident that the situation with the ice melting was changing without stopping, attempts to restore the route began. In 2008, the President of the Russian Federation Dmitry Medvedev approved the first document related to the Arctic area, which is unofficially known as the first Russia's doctrine on the Arctic Region. In the document entitled "*The Foundations of the Russian Federation's State Policy in the Arctic until 2020 and Beyond*", the Russian plan of activities in the Arctic in the next decade is described with exactitude. The content of the foundations is the objectives, priorities, and system of measures for strategic planning of the socio-economic development of the Russian Arctic Zone, as well as the mechanism for the realization of Russian state policy in this area and the provision of the national security of the Russian Federation too. According to the stated document, the Russian national interest in the Arctic is not only the conservation of the Arctic's unique ecosystems but the use of the Northern Sea Route as a national integrated transport-communication system too. The international shipping through the Northern Sea Route was secured within the framework of Russian jurisdiction and besides this in correspondence with its international treaties. The same formulations were incorporated in the Strategy for Development of the Arctic Zone from 2013 (Moe, 2020).

The gradual melting of the polar ice caps and an accident in the Suez Canal in the first months of 2021 have contributed to the discussion about the possibility of the Northern Sea Route. After these occurrences, the route across the Russian Arctic coast is perceived as a potential transit route between Asia and Europe. This possibility to ship the goods, cargo, and energy commodities through the northern transport artery is connected with the question of who indeed has the power over the Northern Sea Route. The Russians claim that the route is in its territorial waters, so, according to international law, it has the exclusive rights to develop the stated area and patrol ships. This claim is not recognized by the USA and other states; this leads to the paradoxical situation in which neither state admits the veracity of who can control this passageway. According to international law, the North Pole or the ocean in its immediate vicinity is owned by no state. The five Arctic states—the

USA, Russia, Canada, Norway, and Greenland—can claim an exclusive economic zone, which is a marine area within a distance of 370 km from the coast. It is essential to say that several contradictions exist, which determine whether other parts of the sea fall within areas belonging to national or international waters. Other Arctic states are Sweden, Finland, and Iceland. They are the part of the area north of the Arctic Circle only with the piece of their territory (David & Hansut, 2019). Regarding the recognition of claims, it is interesting that Canada and Russia recognize the claims each other—Canada admits Russia's claim over the Northern Sea Route, and Russia reciprocates Canada's claim over the Northwest Passage as the sea route between the Atlantic and Pacific oceans through the Arctic Ocean (Gricius, 2021).

The vision of the Russian Federation is to establish the Northern Sea Route as a new global shipping artery, which is, according to many analysts, a courageous plan. We agree with the claim that everything will depend on global warming and whether it will be navigable throughout the whole year or only for a few months. The Northern Sea Route is an alternative to the traditional used southern sea route through the North Pacific Ocean, the South China Sea, the Strait of Malacca, the Indian Ocean, the Bab-el-Mandeb Strait, the Suez Canal, the Mediterranean Sea, and the North Atlantic (David et al., 2021). Nowadays, the route is open for longer periods because of ice melting. According to the observations in the Arctic, the amount of ice was reduced by 40 percent over the last 30 years. Russia considers the NSR as its top priority project (Arctic Bulk, 2021).

The Northern Sea Route offers the possibility of shortening the distance between Asia and Europe and decreasing the transport costs. This relates to the reduction of transportation time, fuel consumption, and environmental emissions. Also, unlike areas of the South China Sea, Indonesia, Strait of Malacca, Indian Ocean, Arabic Sea, and the Gulf of Aden, where they operate, there is no threat of pirates. Talking about threats, it is also necessary to mention the instability in Yemen, Northern Kenya, and Somalia on the sea routes across the coast of the states from Asia to Europe for big tankers (David & Hansut, 2019). The huge advantage of this shipping artery against the southern sea route is the possibility of navigation for various types of vessels (Arctic Bulk, 2021). Considering an alternative, the countries began to think about the Northern Sea Route after Ever Given cargo ship accident in the Suez Canal, which blockaded the passage for some days.

On the other hand, the establishment of a reliable transit route requires investment in infrastructure across the Russian Arctic coast. The Russian Federation launched the state program for socio-economic development in the Arctic. Its budget for the years 2021–2023 is about ₽17.6 billion. Compared with the program for Crimea, whose budget is more than ₽300 billion, the amount of finances is substantially less. The Arctic region is about itself one of the problematic and structurally weak regions across the Russian Federation (Kluge & Paul, 2020). For the needs of the successful development of the stated area, there are necessary finances not only from the domestic entrepreneurs and companies but also from the foreign companies, not only in the close surround but also from, for example, the rich Arabic countries.

In 2020, almost 33 million tons of goods were transported through the Northern Sea Route as transport corridor. According to the figure of the Russian Federal Agency for Maritime and River Transport, in the first 9.5 months of the year 2021 (January until mid-October), the shipping volumes on this route increased by 4.5 percent, as compared with the volume in the same period of the year 2020 (Staalesen, 2021). Most of the total 65 voyages in mid-October were international; only 11 of them were voyages between two Russian ports. The ships transported general and bulk cargo (CHNL Information Office, 2021). The plan of the Russian President Vladimir Putin is to transport the goods through this route to the year 2024 amounting 80 million tons, with the prospect of 150 million tons of goods to the year 2030. The vision of these grandiose and ambitious plans of the Russian Federation is connected with the project of new icebreakers with nuclear power. They consider the continuing global warming and also the subsequent disappearance of the sea ices (Staalesen, 2021).

In the Russian Arctic currently, there is a lack of infrastructure and also poor conditions on this route. The fully-competitive Northern Sea Route to the traditional southern shipping route requires the building of new ports, airports, a network of lifeguard systems, or navigation systems. According to Gricius (2021), the Russian development of the new ports throughout the region belongs to one of the most interesting developments in the Arctic region over the last years. The Russian Federation is the principal mover in this area. It builds two types of ports: hydrocarbon and military-oriented. The reason for the so-called build-up and the militarization of the Arctic by Russia is the defense and survivability of second-strike assets, protection of own commercial interests, and addressing economic challenges in the polar region. This activity from Russia's side is explained as the signal of the historic domination in the region. Russia will need the aforementioned ports because of the longer navigation season on the Northern Sea Route and the increasing hydrocarbon extraction and research (Gricius, 2021). Some observers notice that the military presence of the Russian Federation in the area increased last years through the modernization and reactivation of outmoded Russian military bases in the Arctic. Besides this, the new military bases are being built too.

In the past, the ships traveling through the Arctic Ocean along the coast needed the assistance of icebreakers; these made the route passable for various types of other ships. Nowadays, the situation regarding global warming is slowly changing. The presence of the icebreakers is also desirable, but examples have been seen of traveling through the passage without the assistance of an icebreaker. To gain experience with the journey through the Northern Sea Route, the Danish company Maersk Line sent its ice-class container ship through the route. The ship sailed from the Far East autonomously, and the assistance of an icebreaker was only required on three days. The other example was the voyage of the natural gas tanker Christophe de Margerie by the Russian company Sovcomflot shipping company from the Chinese port of Jiangsu to the Sabetta port in the first months of 2021; this provided proof of possible year-round use of the Northern Sea Route (although it is necessary to point out that such ships have a reinforced boat hull) (Hataya & Huang, 2021).

The Russian Federation invests greatly in the construction of nuclear-powered icebreakers. In 2020, it launched the world's largest icebreaker, "Arktika". From the same series, the other two icebreakers, "Ural" and "Sibir", have also been produced. Also, two nuclear icebreakers from this type of series, "Jakutka" and "Chukotka", are under construction. The Russian Federation has the largest flotilla of icebreakers—including nuclear ones—and still is working on the building of new icebreakers with suitable parameters for use in the Arctic, especially the length and width of these stated icebreakers and also their power and capacity. Under construction, there is also the first Lider-class ship. Russia expects that year-round shipping on the route will be possible as soon as 2023–2024. In this context, the Russian Federation will need the whole fleet of ships, together with approximately 30 new tankers, 40 bulk carriers, and 22 container ships for use in the Arctic area (Humpert, 2020; Odynova, 2021; Staalesen, 2021). Besides these ambitious plans, Russian's Arctic coast could be developed through several industrial projects; examples include the Arctic LNG 2, the Syradasaysky coal project, and Rosneft's Vostok Oil. Their contribution will be visible in near future in the explosive growth in shipping (Humpert, 2020).

Many years ago, extreme climate conditions protected the Arctic coast of the Russian Federation as a natural barrier. In the context of ice melting, the conflict potential is bigger. The Northern Sea Route offers a transport advantage for the fast-growing regions in the Far East. The Northern Sea Route has a huge impact on the countries in the region. One of them is South Korea, which has an interest in the hydrocarbons in the Arctic and created the research base in Svalbard too. The People's Republic of China is dependent on global shipping—it is characterized as a powerful manufacturer. With the Russian Federation, China has a common border, and they cooperate in many developing railway projects. Their mutual relationship underlines the partnerships in the international organizations operating in this area or the post-Soviet area. The Russian Federation and the People's Republic of China are members of the United Nations, the World Trade Organization, Group 20, and in the past also Group of Eight + Five. From the post-Soviet area, there is a membership in the Shanghai Cooperation Organization. Both states develop the relationships in the format BRICS. Together with the other member states of BRICS—the Federative Republic of Brazil, the Republic of India, and the Republic of South Africa—they established the New Development Bank in 2015. The partnership between Russia and the People's Republic of China is devoted especially to answer economic and armed questions. In the context of the Northern Sea Route, the People's Republic of China is building a net of the icebreakers itself; it also launches research and scientific expeditions in the Arctic area, where has ambitions on exploration, mining, and sea fare (Javaid, 2014). China is interested in local mining opportunities because it is one of the world's major importers of raw materials (David et al., 2021). In this context, it is worth emphasizing the investment of China and its ability to supply the technology for exploration in the Barents Sea (Kluge & Paul, 2020). The economic prospects of the Northern Sea Route interest Asia's largest exporters, with China at the head, followed by others like Japan and South Korea. In relation to the Russian efforts to develop the Northern Sea Route, such states have started to invest their finances in ice-capable vessels. China's political interest to be a part of the Arctic action was shown by signing the free-trade agreement with Iceland in 2013. Besides this, the next important symbol of the stated interest was the attainment of observer status in the Arctic Council together with Japan and South Korea (Bekkers et al., 2015).

The Russian Federation recently adopted the new Strategy for Developing the Russian Arctic Zone and Ensuring National Security Strategy through 2035. This aims to advance the development of the region as well as to improve living conditions for the population. The document reflected the hopes and also the threats connected with the warming of the Arctic. From Russia's side, it is counted that the Arctic area, after the increasing ice melting, could be the place for vulnerabilities in the so-called Far North. For this purpose, the complete development program of the Arctic area includes the military presence of Russian troops.

The Russian Federation also wants to preserve the ecological balance of this area, which is the same aim together with the European Union. In the new Arctic strategy, the intention of the Russian Federation to protect the Arctic environment is described because the heavy industries with the thawing permafrost and other global climate effects create a toxic mix for the Arctic ecosystems. The strategy responded to the dangerous situations in the Arctic region over the last years. One of them, from June 2020, was the spillage of 20 thousand tons of diesel into the Ambarnaya River. This situation happened after thawing the permafrost settled under the large storage tank. Another example is the forest fires in the area of the Russian Arctic, which burst in the year 2019 and 2020. According to Kluge and Paul (2020), "the interests of the energy sector and the military will be served, while funding to improve environmental protections and living conditions will remain inadequate" (Kluge & Paul, 2020). From May 2021, for two years, Russia will chair the Arctic Council (Kluge & Paul, 2020; Paul, 2021a). In comparison with the Island's presidency in the Arctic Council in years 2019–2021, its priorities were the Arctic sea environment, the climate changes (including the practical solutions, which offer green energy), people and communities of the Arctic, and the strengthening of the Arctic Council. Most of the Island's presidency took place in the time of COVID-19, but the initiative against the marine micro-plastic pollution started (Paul, 2021a).

The Arctic Council is described as the leading intergovernmental forum promoting cooperation in the Arctic. The main document including the cross-cutting priorities of the Russian Federation in the position of the council's chairman is titled *Responsible Governance for a Sustainable Arctic*. Through this document and planned activities during the years 2021–2023, Russia wants to promote collective approaches to the both the sustainable development and the environmentally, socially, and economically balanced development of the Arctic. Besides this, the important point of the Russian chairmanship will be to enhance synergy, cooperation, and coordination with other regional structures. According to *Responsible Governance for a Sustainable Arctic*, the multilateral cooperation includes four priority areas:

1. Human capital;
2. Environment protection;
3. Socio-economic development (in Russia, focused on the cities and towns in the northern regions and across the Northern Sea Route);
4. Strengthening the Arctic Council.

For the Arctic area of the Russian Federation, it is a typical year loss of the population. The Russian offices state that the loss is in some 18,000 people from 2.4 million residents. According to the aforementioned document, the effort will be concentrated in enhancing sustainability and quality of life of the Arctic inhabitants, including indigenous peoples, as well as ensuring sustainable socio-economic development in the region. According to the new Strategy for Developing the Russian Arctic Zone and Ensuring National Security through 2035, adopted in October 2020 by President Vladimir Putin, the improving of the living conditions in the Arctic is a prominent priority (Kluge & Paul, 2020). Attention will be devoted to the preservation of the linguistic and cultural heritage of indigenous peoples of the Arctic and the youth cooperation across the borders through promoting exchanges (scientific, educational, and cultural), tourism, and contacts between peoples and regions too.

The previous Strategy for Development of the Arctic region of the Russian Federation until 2020 was adopted in 2008. According to this document, the Arctic was described as a key strategic territory of the country. In comparison with today's strategy, the main objectives were the creation of a common transport area, more involvement in the global transport chain, the provision of affordable and quality transport services, and finally, the reduction of the negative environmental impacts in the Arctic (David et al., 2021).

The visible demonstration of climate changes in the Arctic is the degradation of the permafrost and the icy gas hydrides emissions. In this context, it is important to continue in supporting the efforts to mitigate the negative effects of climate change, as well as the preservation and restoration of the environment, the sustainable use of natural resources, and maintaining the health of the Arctic ecosystems. In relation to global climate change, the task for states will be the promotion of the introduction of advanced sustainable innovative technologies in many areas—for example in the transport sector, energy, industry, infrastructure—just as the use of renewable energy sources.

Talking about the socio-economic development of the Arctic, the key is cooperation between the Arctic and non-Arctic states in the region. It is also important to mention that this cooperation includes the areas of developing reliable energy infrastructure, as well as sustainable transport routes, including shipping. The finances are necessary to telecommunication systems and the food production sector. Besides cooperation among the states, the improvement of conditions for sustainable investment flows is crucial for the socio-economic development of the Arctic. It is difficult to say exactly what importance the member states and observers give to the Arctic Council. In general, the Arctic Council introduces the format for international Arctic cooperation, including the projects, programs, and activities not only of the Council but also of the Working and Expert Groups and Secretariat (Russia's Chairmanship Priorities for the Arctic Council 2021–2023, 2021; Paul, 2021b).

There are also elements about international cooperation in the Russian Arctic Strategy too. For Moscow, technologies and investment in the energy sector are important. Foreign companies can help to develop infrastructure projects and tackle environmental problems. In addition, the strategy underlines the possibility of joint international research on ecosystems and the effect of climate change. In this context, Russia started cooperation with Germany. The result is the international Multidisciplinary Drifting Observatory for the Study of Arctic Climate (Kluge & Paul, 2020).

We agree with the opinions of those scientists who say we must perform now and approve the immediate action. According to Osho (2021), "protecting the Arctic requires immediate global action to cut carbon dioxide but also emissions of short-lived climate pollutants such as methane and black carbon". He stresses that if we do not approve any steps, we lose the chance to protect ourselves against climate change (Osho, 2021).

14.5 Conclusions

The importance of the Northern Sea Route will increase as the climate changes and the Earth warms. Building and converting the Northern Sea Route to an almost year-round transportation corridor will require significant investments to the infrastructure along the whole line of the route. From the geopolitical view, the importance of building the Northern Sea Route is connected to several factors. On the one hand, the development of this area will include the development, rejuvenation, and building of old ports, new ports, airports, a new flotilla of icebreakers, and rescue centers (in case of dangerous situations). Besides this, the construction of the net of the automated observation stations and weather monitoring equipment will be important too. The Arctic region is rich in oil and gas reserves and other sources of mineral wealth, with which the possibility of the new production is connected. The large undersea oil and gas findings have attracted the attention of the world's major oil companies. Climate changes, which are causing the polar ice to melt, will provide new and unprecedented opportunities for exploration and resource extraction. In this context, it is possible to discuss the new markets trading with these scarce commodities. It can be predicted that rich oil and gas reserves in the Arctic area and their findings will affect the energy sector worldwide. With global warming and the ice melting, the new zones for shipping, fishing, and researching are opening, and there is also the possibility of Arctic tourism. Shipping through the Arctic Ocean will require new types of ships and tankers.

On the other hand, the Arctic is still a region with long dark winters, low winter temperatures, and limiting water depth in some parts. The melting of Arctic ice will have a negative aspect on life in many areas of Earth. The levels of oceans, seas, and rivers will be increasing, and the life, property, and health of people will be threatened. Some kinds of animals will be extinct. Besides this, after the starting of the regular year-round shipping across the Russian Arctic coast, the danger of ecological accidents will increase. Also, invigorated military presence of the Arctic and non-Arctic states in the area is expected.

We believe that the most attention has to be devoted to the climate changes in the Arctic area, as these have an enormous effect on the climate of Earth. In the context of the effort of great powers and states to reduce the emissions and approve the concrete measures, it is necessary to pay attention to the thawing of permafrost. It is estimated that the thawing of permafrost will affect a minimum of 17 million people living in the permafrost area in Russia, alongside the whole infrastructure including houses, bridges, roads, and the energy infrastructure too. From the Russian Arctic areas, the falloff of the houses and roads is actually being reported now. For the purpose of the control, the system of monitoring infrastructure in Russia was established, which is a measure that we welcome. The thawed permanent frozen ground in Siberia releases carbon dioxide and methane constituted by the decay of the plant and animal remains, which warms the Earth faster and has a negative effect.

The inevitable target of states from all over the world would sustain global warming under the critical level of 1.5 degrees Celsius in comparison with the era before the Industrial Revolution. The reduction of emissions in the fastest and most economical way will be our mutual effort; it is the challenge we face together. The summit G20 in Rome resulting in declarations to maintain the warming under the critical value, as well as the introduction of the global minimum tax and the solution of duties for steel, but the common additional statement is vague without the concrete measures for decreasing emissions. As a disadvantage, we consider the absence of a concrete date for the essential reduction of greenhouse gases. We appreciate that since the approval of the Paris Agreement, proper attention has been devoted to the theme of global warming and climate change. We are convinced that it is necessary to approve concrete measures.

In relation to the importance of the Northern Sea Route and its further building, the key factors are climate changes and global warming. These are the elements on which the success of this ambitious project and the volume of the transported goods through the new alternative trading route will depend. Maybe the changing conditions will enable the year-round use of the route in the year 2030, as many climatologists assume, or maybe earlier, or maybe later.

References

Ač, A. (2021). *In island 21 Celsius degrees, in our country 0: Does the warmer Arctic have a connection with the weather in Europe?* Science Within Reach. https://vedanadosah.cvtisr.sk/priroda/zem/na-islande-21c-u-nas-0-suvisi-teplejsia-arktida-s-pocasim-v-europe/ (in Slovak).

Arctic Bulk. (2021). *Northern Sea Route—The shortcut between Asia and Europe.* http://www.arcticbulk.com/page/223/What_We_Offer

Arctic Council. (2021). *Russia's chairmanship priorities for the Arctic Council 2021–2023.* https://oaarchive.arctic-council.org/bitstream/handle/11374/2646/Арктика%20приоритеты_англ_21.06.2021.pdf

Bekkers, E., et al. (2015). Melting ice caps and the economic impact of opening the Northern Sea Route. *The Economic Journal, 128*(610), 1095–1127. https://doi.org/10.1111/ecoj.12460

Blunden, M. (2012). Geopolitics and Northern Sea Route. *International Affairs, 88*(1), 115–129. https://doi.org/10.1111/j.1468-2346.2012.01060.x

CHNL Information Office. (2021). *Transit voyages on the NSR in 2021: The results as of the current date.* https://arctic-lio.com/transit-voyages-on-the-nsr-in-2021-the-results-as-of-the-current-date/

David, A., & Hansut, L. (2019). *Cargo transport options between Asia and Europe through the Northern Sea Route.* http://www.svetdopravy.sk/moznosti-prepravy-nakladu-medzi-aziou-a-europou-cez-severnu-morsku-cestu/

David, A., et al. (2021). The Northern Sea Route as a new route for maritime transport between the far East and Europe. *Communications, 23*(2), A74–A79. http://komunikacie.uniza.sk/index.php/communications/article/view/1674/1571

Didenko, N. I., & Cherenkov, V. I. (2018). *Economic and geopolitical aspects of developing the Northern Sea Route.* IOP Conference Series: Earth and Environmental Science. https://iopscience.iop.org/article/10.1088/1755-1315/180/1/012012

Gress, M. (2021). Northern Sea Route as an alternative to the Southern Sea Route. In *Economic, Political, and Legal Issues of International Relations 2021: Proceedings from the International Scientific Conference* (pp. 136–141). https://fmv.euba.sk/www_write/files/veda-vyskum/konferencie-virt/2021/zbornik.pdf#page=97

Gricius, G. (2021). *Geopolitical implications of new Arctic shipping lanes.* The Arctic Institute. https://www.thearcticinstitute.org/geopolitical-implications-arctic-shipping-lanes/

Hataya, S., & Huang, M. C. (2021). The opportunity and challenges of the Northern Sea Route (NSR) after the Suez Obstruction of 2021. *OPRI Perspectives, 22.* https://www.spf.org/opri-intl/global-data/report/perspectives/20210615152933629.pdf

Humpert, M. (2020). Cargo volume on Northern Sea Route remains stable at 32 million tons in 2020. *Hugh North News*. https://www.highnorthnews.com/en/cargo-volume-northern-sea-route-remains-stable-32m-tons-2020

Javaid, F. (2014). The use of the Northern Sea Route and its effect on climate change. *International Pollution Issues*. https://intlpollution.commons.gc.cuny.edu/the-use-of-the-northern-sea-route-and-its-effect-on-climate-change/

Kluge, J., & Paul, M. (2020). *Russia's Arctic strategy through 2035: Grand plans and pragmatic constraints*. SWP—German Institute for International and Security Affairs. https://www.swp-berlin.org/10.18449/2020C57/

Labe, Z. (2021). *Arctic temperatures*. https://sites.uci.edu/zlabe/arctic-temperatures/

Lynas, M., et al. (2021). Greater than 99% consensus on human caused climate change in the peer-reviewed scientific literature. *Environmental Research Letters, 16*(11). https://iopscience.iop.org/article/10.1088/1748-9326/ac2966

Moe, A. (2020). A new Russian policy for the Northern Sea Route? State interests, key stakeholders and economic opportunities in changing times. *The Polar Journal, 10*(2), 209–227. https://doi.org/10.1080/2154896X.2020.1799611

Odynova, A. (2021). *Russian tanker cuts a previously impossible path through the warming Arctic*. https://www.cbsnews.com/news/russian-tanker-cuts-a-previously-impossible-path-through-the-warming-arctic/

Osho, Z. (2021). *The Northern Sea Route will shorten trade journeys—And augur climate disaster*. https://science.thewire.in/environment/the-northern-sea-route-will-shorten-trade-journeys-and-devastate-the-arctic/

Paul, M. (2021a). *A new Arctic strategy for the EU: Maritime security and geopolitical signalling*. SWP Berlin—German Institute for International and Security Affairs. https://www.swp-berlin.org/en/publication/a-new-arctic-strategy-for-the-eu

Paul, M. (2021b). *The rush for the North Pole: A furious start to Russia's Arctic Council chairmanship*. SWP Berlin—German Institute for International and Security Affairs. https://www.swp-berlin.org/publikation/the-rush-for-the-north-pole

Solski, J. J., et al. (2020). Introduction: Regulating shipping in Russian Arctic waters: Between international law, national interests and geopolitics. *The Polar Journal, 10*(2), 203–208. https://doi.org/10.1080/2154896X.2020.1818487

Staalesen, A. (2021). *Arctic sea-ice reaches this year's low, but shippers still snub Northern Sea Route*. Barents Observer. https://thebarentsobserver.com/en/arctic/2021/09/arctic-sea-ice-reaches-years-low-shippers-still-snub-northern-sea-route

United Nations. (1992). *United Nations framework convention on climate change*. https://unfccc.int/files/essential_background/background_publications_htmlpdf/application/pdf/conveng.pdf

Vojtilla, S., & Široký, P. (2009). *Slovak climate coalition: Global warming and climate change in the world*. O. Z. ZA MATKU ZEM and Slovenská klimatická koalícia. https://www.minv.sk/?Klimaticka_zmena&subor=202671 (in Slovak).

Wan, Z., et al. (2021). Key barriers to the commercial use of the Northern Sea Route: View from China with a fuzzy DEMATEL approach. *Ocean and Coastal Management, 208*(1), 1–10. https://www.sciencedirect.com/science/article/abs/pii/S0964569121001150?dgcid=rss_sd_all

World Meteorological Organization. (2021). *State of the global climate 2020*. https://library.wmo.int/doc_num.php?explnum_id=10618

CHAPTER 15

Northern Sea Route and the New Energy Agenda

Natalia A. Nevskaya

15.1 Introduction

The struggle for energy resources around the world has exacerbated competition between the countries of the Arctic Zone for influence in the northern territories and increased interest in the Northern Sea Route (Torkunov, 2018). The reasons for such heightened attention are as follows:

- Changing climatic conditions and global warming;
- The growing manifestations of the systemic capitalist crisis, the result of which is the intensification of the sanctions struggle between countries, causing energy and transport problems;

N. A. Nevskaya (✉)
Department of Economic Studies, Institute of Europe of the Russian Academy of Sciences (IE RAS), Moscow, Russia
e-mail: nnevskaya@instituteofeurope.ru

V. I. Salygin (ed.), *Energy of the Russian Arctic*,
https://doi.org/10.1007/978-981-19-2817-8_15

- The outbreak of the COVID-19 pandemic, which cut off established production chains and changed transaction costs;
- Traffic problems on the Suez Canal.

These reasons have caused increased attention to the Northern Sea Route not only in the Arctic countries but also in non-Arctic countries participating in international trade. One of the main reasons for this increased attention is the growing importance of the energy factor in the Arctic. Interests in the Arctic shelf are associated with the large-scale development of hydrocarbon deposits and their transportation. Global warming makes it possible to increase the navigability of the Northern Sea Route.

The priorities for action here are quite contradictory, due to the lack of common approaches to the demarcation of shelf boundaries beyond the 200-mile zone. Icebreaker, transport, and rescue fleets, as well as ports, navigation, hydraulic, and hydrometeorological services are in a grave condition. The main development of the Northern Sea Route took place during the Soviet era when there was a planned economy with directive planning and management. The country's transition to market relations harmed the development of the Northern Sea Route, but the aforementioned reasons for the increased attention to the energy problem recently influenced the change in the state's attitude toward the development of the Arctic Zone. Strategies and plans for the development of the Northern Sea Route have been developed, which include a very high share of energy interests associated with the export-oriented development model of Russia.

15.2 Materials and Methods

The energy problems of the Russian Arctic are considered in this article through solving the problems of the development of the Northern Sea Route. On the one hand, the Northern Sea Route participates in the product value chain in terms of the provision of transport services for the movement of energy and non-energy goods. The development of the Northern Sea Route is closely related to the production of energy products on the Arctic shelf. On the other hand, the Northern Sea Route is a consumer of energy products within the framework of the industrial consumption of nuclear engineering—the nuclear icebreaker fleet. To consider the development of the Northern Sea Route, an analysis of Russian regulatory legal acts has been made, ranging from the strategy for the development of the Northern Sea Route to specific plans for its implementation. The article deals with the history of the formation of the Northern Sea Route and the problems of its development at present.

Historical matters about the development of the Northern Sea Route can be found in the works by Andreeva et al. (2020) and Zhukov (2019), as well as official archive documents of Foreign Affair Ministry of Russian Federation.

The articles by Aksenov et al. (2017) and Melia et al. (2016) show the future navigability of Arctic Sea routes. Furuichi and Otsuka (2015) propose a common platform of a shipping cost analysis of the Northern Sea Route and the Suez Canal route. Artamonov (2021) and Blanutsa (2021) tested several hypotheses about the spatial development of the Russian Arctic Zone. Leonov and Zaostrovskikh (2021) analyze the propulsive potential of the ports of the eastern section of the Northern Sea Route and indicate the influence of the Northern Sea Route on increasing the efficiency of the key zone's formation for the development of the Far Eastern Arctic.

15.3 Results

15.3.1 The History of the Development of the Northern Sea Route and Energy Industry

State participation in the development of the Northern Sea Route is closely related to the development of world science, experience in operating various types of transport, and access to new resources. The territories of the Arctic Ocean attracted the attention of researchers and sailors even in the era of Ivan the Terrible, but the Russian systemic state participation in the development of the Northern Sea Route dates back to the period of the formation of a new socialist state. Planning was one of the leading management functions in the new country; therefore, the State Planning Committee was also involved in the development of transport. The development of the Arctic route had objective grounds in connection with the growing volumes of international trade, primarily from the developed European countries and the United States.

The Presidium of the USSR State Planning Committee approved the conclusion of the Bureau of Congresses on January 29, 1925, which reflected 3 main provisions (archive of foreign affairs Ministry of the Russian Federation, 04/ 59/57704/437/13):

1. On scientific issues of metrology and aerology—"the scientific significance of specially designed airships for research purposes is limited mainly to the field of meteorology and aerology; but the data obtained here will to a large extent ensure both correct voyages along the Northern Sea Route and the implementation of hydrographic and field studies, being an aid in explaining the main processes affecting the weather intemperate, and maybe even in the southern latitudes of the USSR";
2. On economic issues—"the organization of aeronautics could have only based on a solid administrative and economic North and in connection with the resolution of several other issues included in the problem of overcoming the spaces of the North, such as the Northern Sea Route, the development of a coastal communication network, the use of aircraft, tractors, and other things. All these forms of transport should be combined with aeronautics into a commonly agreed production plan, for which it is necessary that the routes of air flights comply with the economic needs of the regions of our North, and not be dictated by the requirements for the speed of transporting passengers and transit cargo through the Arctic";
3. Regarding the defense of the country—"in the present international situation of the USSR, the organization of even international transpolar aeronautics, using our northern coasts, will require large expenditures for the protection of airbases, protection, which will still not be able to provide sufficiently reliable control over possible aggressive actions from hostile states. On the contrary, being in our own hands and having our base for the production and restoration of airships, the latter will be an incomparably effective means of protecting the coasts, estuaries of rivers and islands of the North and Northeast, an instrument of counteracting the predatory export of furs and other riches of our North".

As a result, 6 months was approved for the development of a plan for the development of aeronautics, the arrangement of construction shipyards (airship base), and the training of the necessary personnel of specialists of various qualifications, to find out possible sources of funding (Archive of foreign affairs Ministry of the Russian Federation 04/59/57704/437/13).

On February 21, 1925, the commission on transpolar aeronautics was established under the Council of People's Commissars of the USSR under the leadership of Gorbunov. It was equal in its powers to the Administrative and Financial Commission (AFK) under the USSR Council of People's Commissars, the Labor and Defense Council (STO), and the Main Concession Committee.

The aeronautics project was adopted since the USSR did not have the necessary ice-class fleet or air transport necessary for the "effective occupation" of the Arctic coast and adjacent territories. Otherwise, they could have been lost very soon, which would have been forever.

On August 20, 1926, the Council of People's Commissars of the USSR adopted a resolution "On the organization of a trans-Siberian air route in airships", signed by the Chairman of the Council of People's Commissars of the USSR Kuibyshev. The project was implemented in the form of a joint-stock company with the participation of foreign capital with a cash contribution of 50 percent of the share capital. At the same time, the joint-stock company was granted the right to transit cargo through the territory of the USSR, which ensured the interests of the USSR, privileges in the acquisition of building materials and the use of land for the construction of the company, and tax benefits and guarantees of the government of the USSR in the first years of operation of the line.

It was assumed that the project would be of interest in Japan, which has shown an active interest in the development of northern transport arteries, and Germany, which was leading development in the field of various types of vehicles.

In the end, however, foreign states did not show interest in this project, and on January 11, 1927, the commission on the organization of the Trans-Siberian air route on airships was liquidated.

Dissolving the commission for organizing a trans-Siberian air route on airships, the USSR did not abandon its long-term plan to establish reliable control over its sector of the Arctic using transport.

Thanks to the development of aviation, shipbuilding, and science and technology, Great Britain, Canada, and the United States were firmly established in their sectors of the Arctic.

Industrialization was actively developing in the USSR based on the first five-year plans. The economic issues of the Far North were handled by joint-stock companies subordinate to the People's Commissariat of Trade "Komseveroput" and "AKO".

In the summer of 1932, the icebreaking steamer Sibiryakov (a Sovcomflot vessel) left Arkhangelsk in one navigation of the Northern Sea Route on July 28, passing the Vilkitsky Strait, the Laptev Sea, the East Siberian Sea, and the Bering Strait to reach Petropavlovsk-Kamchatsky. In connection with the successful experience of overcoming the Northern Sea Route in one navigation, on December 17, 1932, a decree was adopted on the establishment of the Main Directorate of the Northern Sea Route (Filin, 2020). This was transferred to the management of meteorological and radio stations (*Pravda*, 1932, December 21), while icebreakers and non-icebreaking ships were transferred to the jurisdiction of the Arctic Institute and assigned the responsibility of creating coal bases and the production of port equipment. In 1933, Komsevput was liquidated.

The main stages of state management of the Northern Sea Route were:

- Main Directorate of the Northern Sea Route (1932–1964);
- State Committee of the RSFSR (Russian Federation) for the development of the North;
- State Committee of the Russian Federation for Northern Affairs (Goskomsever of Russia).

The development of the ports of the Northern Sea Route is associated with harsh climatic conditions, problems of waterway depth, and the navigability of ships. This contributed to the development of rail transport in the Norilsk industrial region and the northern regions of the Yamal-Nenets Autonomous District. It was along the route that the largest gas fields in Russia and the world were discovered (Urengoyskoye, Zapolyarnoye, and Medvezhye). Railroad lines planned and under construction will connect major Arctic ports and places of extraction of minerals, including oil and gas. There is active development and exploitation of oil and gas deposits in the Yamal regions. Forecasts of melting Arctic Sea ice increase interest in the development of the Northern Sea Route as a shorter and cheaper route from Europe to Asia. China is particularly interested.

The volume of cargo shipments in the waters of the Northern Sea Route is growing and in 2018 amounted to 20,180.2 million tons; in 2019—already 31,531.3 million tons; in 2020—32,978.9 million tons; cargo turnover of Russian seaports in 2021 increased by 1.7 percent compared to the previous year and amounted to 835.2 million tons, but the cargo turnover of ports of the Arctic basin in 2021 decreased by 1.9 percent, to 94.3 million tons on the statistical data base of the Federal Agency for Sea and Inland Water Transport of the Russian Government.

Russia belongs in the group of countries with a large gross cargo turnover of seaports; Russian statistical data is comparable with Great Britain, Japan, and Australia. China's attention to the use of the Northern Sea Route will attract additional investments in the development of energy infrastructure of the Russian Arctic region and may contribute to the growth of production of the energy industry. The Northern Sea Route is not only a transit route for goods sent from China to Europe and turnout also a link for producers of energy products in the North of Russia with consumers in Europe.

The high capital intensity of Arctic projects requires innovative approaches to the development of the Arctic artery of goods supply and the development of infrastructure for the production of energy commodities. Investment is increased in the infrastructure; it will make cheaper the unit weight of one transportation and can compete with other ways of delivery of transit and energy goods in the future. European consumers need gas, the problem of which is very acute in Europe in the winter of 2021–2022, and Asian countries have developed at a high rate and need affordable oil and energy resources.

About 60 percent of the explored hydrocarbon reserves are concentrated in the Arctic, and the extractive energy industry is actively developing. The major projects of cargo traffic on the Arctic routes related to the fossil fuel industries, which include petroleum industries (oil companies, petroleum refiners, fuel transport, and end-user sales at gas stations), coal industries (extraction and processing), and the natural gas industries (gas condensate, oil, and coal deposits), can be distinguished in the civil sector at present:

1. Yamal LNG—Yuzhno-Tambeyskoye gas condensate field. This is exported through the Sabetta port terminal in the Gulf of Ob in the Kara Sea.

2. Arctic LNG-2—oil and gas condensate fields on the Gydan Peninsula. Sabetta port terminal in the Gulf of Ob, Kara Sea.
3. The New Port—Gateway to the Arctic—oil fields in the western coast of the Gulf of Ob. It is exported through the "Gateway of the Arctic" sea terminal.
4. Tanalau oil terminal—oil fields of "Payakhskaya" group. It is an export terminal at the mouth of the Yenisei River.
5. Prirazlomnoye field—oil fields at the offshore ice-resistant stationary platform Prirazlomnaya. This is exported by large-tonnage ice-class tankers.
6. VostokUgol—development of anthracite of Taimyr coal basin. Dikson in the mouth of the Yenisei River of two deep-water port terminals "Chaika" and "Sever".

For an example of the importance of the development of the Northern Sea Route in the export of energy resources, consider the first project in this list.

In the Yamal-Nenets Autonomous District, the reduced natural gas plant was launched in several stages (2017, 2018, and 2019). This major energy project crossed the interests of the main participants in the energy and transport markets—Russia, France, and China—as Yamal LNG, Ltd, is a joint venture of Novatek (50.1 percent), Total (20 percent), CNPC (20 percent), and the Silk Road Fund (9.9 percent). Construction of the liquefied gas storage tanks and work at the port, from which tankers with gas will be sent to customers in Europe and Asian countries, is currently being completed. Plans call for year-round navigation without icebreaker escort in the western direction and during the Arctic summer in the eastern direction along the Northern Sea Route. 255 shipments or 18.6 million tons of LNG and 24 shipments or 1 million tons of stable gas condensate were shipped through the port in 2020. Liquefied natural gas (LNG) has been export from the sea border crossing Sabetta (Yamalo-Nenets Autonomous Okrug) more than 18.7 million tons and over according to the customs department in January–December 2021.

15.3.2 *Modern Problems of the Development of the Northern Sea Route for Energy Industry*

Currently, the growing attention given to the Arctic raises the question not only of economic competition but regional and global conflict, catalyzed by competition for access to northern energy resources. The elements of danger created by the development of resources form not only localized political and social risks, but also strategic risks. Opponents of Russia in the Arctic region note the strategic importance of the Arctic for the economy and strengthening the integrity of the Russian state, while for the Western countries, the Arctic serves to derive profit. Therefore, for the autonomous territory of Greenland, resource development is not only central for the development of the island's economy, but also appears to be a decisive factor in long-term political development plans independent of Denmark. In addition, such a decision will significantly weaken China's position as one of the main participants in the Greenlandic investment processes, especially in the extraction of rare earth metals—20 percent of the world's reserves of strategically important elements used in the latest weapons and equipment. In 2020, China's market share of the rare earth metals market was nearly 80 percent (Artamonov, 2021).

The number of permits for navigation in the water area of the Northern Sea Route, including under a foreign flag, is increasing. The volume of cargo transportation increased from 10 and 7 million to 19 and 7 million tons (Andreeva et al., 2020: 194–197). The increase was obtained due to gas condensate and liquefied natural gas and the transportation of hydrocarbons. This creates risks and provokes protests by the greens being used in unscrupulous competition on a global scale. Since 2019, a new procedure has been introduced for the passage of foreign ships along the Northern Sea Route. Foreign states must submit a notification of their intention to sail along the Northern Sea Route to the administration of the Northern Sea Route 45 days in advance required for issuing a permit, explaining the ship's name on the navigation routes, the main parameters of the floating craft (in particular displacement), and engine characteristics. A Russian pilot must be present onboard the warship. All this testifies to the progressive security risks in the Arctic geopolitical or economic region and the need to strengthen measures of state regulation of economic and political processes.

Interest in the development of the Arctic is associated with three important circumstances:

1. Climate change. Warming will entail a decrease in the length and thickness of the ice, which will make it possible to use the Northern Sea Route by non-ice-class vessels. Scientists estimate this will happen by 2050 (Aksenov et al., 2017).

At the same time, the preservation of Arctic ice off the coast of Canada and Greenland limits the possibility of sea vessels in the Northwest Passage through the Canadian Arctic.
2. The sanctions policy of the states-trade partners of Russia and the threats to national security associated with it.
3. High risks of operation in the Suez Canal on the example of the transport problem of the bulk carrier Evergrande.

15.3.3 State Regulation of the Development of the Northern Sea Route and Energy Infrastructure: The System of Normative Legal Acts and Meaningful Analysis

As of today, a system of normative legal acts has been formed that provides the basis of state management of the Arctic, including the Northern Sea Route.

The basic document of state planning is the Federal Law of June 28, 2014, No. 172-FZ "On strategic planning in the Russian Federation".

The system of by-laws is formed by two decrees of the President of the Russian Federation:

1. Following article 17 of the federal law "On strategic planning", the Decree of the President of the Russian Federation of March 5, 2020, No. 164 "On the foundations of the state policy of the Russian Federation in the Arctic for the period up to 2035" was adopted. It defines the goals, main directions, and tasks of the state policy of the Russian Federation in the Arctic, as well as the main mechanisms and indicators of the effectiveness of the implementation of the state policy of the Russian Federation in the Arctic.
2. Decree of the President of the Russian Federation of October 26, 2020, No. 645 approved the "Strategy for the development of the Arctic Zone of the Russian Federation and ensuring national security for the period until 2035". The strategy defines the main directions, tasks, and measures for the development of the Arctic Zone, as well as mechanisms, stages, and expected results of their implementation. The implementation of this strategy is designed in three stages: the first stage (2020–2024), the second stage (2025–2030), and the third stage (2031–2035).

The provisions of the document will be ensured by amending the state program "Socio-economic development of the Arctic Zone of the Russian Federation", regional state programs, and the implementation of measures for the development of the infrastructure of the Northern Sea Route. We will consider the analysis of the content of the document in essence below, so as not to deviate from the structural logic and sequence of documents.

To implement two decrees of the President of the Russian Federation, the order of the Government of the Russian Federation of April 15, 2021, No. 996-r "Unified plan of measures for the implementation of the foundations of state policy of the Russian Federation in the Arctic for the period up to 2035 and the strategy for the development of the Arctic Zone of the Russian Federation and ensuring national security for the period up to 2035" was ratified.

The unified plan is a consolidated practical document reflecting the social and economic development of the Arctic Zone of the Russian Federation, the development of science and technology in the interests of the development of the Arctic, environmental protection and environmental safety, and the development of international cooperation. The plan implies the implementation of the Fundamentals and the Strategy in individual constituent entities of the Russian Federation and municipalities.

The transport problems of the Arctic are systematically considered within the framework of the order of the Government of the Russian Federation dated November 27, 2021, No. 3363-r "On the approval of the Transport Strategy of the Russian Federation until 2030 with a forecast for the period up to 2035". Following the Transport Strategy, "the key transport project of the Arctic Zone of the Russian Federation will be the Northern Latitudinal Railway, within the framework of which a direct railway connection of Yamal and Eastern Siberia with the ports of the North-West will be provided, as well as the development of navigation on the rivers of Siberia to ensure an approach to the Northern Latitudinal Railway. The basis of the cargo base of the Northern Latitudinal Railway will be liquid hydrocarbons and gas condensate, cargo for the development and supply of promising raw material Arctic projects".

Following the previous versions of the Arctic Development Strategy, documents to clarify the development of the Northern Sea Route were adopted. Following the Order of the Government "On approval of the attached plan for the development of the infrastructure of the Northern

Sea Route for the period up to 2035" dated December 21, 2019, cargo transportation should increase to 160 million tons by 2035. Following the Decree of the President of Russia dated May 7, 2018, No. 204 "On national goals and strategic objectives of the development of the Russian Federation for the period up to 2024", by 2024, freight traffic along the Northern Sea Route should increase to 80 million tons. However, in 2020, these indicators corrected downward, due to the economic crisis and the COVID-19 pandemic.

15.4 Discussion

The volumes of goods turnover set in the decree presuppose the development of the transport infrastructure of the Russian Arctic. On preferential terms, it is planned to create eight support zones based on the Arctic ports: (1) Kola (Murmansk and Teriberka); (2) Arkhangelsk (Arkhangelsk); (3) Nenets (Naryan-Mar, Indiga); (4) Vorkutinskaya (Varandey, Amderma); (5) Yamalo-Nenetskaya (Sabetta Kharasavey, Novy Port); (6) Taimyro-Turukhanskaya (Dikson, Dudinka); (7) North Yakutsk (Tiksi); and (8) Chukotka (Pevek and Berengovsky). Large hubs are supposed to be created in Murmansk and Petropavlovsk-Kamchatsky—at the entrance and exit of the Northern Sea Route.

The development of transport infrastructure should remove the technical and organizational limitations of logistics, which will attract additional transit cargo to the Northern Sea Route and ensure the loading of transport with a national cargo base (cargo produced in this country and transported by its sea fleet) (Lipina, 2019).

The implementation of these projects will ensure the territorial connectivity of the country envisaged in the draft transport strategy. The increase in capacity entails the need to develop icebreaker transport for the transport of goods.

Within the framework of the federal project "Northern Sea Route" adopted by the Order of the Government of the Russian Federation No. 3120-r dated December 21, 2019, the operation of the lead universal nuclear icebreaker of project 22220 "Arktika" with a capacity of 60 megawatt (on shafts) and the construction of a 25-megawatt linear diesel icebreaker "Viktor Chernomyrdin" were completed. Following the report on the implementation of the federal project "Northern Sea Route", budgetary financing in the field of transport in 2020 amounted to ₽1447.4 million.

Within the framework of the existing specialization of Russia in the export of natural resources—in particular energy—the extractive industry has the greatest investment potential, but unlike the late Soviet period, most of the new projects rely on maritime logistics.

In January 2020, a plan for the development of the infrastructure of the Northern Sea Route (NSR) until 2035 was approved, concretizing the activities of the federal project "Northern Sea Route". Freight turnover involves the transportation of energy resources. An increase in cargo turnover due to the development of minerals of ore and coal deposits in the Far Eastern Arctic is expected to amount to 10.5 million tons. The growth of cargo turnover in the port of Petropavlovsk-Kamchatsky is about 22.5 million tons. The "Strategy for the development of the port of Petropavlovsk-Kamchatsky" contains a more optimistic option, which implies the construction of a deep-water container terminal in 2024 with a capacity of 160 million tons.

The construction of large infrastructure facilities will lead to an increase in coastal cargo, as it will require the delivery of construction and industrial goods, equipment, and machinery.

However, since the Northern Sea Route passes beyond the Arctic Circle, one should not count on a large synergistic effect from the development of the mining and manufacturing industries, due to the lack of technical ability to implement large projects in difficult weather conditions. Unlike the southern sea routes, there is no way to count on the development of medium and small businesses in the reference points of the development of the infrastructure of the Northern Sea Route.

Support for the development of the Northern Sea Route is reflected in the system of strategic planning documents in the Russian Federation, indicated first in the hierarchy of management documents.

Decree of the President of the Russian Federation of 05.03.2020, No. 164 "On the Fundamentals of the State Policy of the Russian Federation in the Arctic for the Period up to 2035" confirms the development of the Northern Sea Route as a competitive national transport communication of the Russian Federation in the world market and as one of the main national interests of the Russian Federation in the Arctic, which is implemented through infrastructure projects.

In particular, following clause 13, the main tasks in the field of infrastructure development in the Arctic Zone of the Russian Federation are the formation of icebreaker, rescue, and auxiliary fleets; the creation of a control system for ensuring the safety of navigation and traffic management in areas of heavy traffic in the Arctic Zone of the Russian Federation; constructing and modernizing seaports and expanding the possibilities of navigation on the rivers of the Arctic Zone of the Russian Federation, constructing railways and highways, and ensuring the operation of these infrastructure facilities at the modern level of development of technology and technology, taking into account international environmental safety requirements.

Within the framework of this document (clause 5), "Key indicators of the effectiveness of the implementation of the state policy of the Russian Federation in the Arctic" have been developed. Of the direct indicators of the effectiveness of the implementation of the development plan for the Northern Sea Route, one particularly noteworthy item is "o) the volume of cargo transportation in the water area of the Northern Sea Route, including transit traffic". The rest of the indicators are indirect and affect the general socio-economic indicators of regional development.

Decree of the President of the Russian Federation dated October 26, 2020, No. 645 "On the Strategy for the Development of the Arctic Zone of the Russian Federation and Ensuring National Security for the Period up to 2035" states that "the Arctic Zone produces more than 80 percent of combustible natural gas and 17 percent of oil (including gas condensate) in the Russian Federation", and that "the continental shelf of the Russian Federation in the Arctic, according to experts, contains more than 85.1 trillion cubic meters. meters of combustible natural gas, 17.3 billion tons of oil (including gas condensate) and is a strategic reserve for the development of the mineral resource base of the Russian Federation". This decree envisages the construction of five new icebreakers and the creation of the infrastructure of the Northern Sea Route.

The system of benefits for the development of the Russian Arctic is in its infancy.

Historically, the first group of benefits is associated with preferences in attracting labor and capital by residents of the Advanced Development Territories, as well as residents of the Free Port of Vladivostok in the Arctic ports of the Russian Far East.

In July 2020, a package of federal laws on state support for entrepreneurial activity in the Arctic Zone of the Russian Federation was adopted to the existing system of benefits. Federal Law of July 13, 2020, No. 193-FZ "On state support for entrepreneurial activity in the Arctic Zone of the Russian Federation" has led to the introduction of amendments to the Tax Code of the Russian Federation. Following these innovations, an entrepreneur registered in the Arctic who is ready to invest at least ₽1 million in the implementation of a new investment project can be recognized as a resident of the Arctic Zone. Residents are offered:

- Zero rate of income tax to the federal budget with subsidiary responsibility of the region (provided that the law of the subject of the federation will establish a reduced rate of income tax in terms of crediting to the regional budget) for 10 years;
- Zero VAT rate when performing certain works and services (transportation of goods by sea vessels, their wiring, and ensuring the safety of navigation);
- 50 percent tax deduction for the extraction of minerals in new subsoil areas (excluding coal and hydrocarbons) for 12 years;
- Several other benefits associated with the reimbursement of part of the cost of paying insurance contributions to state extra-budgetary funds.

It should be expected that the developed system of preferential taxation will be successful mainly in the western part of the Northern Sea Route, rather than in its eastern part, since it is designed to stimulate the development of the extractive industry.

With a detailed examination of the planning documents, the development of the eastern part of the Northern Sea Route refers to the end of the planning period.

In turn, the development of the Far Eastern Arctic Zones of the Northern Sea Route is an integrating backbone project, forming a sustainable spatial formation. At the same time, though, the development of the port economy is not decisive in the development of the Far Eastern Arctic.

The disadvantage is the small number of ports along the long length of the Northern Sea Route (reliance on Murmansk, Arkhangelsk, Norilsk, and Novy Urengoy), which cannot provide a large network effect of growth and the faster economic development of the Arctic Zone with non-Arctic territories. At the same time, there is no significant digitalization in the programs of the new Northern Sea Route infrastructure project. Currently, even on tourist ships, there are problems with the provision of Internet services. It is possible to implement capital-intensive projects with a high rate of replacement of the labor force with capital, but this point is not provided for in the development programs with a small population of the northern territories.

It is necessary to carry out a faster development of small towns and villages—such as Pevek and Tiksi—due to transport, logistics, and digital factors of economic growth.

The development of the Northern Sea Route involves several large investment projects. The implementation of such a project is possible only by a major actor in economic relations.

In November 2021, a Transport Strategy was adopted, which takes into account the importance of the Northern Sea Route in the transport of energy resources. Following the data of the Transport Strategy (Portnews, 2021):

> in 2019, the volume of transshipment through the ports of the North-Western and Arctic basins, excluding land checkpoints, amounted to 361 million tons of cargo. In the future, until 2035, the volume of transshipment will increase to 498 million tons, while the key factors affecting the change in the volume of transshipment through the ports of the North-Western Basin relative to 2019 will be the following energy products:
>
> - *oil*: a decrease by 21 percent from 126 to 99 million tons due to a decrease in consumption in Europe;
> - *oil products*: a slight increase by 3 percent from 72 to 74 million tons due to a decrease in consumption in Europe and outstripping growth in supplies to the Middle East and Africa;
> - *coal*: an increase of 19 percent from 61 to 82 million tons, depending on the rate of decline in coal consumption in Europe and logistics costs in case of reorientation of flows to other directions (the People's Republic of China, Republic of India), as well as the reorientation of Russian flows from the ports of the Baltic States;
> - *liquefied natural gas*: an increase of 3.8 times from 21 to 79 million tons due to the implementation of the announced Arctic projects and the export of their products through the north-western direction.
>
> Forecasted volumes of foreign trade freight traffic in the ports of the Far Eastern basin. In 2019, the volume of transshipment through the ports of the Far Eastern Basin amounted to 214 million tons of cargo. In the future, until 2035, the volume of transshipment will increase to 397 million tons, while the key factors affecting the change in the volume of transshipment through the ports of the Far East basin relative to 2019 will be:

- *oil*: a decrease of 27 percent from 52 to 38 million tons due to the global decline in consumption;
- *oil products*: an increase in volumes by 19 percent from 16 million tons to 19 million tons, an increase in domestic consumption will be offset by the export of oil products from the Siberian and Ural federal districts;
- *coal*: growth by 2.5 times from 99 to 241 million tons due to the replacement of Australian coal in the markets of the People's Republic of China and an increase in demand in the countries of Southeast Asia;
- *liquefied natural gas*: a fourfold increase from 11 to 40 million tons due to the implementation of the announced Arctic projects, projects on Sakhalin, and the export of their products using the transport infrastructure of the Far East.

15.5 Conclusions

The development of the Northern Sea Route is an important stage in increasing the income of the export-oriented economy of Russia from world trade in energy resources (Salygin & Lobov, 2021). In this case, the emphasis is on fossil energy resources and their export abroad. However, as a result of the adoption of several documents of state regulation of the development of the Arctic, it is planned to create five nuclear icebreakers, which will contribute to the development of nuclear engineering and nuclear energy. In addition, the documents suggest the construction of infrastructure, including digital. A distinctive feature of the recently adopted strategies for the development of the NSR involves measures for social support of the local population, which implies the creation and development of new jobs in education and medicine to meet the needs of employees of the NSR. To attract private investment, more favorable tax conditions have been created.

The disadvantage of the system of state regulation of the development of the Northern Sea Route is the systemic underfunding of programs, which in some cases jeopardizes the success of the entire system of measures taken. The presence of private capital in highly profitable economic activities in the Arctic Zone against the backdrop of underfinancing of government obligations may lead to a shift of resources to a more profitable sphere and a disruption of work in less paid divisions serving the Northern Sea Route. This can lead to operational disruptions and the potential loss of foreign clients. The sanctions pressure jeopardizes several energy projects in the extractive industry and requires the ability to diversify activities to protect against the risks of investment loss. The transport strategy assumes a planned reduction in the consumption of fossil energy resources in connection with new environmental standards. However, the use of the Northern Sea Route can make it possible to minimize transport costs, affecting the price reduction, which can help maintain the same volumes of consumption of raw materials, but with the replacement of consumers from developed countries toward developing economies. An important feature of the Northern Sea Route is the operation of nuclear icebreaking ships, which minimizes the emission of carbon dioxide into the air.

References

Aksenov, Y., Popova, E., Yool, A., Nurser, A. J. G., Williams, T., Bertino, L., & Bergh, J. (2017). On the future navigability of Arctic sea routes: High-resolution projections of the Arctic ocean and sea ice. *Marine Policy, 75*, 300–317.

Andreeva, E. N., Voronina, E. P., Zaydfudim, P. H., et al. (2020). *The Arctic in the spatial development of the Russian Federation: Management problems.* Northwestern Institute of Management RANEPA (in Russian).

Artamonov, V. S. (Ed.). (2021). *Geoeconomics of the Arctic: Problems of ensuring economic security*. Science (in Russian).

Blanutsa, V. I. (2021). Spatial development of the Russian Arctic zone: Analysis of two strategies. *Arctic: Ecology and Economy, 11*(1), 111–121.

Filin, P. A. (Ed.). (2020). Polar readings—2019: The Arctic: Management issues. In *Materials of the Seventh International Conference "Polar Readings – 2019. State and Public Organizations in the Management of the Arctic: Past, Present, Future". St. Petersburg, April 29–30, 2019*. Paulsen.

Furuichi, M., & Otsuka, N. (2015). Proposing a common platform of shipping cost analysis of the Northern Sea Route and the Suez Canal Route. *Maritime Economics and Logistics, 17*(1), 9–31.

Leonov, S. N., & Zaostrovskikh, E. A. (2021). Influence of the ports of the Northern Sea Route on the formation of focal zones for the development of the Eastern Arctic. *Arctic: Ecology and Economy, 11*(1), 6–18.

Lipina, S. A., Smirnova, O. O., & Kudryashova, E. V. (2019). *The Arctic: Development strategy*. NArFU (in Russian).

Melia, N., Haines, K., & Hawkins, E. (2016). Sea ice decline and 21st century trans-Arctic shipping routes. *Geophysical Research Letters, 43*(18), 9720–9728.

Portnews. (2021). *Transport strategy of Russia until 2030 with forecast until 2035 published on official portal*. https://portnews.ru/upload/basefiles/2510_0001202112030006.pdf

Salygin, V. I., & Lobov, D. S. (2021). Defining major oil and gas companies' development strategies in the era of energy transition. *MGIMO Review of International Relations, 14*(5), 149–166. https://doi.org/10.24833/2071-8160-2021-5-80-149-166

Torkunov, A. V. (2018). Great Eurasian power in the stream of changes. *International Trends, 1*(52), 6–13 (in Russian).

Zhukov, Y. N. (2019). *Stalin: Project "Arctic"*. Conceptual (in Russian).

CHAPTER 16

Energy Development of the Russian Arctic Within the Framework of the Belt and Road Initiative and Greater Eurasia: A View from China and India

Kobilzhon Kh. Zoidov, Alexei A. Medkov, and Zarina A. Dadabayeva

16.1 Introduction

Climatic changes and global warming (an increase in the average annual air temperature on Earth) make the ice melt and significant areas of the water surface suitable for unhindered navigation by the shortest

K. Kh. Zoidov (✉) · A. A. Medkov · Z. A. Dadabayeva
Market Economy Institute, Russian Academy of Sciences, Moscow, Russia
e-mail: kobiljonz@mail.ru

A. A. Medkov
e-mail: medkov71@mail.ru

Z. A. Dadabayeva
e-mail: zarina.17.06@list.ru

Institute of Economics, Russian Academy of Sciences, Moscow, Russia

V. I. Salygin (ed.), *Energy of the Russian Arctic*,
https://doi.org/10.1007/978-981-19-2817-8_16

routes; hydrocarbons and other mineral resources can be developed in the Russian Arctic.

A new global maritime transport network is being formed, transportation distances between Asia and Europe, Asia and America, and Europe and America are being reduced, and large vessels that were not originally intended to pass the Suez and Panama Canals (as well as other narrow points of the World Ocean) can be built and used.

The development of the Northern Sea Route (NSR) is a priority of the Russian state economic and transport policy. The country should make full use of the emerging opportunity to show a synergistic effect from the export–import, domestic ("northern delivery", transportation of fish from the Far East to Western Russia, etc.), and transit cargo transportation.

According to the estimates of the US Geological Survey, the Arctic contains 30 percent of the world's natural gas reserves, 13 percent of oil, and 9 percent of coal, as well as large reserves of metals, including rare earth metals, and minerals (diamonds, graphite). Significant reserves of natural resources in the Russian segment of the Arctic, alongside the emergence of production, technological, organizational, and economic opportunities for extracting natural resources on the continental shelf of the northern seas, open up tempting prospects for the spatial and economic development of the Russian Far North.

The program document "China's Arctic Policy" adopted by the State Council of the People's Republic of China (PRC) in January 2018 outlined the main directions of China's Arctic policy. They are related to the prospects for the development of mineral resources, as well as the use of the transport and transit potential of the region to reduce dependence on the bottlenecks of the global maritime transport system, primarily the Strait of Malacca and the Suez and Panama Canals.

Chinese companies and the Silk Road Foundation are known to participate in the project for the creation and operation of the Russian enterprise Yamal LNG. In 2021, PJSC Novatek (Russia) and Zhejiang Energy (China) signed an agreement on long-term supplies of liquefied natural gas (LNG) from the new Arctic LNG-2 project for 15 years in the amount of up to 1 million tons per year at the St. Petersburg International Economic Forum (RIA, 2021).

At the same time, it should be borne in mind that China's foreign policy is aimed at internationalizing the Arctic space, which is under the jurisdiction of basic agreements of international law (the Charter of the United Nations [UN] and the UN Convention on the Law of the Sea). In this regard, Russia is aimed at ensuring national security, countering the transformation of the NSR into an international highway, and neutralizing the claims of other states to the Russian part of the continental shelf.

On the other hand, the implementation of joint Russian-Chinese oil and gas projects in the Arctic is a way to counter the expansion into the United States, Canada, and the Nordic countries (Denmark, Norway, Sweden, and Finland). Russia can take advantage of the geopolitical and geo-economic rivalry between China and India when it comes to access to natural resources during the internationalization of the Arctic space.

This paper hypothesizes that the energy-resource, mineral-raw materials, and transport-transit potential of the Russian Arctic can be effectively used only through nominating and implementing large infrastructure-integration projects: these are associated with the Chinese Initiative "Belt and Road", India's program for the development of innovative industrial belts of the transport corridor "North-South", and implementing Russia's initiative to form a Large Eurasian Partnership (Tsvetkov et al., 2014, 2019; Zoidov et al., 2017, 2021a, 2021b).

16.2 Review of Scientific and Practical Approaches

Many scientific monographs, articles in journals, and collections have been published on the research topic, dissertations have been defended, and thematic reports have been prepared at specialized conferences. We should distinguish the works of those authors whose information, ideas, and arguments contribute to proving the hypothesis put forward in this article.

For example, the monograph by Indzhiev (2010) indicated that the full-fledged development of offshore hydrocarbon deposits in Russia requires the following:

- Experience in the construction of floating drilling platforms capable of operating in the harsh Arctic climate;
- Large-tonnage tankers designed to work in ice;
- Deep-water terminals for their moorage;
- Pipeline systems resistant to low temperatures and their fluctuations;
- Trained personnel, developed social and household, transport, and road infrastructure.

The author came to the disappointing conclusion that "while Russia does not have all the necessary instruments, it can receive no more than 0.2 per cent of total oil and gas production on the shelf, and everything else has to be left to foreign corporations" (Indzhiev, 2010: 51).

According to the former US Deputy Secretary of State for Oceans and International Environmental and Scientific Affairs Claudia McMurray, the natural resources of the Arctic should belong to transnational corporations (Indzhiev, 2010: 112). The United States can oppose Russian expansion under the guise of developing and implementing Arctic assistance programs aimed at preserving the traditions of the indigenous population, adjusting the environmental agenda, developing education, etc.

One cannot but mention the opinion of Sergey Sobyanin, who was the Deputy Prime Minister of the Russian government, cited by the same author, that further development of the Arctic and North-Eastern Siberia can be carried out only in partnership with the largest Western and East Asian corporations (Indzhiev, 2010: 62). The creation of trilateral joint ventures for the extraction of hydrocarbons in the Russian Arctic using Chinese capital and Western technologies is quite promising.

The closed process of global warming in the Arctic is associated with its acceleration due to the increasing production of hydrocarbons and the burning of associated gas, the melting of snow and ice reflecting sunlight, and the release of the dark surface of the earth and water, which contributes to even greater atmosphere warming.

Dodin points out that the long-term socio-economic interests of Russia in the Arctic correspond to the connection of two unique transport routes into a single whole: the North-Western Sea and the Northern Sea Routes, which will ensure the creation of the shortest of all existing international water transport corridors: Pacific Ocean—Atlantic Ocean—Arctic Ocean—Pacific Ocean (Dodin, 2005: 209).

One cannot but agree with the opinion of Zbarshchenko that "the analysis of the existing cargo flows within the framework of the world's most powerful East-West transport corridor through the Suez Canal between the countries of Southeast Asia and Western Europe from the point of view of the expediency of switching part of the cargo to the Northern Sea Route showed unambiguously negative results mainly due to the higher cost of transportation along the NSR" (Zbarshchenko, 2016: 72). The experience of operating national transport vessels of reinforced ice-class, such as the Norilsk motor ship and the Sevmorput nuclear-powered vessel, proved the high efficiency of such vessels in the Arctic, but also demonstrated the high cost of cargo transportation on these vessels in the open water of southern routes.

The proposed formation of a trunk-feeder transport and transit system (TTS) forms an economic interest in the development of transit traffic along the NSR. The use of a hull device of the same type for all types of transportation should ease the creation of the Arctic fleet and significantly reduce the construction cost of reinforced ice-class shuttle ships.

Zbarshchenko warns that "due to the current principles of freedom of navigation, the transportation of transit cargo by foreign charterers along our Northern Sea Route in conditions of a shortage of icebreaking fleet will never, under any circumstances, provide the Russian economy with any significant income, but threatens to forever pollute our coastal waters in the Arctic" (Zbarshchenko, 2016: 83).

Another article (Strek et al., 2016: 122–127) presents the results of the development of projects for the construction and use of ice-class vessels Arc7 with a container capacity of 2000, 3000, and 4000 TEU,[1] as well as container ships of the highest ice-class Arc9, which do not require icebreaking support throughout the year. It is noted that the transport schemes for the use of container ships are closely related to their size: for example, the traditional route through the Sannikov Strait can be used with a capacity of up to 3000 TEU, whereas a draft of no more than 12.2 m, with a larger draft and capacity, requires only a high-latitude path above the Novosibirsk Islands.

According to the results of mathematical modeling and economic calculations, the ice-class container ship Arc7 with a capacity of 3000 TEU was recognized as optimal. It is characterized by economic efficiency and a minimal need for icebreaking support. After the authors studied foreign shipyards capable of bidding for the construction of Arctic-class vessels, three of them were selected, among which there is a COSCO shipyard (China).

The weaknesses and difficulties in organizing the internal export–import and transit transportation of goods using the SPM raises more interest in the projects of the North-South ITC on the route Russian Arctic—South Asia. The Indo-Siberian Continental Railway project may include several routes from the Eurasian Economic Union (EAEU) to

[1] Twenty-foot Equivalent Unit.

Afghanistan, Pakistan, Iran, and India (Pak, 2021; Pak & Sarkisov, 2014). According to Alexander Sobyanin (2012), the gauge of Russian meridional railways should be 1520 mm, and new latitudinal China–Europe routes should probably have a gauge of 1435 mm.

He notes that the growth of the Islamic population in the post-Soviet countries is restrained by Europeanized thinking, in particular by overcoming the tradition of polygamy. However, shortly, a lot of people may return to this centuries-old tradition. Population growth in Central, Western, and South Asia increases the export of grain-producing regions of Siberia, as well as the demand for Russian energy and food products, and exacerbates the problem of water scarcity.

Demographic growth stimulates the construction of industrial, agricultural, transport, housing, and socio-cultural facilities, and, as a result, increases the demand for Russian construction materials and equipment. In turn, Russia should become a capacious market for textile goods, furniture, household appliances, and cars produced in the countries of the Indian Ocean.

The railway network of Siberia and Central Asia is separated from India and Pakistan by the high mountain ranges of the Pamirs, Hindu Kush, and the Karakoram. However, modern construction technologies applied in the Andes and Tibet allow us to overcome these problems. According to Sobyanin, "it is in the highlands that the most straightened aboveground rail tracks can be laid using the modules produced in factory conditions and transported to the site by the most energy-efficient mode of transport – an airship" (Sobyanin, 2012).

As noted by Sidorov (2018), China considers Denmark, Sweden, Norway, Finland, and Iceland as door openers to the Arctic region and at the same time to Europe, as well as the owners of the technologies China needs to develop the Arctic. The objectives of China in Northern Europe are:

- Diversification of cargo transportation routes by sea;
- The formation of an alternative to the European hub port Rotterdam on the Scandinavian Peninsula;
- The creation of nodal points of sea cargo transportation on transaction routes.

Finland serves as an intermediary in China-EU relations. The cooperation between the two countries in the field of scientific research into the Arctic continues and is gaining momentum. For example, the second Chinese icebreaker Xuelong-2 was built with the participation of Finnish experts. There was further cooperation in the field of the construction and modernization of transport and communication infrastructure: in particular, in the construction of the Rivne-Kirkenes railway, and the laying of a tunnel between Helsinki and Tallinn under the Baltic Sea within the framework of the Rail Baltica project.

However, the Arctic states treat the Chinese initiatives with a high degree of caution. In particular, Norway refused to create a free trade zone with China. The government of Greenland took into account the opinion of the EU supranational bodies and did not grant China the right to extract rare earth metals. The Icelandic authorities did not allow Chinese entrepreneurs to buy a land plot in the area of Grímsstaðir (300 square kilometers) for building a tourist complex.

The Nordic countries fear that China's commercial projects hide its large and far-reaching geopolitical and military-strategic plans. Besides, European environmentalists know the methods of spatial and economic development of foreign territories that Chinese companies use. That is why they are concerned about the consequences of China's expansion into the Arctic, as it can affect the fragile nature and life quality of the indigenous peoples of the Far North.

Comparing the strengths and weaknesses of the two available sea routes through the Arctic—the NSR and the Northwest Passage along the US coast (Alaska) and the islands of Canada—Chinese experts concluded that "the last available option for organizing the Arctic transport corridor was the Russian Northern Sea Route" (Sidorov, 2018).

In 2020, the Chinese shipping company COSCO Shipping performed 11 flights along the NSR. Yeranosyan, Expert of the Bureau of Military-Political Analysis, believes that the development of the NSR creates conditions for increasing the cargo turnover of the ports of Northeast China (Dalian and Qingdao). The author draws attention to the unofficial statements of Rear Admiral Yin Zhuo and researcher at the Shanghai Institute of International Studies Chen Baozhi that "the Arctic belongs to the whole world, so no nation has sole power over it" (Yeranosyan, 2021).

Developing cargo transportation in the Arctic, China will inevitably strive to independently perform ice wiring and refuse the services of foreign pilots as the national icebreaking fleet is created. In June 2018, China laid the keel of its first nuclear icebreaker.

Due to the high competition in the market of global cargo transportation by sea, the development of TTS in the Russian Arctic should be based on the undeniable advantages of Russian companies and scientific organizations in the development and implementation of advanced transportation processes. For example, there will be the development, improvement, implementation, and wide dissemination (scaling) of magnet levitation technologies for high-speed cargo transportation over long distances.

Zanin, General Director of JSC "National High-speed Roads", highlights the following competitive advantages of magnet levitation technologies (Maglev) in comparison with the existing wheel-rail transport systems for cargo transportation (Zanin, 2017: 80–82):

- The Maglev highway is located above the ground on overpasses, which eliminates or minimizes the threats and consequences of man-made accidents and environmental disasters;
- There are no carbon and other emissions into the environment;
- Full automation, robotization, and digitalization of transport system management based on the use of unpopulated technologies is provided;
- Each unit of rolling stock (platform) can be sent either separately or as part of any number of platforms, limited only by the capacity and specifics of loading and unloading operations;
- The platforms can reach the speed of 500–600 km/h, which is facilitated by the optimized aerodynamic shapes of rolling stock and transported goods.

Zanin notes a reduction in operating costs (except energy) by 3–4 times compared to wheel-rail high-speed roads due to the absence of rubbing and vibrating structures and the automation of monitoring the condition of the overpass. In this regard, the transportation of coal, timber cargo, and woodworking products will not require reducing tariffs and providing subsidies, "which will make Siberian resources more competitive on world markets" (Zanin, 2017: 81).

He says that magnet levitation technologies using superconductivity should appear, which significantly reduces energy consumption and increases speed to 1000 km per hour or more. The conclusion is unambiguous: "the future of Arctic transport is not for wheeled roads, but 'flying' trains of today and the future" (Zanin, 2017: 82).

Haidarov cites a system of arguments in favor of the transportation of liquefied natural gas (LNG) in LNG containers, "which allow realizing the potential of the Northern Sea Route with minimal investment and optimizing logistics on the Arctic Container Line" (Haidarov, 2017: 61–65), namely:

1. The liquefaction of natural gas provides for its additional purification from hydrogen sulfide and carbon dioxide impurities, which increases the efficiency of electricity production and heating and reduces the number of harmful emissions into the atmosphere;
2. Containers do not require an additional regasification system: gas is discharged directly from the device;
3. There is no need for the construction and purchase of large-sized and expensive gas tankers—conventional container ships can be used;
4. Compliance of LNG containers with the ISO standard ensures optimal load capacity and transport accessibility: the container is suitable for organizing multimodal transportation by rail, road, and air.

Particular attention is paid to the absence of economic risks when transporting gas in LNG containers, "taking into account the annual climate warming, the coastal infrastructure should be influenced by a thermostat, so the soil subsidence can destroy production facilities and gas pipelines" (Haidarov, 2017: 65).

Dosenko et al. (2017: 67) point out that "when operating vehicles in the Arctic, one has to strictly regulate the driver's shift, taking into account natural illumination or its absence".

The need for adaptation therapy is also associated with the consequences of high-latitude conditions and changing climatic zones that disrupt the rhythms of the human body. An increase in the volume of cargo flows in the Arctic can provoke a surge in the morbidity of the transport and road employees, as well as an increase in the number of accidents.

These considerations should be taken into account when scaling up the practice of shift work in the Russian Arctic, attracting foreign labor (including from China, India, Central Asian countries), and making further changes.

16.3 Results and Discussion

16.3.1 Weaknesses of the Arctic Transport and Transit System of Russia and Its Main Transport Artery—the NSR

There are weaknesses of the Arctic marine transport and transit system (TTS), the most important component of which is the NSR.

1. Nature and climate

   1.1. Low air and water temperatures; high probability of severe ice conditions, ice bridges, ice storms; mobility of Arctic ice fields, impassable areas, hummocks, the need for ice exploration.
   1.2. Prolonged polar night (significant duration of the dark time of day in winter).
   1.3. Relatively shallow depths (on the shelf on average 12 m), numerous shoals.
   1.4. The NSR not being a separate route, but a set of navigable routes that differ in navigational and hydrographic terms.

2. Production and equipment

   2.1. Delays in the construction and commissioning of new icebreakers, their high cost is increasing over time.
   2.2. Regular employment of the icebreaker fleet in shipping raw materials for export, providing "Northern delivery", and escorting ships of the Russian Navy.
   2.3. Low speed of icebreaking (5–7 knots), the need for the repeated splitting of thick ice after gaining momentum.
   2.4. Insufficient width of Russian icebreakers, which does not correspond to the size of modern oil tankers, gas carriers, container ships, and bulk carriers.
   2.5. Inefficient use of the icebreaker fleet when escorting ships (caravans) in open water.
   2.6. Transfer of the functions of ice wiring of ships from nuclear and mainline icebreakers to diesel-electric ones at the mouths of

northern rivers during the implementation of river-sea transport and logistics technologies, as well as to port icebreakers.

2.7. The need to use reinforced ice-class vessels adapted to sailing under low temperatures and icing.

2.8. The special character of marine and man-made disasters in the Arctic, especially when the ships carrying hydrocarbon raw materials (as well as coastal reservoirs) are involved.

3. Finance and economy

3.1. Doubts of cargo owners, consignees, and transport and logistics companies about the high profitability of cargo transportation through the Northeast Passage.

3.2. The contradiction between the need for icebreaking and the development of corporate fleets, including reinforced ice-class vessels.

3.3. The absence of or low revenues from the passage of transit vessels under the flag of foreign states moving along high-latitude routes, open water spaces or ice fields without icebreaking and pilotage escort.

3.4. Significant costs for accident and man-made disaster management.

4. Transport and logistics

4.1. The need to form not only large ship parties but also caravans of ships to increase the efficiency of icebreaking support for transit cargo transportation, low efficiency of one or even two icebreakers.

4.2. The need to use two icebreakers for working with wide large-tonnage merchant ships and warships.

4.3. The expediency of transshipment at the initial and final points of the cargo route from reinforced ice-class vessels to conventional vessels using high-speed technologies of cargo operations and the organization of clear transportation and transshipment.

4.4. Competition of the North-Eastern (including the NSR) and North-Western Arctic passages.

4.5. Competition between the ports of Petropavlovsk-Kamchatsky and Korsakov on Sakhalin for the right to become the eastern nodal point (hub) of the NSR.

4.6. The need for isothermal protection of temperature-sensitive goods transported along the NSR.
4.7. The underdeveloped coastal infrastructure that does not meet the requirements of ensuring the safety of navigation, especially in the event of a technical breakdown or the illness of crew members.

5. Institutes

5.1. Attempts by the individual Arctic and non-Arctic States to internationalize navigation along the NSR.
5.2. The relevance of geoclimatic problems that determine the positions and actions of the main players of Arctic policy, insisting on granting the right of free navigation in the water areas freed from ice due to global warming.
5.3. Unfavorable socio-cultural situation in many regions of the Russian Far North, the demographic situation, natural, and migration population decline.

Today, five nuclear icebreakers carry out icebreaking in the seas of the Arctic Ocean. Four nuclear icebreakers of the Arctic type will be built and put into operation by 2026, and the first icebreaker of the Leader project will be built in 2027. By 2030, the ice-class cargo fleet should be increased by more than three times, for which there is no need to build more than 30 tankers, 40 bulk carriers, and 22 container ships (Belov, 2021).

These ambitious plans require significant funding from the state budget, the National Welfare Fund, attracting investments from state corporations, and private businesses. For example, the construction of the main icebreaker of the project LK-60YA "Arktika" required ₽47 billion, while the initial cost had been about ₽37.8 billion. Today, the cost of the new icebreaker LK-120YA "Leader" is estimated at ₽130 billion, but the costs should grow and exceed ₽150 billion with a corresponding increase in its payback period (Alyoshina, 2021b).

The unified transport and logistics operator of the NSR of the Rosatom State Corporation plans to create and develop a container corridor in the Arctic transit waterway using Russian container ships of reinforced ice-class; the first of these should sail in 2025 (RZD-Partner, 2021).

The main barrier to implementing these plans may be the lack of icebreaking support (Chernyshevskaya, 2021). The situation is aggravated by the fact that the icebreaker fleet currently in use is already worn-out and there are parameters that do not meet the requirements of the ice channeling of large merchant ships.

The most advanced nuclear icebreaker "50 Let Pobedy" has a width of 30 m and can create a corridor in the ice with a width of 33–35 m, while the width of Panamax-class vessels is 40 m, and gas carriers is 50 m. In this regard, the efforts of two icebreakers are required for channeling modern merchant ships along the NSR.

According to A. Afonin, the dean of the Arctic Faculty of the Admiral Makarov State University of Maritime and Inland Shipping, the draft of the main container ship with a width of 59 m is 12–15 m, and the guaranteed depths on the main routes of the NSR are no more than 12 m. At the same time, the drifting nature of the ice fields forces the captains to constantly deviate from the planned routes. In addition, the routes themselves should be built only after continuous hydrographic studies, obtaining the information about the bottom relief, identifying safe water areas and routes for large-tonnage vessels (Alyoshina, 2021b).

In 2019, about 700 thousand tons of cargo were transported through the NSR in transit. In 2020, the volume of transit cargo transportation increased by 71 percent and amounted to slightly less than 1.2 million tons, which accounts for less than 4 percent of the total volume of cargo transportation along the NSR (32.9 million tons). At the same time, about 1.17 billion tons of cargo were passed through the Suez Canal in 2020 (Gusachenko, 2021b).

It should be borne in mind that the southern sea trade routes (Asia–Europe) have long been worked out, provided with port infrastructure, can be scaled, are influenced by global container services and states defending the interests of national participants in the transport and logistics chain.

The vessels following the southern sea routes do not have an ice-class, and therefore carry a large payload per unit of their weight and have greater commercial efficiency. It is not the speed of goods delivery that is important for shippers and consignees, but the exact time of their departure and receipt, which was guaranteed by global sea container services during the pre-pandemic period.

The instability of the Arctic ice fields can negate the economic efficiency of the NSR as a shorter Asia-Europe route compared to the Suez Canal: it is difficult, if not impossible, to calculate and maintain the schedule of cargo delivery and their transshipment time (Alyoshina, 2021a).

A rescue and assistance network should be established along the entire route of the NSR. Rescue stations equipped with ice reconnaissance and rescue planes and helicopters should be created in the ports of Pevek, Tiksi, Sabetta, and Dixon, and sanitary aviation should be recreated.

The ports of Petropavlovsk-Kamchatsky and Korsakov on Sakhalin compete for the right to become the eastern nodal point (hub) of the NSR. A multifunctional deep-water hub port should also be created based on the Korsakov port. The capacity of the container terminal of the port should reach 4 million TEU per year after the expansion and modernization (Gusachenko, 2021a).

According to Yuri Chizhkov, member of the Presidium of the Arctic Public Academy of Sciences and the head of the North-Western Branch of the International Academy of Transport (IACT), "if the ice melts or 'stands' for less than 6 months a year, Russia will lose the rights to special regulation of navigation in the waters of the Northern Sea Route and the ability to grant privileges or set any restrictions to any carrier with any cargo on board. The navigation regime in the Arctic will be approximately the same as it is now in the Baltic Sea" (Alyoshina, 2021b).

According to experts from the International Institute for Strategic Studies (London), "despite the shorter distance, the commercial viability of the route through the Northeast Passage to Europe is declining due to the need to use ships of lower payload capacity, the seasonality of the route and the extreme conditions in which the trip is made" (Dobrynin, 2018).

Besides, the authorities of Canada and other countries are interested in Chinese vessels using, first, the Northwest Passage, not the Northeast.[2]

16.3.2 Development of the TTS of the Russian Segment of the Arctic and Implementation of Energy Projects in the Far North of Russia: India's Position and Participation

The development of the Russian Arctic strategy requires a clear and precise definition of the policy of China and India in terms of presence, development, and use of the Arctic. China and India position themselves

[2] The Northeast Passage connects Europe and Asia via the NSR and waterways around the Scandinavian Peninsula.

as world powers, and therefore they should show interest and actively participate in solving global problems, which include the following:

- Fighting against climate change;
- Stimulating the decarbonization of the world economy;
- Developing natural resources of the Arctic region;
- Taking part in the work of international organizations (Arctic Council);
- Defining and clarifying the legal regime of the Arctic regions, complying with the implementation of the UN Convention on the Law of the Sea;
- Ensuring the global military-strategic and political-economic balance, as well as the security of the main global cargo flows;
- Forming and developing new trade routes and transcontinental transportation routes;
- Developing and implementing innovative transport and transit systems and advanced transportation technologies.

In 2019, China and India became permanent observers in the Arctic Council. India's main interests in the Arctic, and Russia's possible reaction to them, are shown in Table 16.1.

The concern about the effects of global warming is far from accidental for India. The course and intensity of climate change are strongly influenced by the state of the Arctic space, the ice fields of which capture hydrocarbon. Rising sea levels threaten to flood the coastline of the Indian peninsula.

The Arctic has an impact on the course and intensity of monsoon processes, on which the state of agriculture in India depends. That is why the issues of food security and problems of water scarcity are of paramount importance for them. However, the mechanism of this dependence has not yet been fully investigated.

India expresses its readiness to create a scientific group to monitor climate changes in the Arctic related to global warming. The country has a scientific background and research experience in conditions of low air temperatures in the Himalayas and Antarctica, a Himadri research base in Nu-Aalesund (Norway), as well as surveillance systems in Svalbard.

Table 16.1 India's interests and Russia's positioning opportunities (methods)

№	*India's interests*	*Features and methods positioning of Russia*
1.	Concern about the effects of global warming	Involvement of Indian companies and research structures in monitoring climate change in the Arctic
2.	Demographic growth, large-scale industrial, infrastructure and housing construction, motorization	The formation of new transport and logistics infrastructure for supplying energy resources from the Arctic to India, the involvement of Indian specialists and labor
3.	Availability of implemented and promising Russian-Indian energy projects	Development of the transit economy in the North-South direction, with an emphasis on the transfer of high technologies in the energy sector and the transport and logistics complex
4	An increase in the volume of Russian-Indian bilateral trade	Creating a free trade zone between the EAEU and India, organizing joint ventures and alliances in the field of high-tech industries, implementing innovative infrastructure projects in the direction of "North-South"
5.	The formation and development of nodal points (hub ports) of global maritime communications in the Indian Ocean, for which the routes and ports of the NSR can become competitors	Creating and developing the cargo base of the land transport corridor "Arctic Ocean - Indian Ocean" based on the use of advanced railway and magnetic levitation technologies
6.	Internationalization of shipping, the use of TTS and development of natural resources of the Arctic region	Promoting and implementing global infrastructure and integration projects with the participation of India, corresponding to the interests of Russia based on interstate-corporate partnership
7	The formation of the Indo-Pacific region, geopolitical and geo-economic confrontation with China, the promotion of projects alternative to the Chinese "Belt and Road" Initiative	Creating the concept and development strategy and implementing the project to create the Indo-Siberian-Arctic trade route of the twenty-first century alongside its innovation and industrial belt

Source Compiled by the authors

Demographic growth, an increasing share of the urban population, accelerated industrialization, the development of transport and communication infrastructure, mass motorization, and other factors should increase energy consumption and imports, and, consequently, arouse India's interest in participating in natural resource development projects in the Arctic.

However, experts note that the country does not yet have sufficient financial and technological resources for large-scale participation in the development of Arctic deposits. India is also interested in the supply of materials for industrial and residential construction, as a reaction to the increase in population and economic growth.

India is interested in the supply of coking coal and anthracite from the northern regions of Russia. India is the second-largest steel producer in the world after China. The demand for coking coal in India is growing faster than for oil and gas due to the growing steel production. They are also going to implement a project for the construction of a metallurgical plant in the Indian state of Maharashtra, developed by NLMK Group, a Russian company.

Of particular importance is the presence of implemented and promising Russian-Indian energy projects. At the moment, India has already invested $15 billion in Russian oil and gas projects. For example, in 2001, the Indian state oil and gas corporation "Oil and Natural Gas Corporation Limited" ("ONGC"), took part in the Sakhalin-1 oil and gas project through its subsidiary ONGC Videsh Limited ("OVL"), 20 percent of whose shares it owns.

A consortium of Indian companies (ONGC Videsh Ltd., Indian Oil Corp Ltd., and Oil India Ltd.) is participating in the Vostok Oil project implemented by Rosneft. The project involves the development of oil and gas fields located in the north of the Krasnoyarsk Territory: the Vankor group of fields, the Zapadno-Irkinsky site, the Payakhskoye field, and the Vostochno-Taimyr site. $70–100 billion are needed to implement the project, since deposits in remote regions of the Far North, where there is no infrastructure at all, are going to be developed.

Oil is transported from the Vankor deposits via an oil pipeline to Taimyr and then transshipped to oil tankers plying the NSR.

In turn, Rosneft is a major shareholder of the Vadinar Indian oil refinery, which increases the interest of the large Russian company with significant lobbying capabilities in reliable and efficient transport and communication support for the functioning and development of oil production facilities and the export of their products.

Back in 2010, Reliance Industries Limited ("RIL"), the largest private company in India, and SIBUR, a Russian petrochemical company, signed a Memorandum of Understanding on the establishment of a joint venture for the production of butyl rubber in the Indian state of Jamnagar. Today, the possibility of Indian partners joining the projects of PJSC Novatek and PJSC Gazpromneft is being discussed. In particular, a consortium of Indian companies ONGC and Petronet LNG is negotiating with PJSC Novatek to purchase a 9.9 percent stake in the new Arctic LNG-2 project (Kommersant, 2021).

Transport and logistics schemes for the export of coal from the north of Russia have been developed and planned for implementation by organizing shuttle flights of large-tonnage vessels on the route of the Chennai Port (in the east coast of India) to the Far Eastern ports of Russia. At the same time, it should be taken into account that coal transshipment requires significant material and time costs, compliance with the total tonnage of Arctic vessels, and conventional large-tonnage bulk carriers, accurate work with transport and logistics.

The volume of bilateral trade between Russia and India is very low (Table 16.2). According to the Federal Customs Service of Russia, the volume of Russian-Indian trade turnover amounted to $9.26 billion in 2020 (a decrease of 17.6 percent compared to 2019), while Russian exports amounted to $5.80 billion (a decrease of 20.7 percent) and imports to $3.46 billion (a decrease of 11.8 percent). The positive trade balance for Russia amounted to $2.34 billion.

The value of bilateral trade turnover is going to be increased to $30 billion by 2025 alone (Kudashev, 2020). The creation of a free trade zone between the EAEU and India will be the organizational and institutional support for implementing these plans.

Table 16.2 Trade turnover between Russia and India in 2006–2020 ($billion)

	2006	*2009*	*2010*	*2014*	*2015*	*2018*	*2019*	*2020*
Turnover	**3.9**	**7.5**	**8.5**	**9.5**	**7.8**	**10.97**	**11.23**	**9.26**
dynamics in percent	125.8	107.6	114.4	94.4	82.3	117.3	102.3	82.4
Exports	**2.9**	**5.9**	**6.3**	**6.3**	**5.6**	**7.75**	**7.31**	**5.80**
dynamics in percent	126.4	113.5	107.7	90.8	87.2	120.1	94.3	79.3
Imports	**1.0**	**1.5**	**2.1**	**3.2**	**2.2**	**3.22**	**3.92**	**3.46**
dynamics in percent	124.1	89.3	140.5	102.6	71.2	111.1	121.6	88.2
Balance	2.0	4.4	4.2	3.1	3.4	4.53	3.39	**2.34**

Source Compiled by the authors based on Federal Customs Service's official data. URL: https://customs.gov.ru

According to Russian Foreign Minister Sergey Lavrov, "the parties are working on changing the structure of bilateral trade to move away from trade in raw materials to products with high added value. To do this, one has to develop joint projects and alliances in high-tech industries" (Gerdo, 2020).

The development of a transit economy in the Russian Arctic may hinder the implementation of India's plans to form nodal points of global maritime communications in the Indian Ocean. All other things being equal, India is not interested in the development of transit traffic through the NSR. The formation of the Arctic Ocean-Indian Ocean Railway transport corridor (SLO-IO) should remove these contradictions.

The cooperation between Russia and India in the Arctic is opposed by the United States. However, according to Karabyants, "the inability of the United States to meet growing energy needs, competition with China, and the threat of economic stagnation due to the lack of resources have made India change its attitude to projects in the Russian Arctic" (Karabyants, 2021).

India, like China, strongly supports the internationalization of Arctic development. Besides, the country is a member of the quadrilateral security dialogue group (USA, Japan, India, and Australia), as well as an active supporter of the concept of the Indo-Pacific region.

India, unlike China, does not have a single icebreaker, although it has a Sagar Kanya research vessel. According to experts, India's Arctic policy should be aimed at intensifying polar research, building icebreakers, joint development, and hydrocarbon extraction. The main partners of Indian companies in the Arctic should be Russia and Norway (Mikhailichenko, 2020), as they are interested in developing the deposits of copper, platinum, rare earth metals, and phosphorus, as well as the huge biological resources of the region.

Russian companies should step up their participation in the "Made in India" program. Given the demographic situation in India, where there is no birth control, there may be an increase in labor migration of Indians to the Arctic countries.

To give greater clarity to the Indian strategy in the Russian Arctic, it should be integrated with another global geopolitical and geo-economic project of India—the North–South transport corridor. Moreover, the development of this project was influenced by the nomination of the Chinese Belt and Road initiative in 2013.

According to experts, the main emphasis in the Russian-Indian investment cooperation should be placed on promoting Russian investments in such Indian infrastructure projects as the creation of an industrial corridor and the construction of a dedicated Delhi-Mumbai freight railway, as well as in the development of telecommunications, the energy industry and road infrastructure (Konovalova, 2015).

16.3.3 *The Development of Land Communications in the Arctic Is the Most Effective Way to Increase Russia's Transport and Transit Potential and Infrastructure Support for Implementing Energy Projects*

The uncertain status of the NSR and the Arctic Ice-Water Space determines three possible options for developing a transit economy and infrastructure support to extract natural resources in the region.

1. Strengthening Russia's military power, toughening its position on the legal status of the Arctic, which is a counterproductive option in the conditions of economic sanctions and critical dependence on foreign equipment, technologies, and competencies.
2. Developing the Arctic Ocean by transnational corporations or one company operating on the principles of interstate-corporate partnership, maximum internationalization of production, and technological, transport, and logistical, organizational and institutional processes.
3. Developing land communications passing through the Russian territory.

In our opinion, the third option is the most effective and optimal. It involves the implementation of the following projects for developing the TTS of the Arctic region:

A. Construction of the Circumpolar railway from the northern Urals to Chukotka.
B. Formation of the Arctic Ocean-Indian Ocean Transport corridor (AO-IO) through Western Siberia, Southern Siberia, Central Asian countries, Afghanistan, and Pakistan/Iran.

C. Construction of a railway through Yakutia, Chukotka, and a tunnel under the Bering Strait to Alaska.
D. Formation of two transport crosses in the Arctic: (1) in the area of Yamal and the port of Sabetta (the intersection of the Circumpolar Highway and the AO-IO corridor) and (2) in Chukotka (the intersection of the Circumpolar Highway and the Yakutia railway—the tunnel under the Bering Strait—Alaska).

Global warming has a significant impact on the functioning and maintenance of extended linear transport infrastructure (pipelines, highways). The condition of the track infrastructure is constantly monitored by special self-propelled rolling stock, mobile laboratory wagons, track stations, and repair crews. Such works are an integral part of the production and transport process, which distinguishes it favorably from road and pipeline production.

The development of land communications in the Russian Arctic is more in line with the objectives of ensuring inclusive growth of the transit economy than the functioning of the NSR. To intensify these processes, one has to involve more shift workers, migrant workers, students, military personnel of railway troops and other engineering and construction units, the local population, and prisoners in the construction of transport communications as well as the industrial and social infrastructure.

On the other hand, low population density, significant distances between cities, settlements, and shift settlements, and harsh climatic conditions require the development and implementation of advanced technologies for automating, robotizing, and digitalizing transportation by rail and port facilities.

The prolonged polar night is another argument in favor of automating the transportation process in the Arctic, constructing a transport infrastructure that would not require much maintenance, and using unmanned rolling stock. Rolling stock automation eliminates the impact on the physical condition and attentiveness of train drivers which provokes failures in physiological "day-night" rhythms, especially for those who have recently arrived at the Far North. Automation compensates for the fatigue of motor vehicle drivers tired from the monotony of the travel situation and the surrounding landscape inherent in the Arctic region.

An innovative transport and logistics technology that facilitates the complexity and reduces the time of transshipment from rail transport to sea vessels, as well as overcoming the difference in the width of the railway track, implies large-scale containerization of cargo flows along the AO-IO corridor, including coal transportation in open containers of the Open-Top type.

The prospective cargo base of the AO-IO railway corridor between Russia and India, as well as the types of rolling stock, are presented in Table 16.3.

When constructing magnetic levitation transport systems in the Arctic, special attention should be paid to protecting lines from icing, ensuring their work under low temperatures, strong winds, prolonged polar nights, and seismic activity. The Rosatom State Corporation should be involved in the power supply of the energy-intensive transportation based on magnet levitation transport systems in the direction of AO-IO, and renewable energy sources should be widely used, especially in the countries of Central Asia (Salygin & Mustafinov, 2018).

Table 16.3 Promising cargo base of the AO-IO railway corridor between Russia and India and the type of rolling stock

Cargo type	*Packaging and rolling stock type*	*Cargo formation region*
Oil, petroleum products, LNG	Tank containers, fitting platforms[a]	Arctic region, Western Siberia
Coal	Open Top containers, fitting platforms	Arctic region, Western Siberia, Southern Siberia, Kyrgyzstan
Mineral fertilizers	Universal containers, fitting platforms, mineral hoppers	NSR ports
Grain and leguminous crops	Universal containers, fitting platforms	Western Siberia, Southern Siberia
Bottled drinking water	Universal containers, fitting platforms	Arctic region, Western Siberia, Southern Siberia
Sunflower and rapeseed oil	Universal containers with elastic liners (flexitanks), fitting platforms	Southern Siberia, Southern Russia (multimodal transportation)
Perishable foodstuffs (meat, milk, vegetables, fruits)	Ref containers, fitting platforms	Western Siberia, Southern Siberia, Central Asian countries

Source Compiled by the authors

[a]If high-speed magnetolevitation transportation technologies are used, the fitting platforms are replaced by magnetolevitation platforms

The AO-IO transport corridor should become the infrastructure and integration basis for the formation of the Indo-Siberian-Arctic Trade Route of the twenty-first century (hereinafter—ISA TR21) along the route: Russian Arctic—Western Siberia (Russia)—Southern Siberia (Russia)—Kazakhstan—Kyrgyzstan—Tajikistan—Afghanistan—Iran/Pakistan—India.

The political and economic significance of ISA TR21 lies in the fact that this is a major infrastructure and integration project:

- It will complement and become a worthy alternative to the Chinese Belt and Road Initiative;
- It will ensure reliable, regular, and fast supply of India with energy resources, primarily coal and LNG;
- It will contribute to the expansion of trade, economic, financial, and investment ties between Russia, India, and the countries of Western and Central Asia, as well as the transfer of high technologies from the North to the South;
- It will provide an infrastructural basis for attracting labor from the labor-surplus regions of South, West, and Central Asia to the Russian Arctic;
- It will strengthen the transport and transit basis for the functioning of the EAEU, the EAEU + format, and further expansion of this organization;
- It will contribute to normalizing the military-political and socio-economic situation in Afghanistan, ensuring inclusive growth in the country;
- It will normalize relations between India and Pakistan;
- It will oppose the ideas of turning the Siberian regions of Russia—rich in hydrocarbons and other minerals—into an enclave (such ideas may become relevant as the economic crisis is growing deeper).

In general, innovative infrastructure and transport and transit support for sustainable development, inclusive growth, and the implementation of energy projects in the Russian Arctic require the formation and development of three major trade routes of the twenty-first century and their economic belts. They include:

1. The Arctic (Cold) Silk Road.
2. Transport route Mohe (China)—Transbaikalia (Russia)—Yakutia (Russia)—Chukotka (Russia)—Bering Strait—Alaska (USA).
3. The Indo-Siberian-Arctic trade route of the twenty-first century.

16.4 Conclusions

To develop the Russian national TTS, transport and communication support for the functioning and expansion of the EAEU, interface with the Chinese Belt and Road Initiative, and formation of the infrastructure basis of the Greater Eurasian Partnership, Russia needs to create a network of new meridional railways and to develop internal transit by water transport via the SPM as part of the big cabotage.

When determining the prospects for implementing energy projects in the Russian Arctic and developing its TFC, one should not overestimate the flow intensity and the consequences of global warming to avoid incorrect forecasts of the situation with ice, as happened in November 2021 in the eastern sector of the NSR.

The Russian LNG should form the basis of the prospective NSR cargo base, but coal should be soon supplied to India from the deposits located on the Taimyr Peninsula. The widespread use of rare earth metals in making high-tech products of Industry 4.0 underlines the relevance of transport and communication support for their extraction and export to the Russian Arctic and Afghanistan through the formation of ISA TR21 (transport corridor AO-IO).

When developing the Russian Arctic, the military-strategic and political-economic contradictions of the United States, China, India, Pakistan, Iran, Saudi Arabia, and other countries, integration associations and regional structures should be used to attract foreign investment, technologies, and competencies.

The rivalry with the geopolitical opponent in the face of China is an effective incentive to intensify India's participation in the development of the Russian Arctic. In this regard, the interests of Russia, the United States, and India may coincide. In any case, the United States should not object if the route of ISA TR21 (AO-IO transport corridor) passes through Afghanistan and Pakistan bypassing Iran, which is under economic sanctions and controls the Strait of Hormuz.

It turns out that the Arctic is one of the few regions of the world where the interests of Russia, on the one hand, and the United States,

Canada, Norway, and Denmark, on the other, can coincide. According to Dmitri Trenin, the head of the Carnegie Moscow Center, "some of the most important provisions of the Chinese and Russian strategies in the Arctic directly contradict each other, and Russia's basic interests in the Arctic tend to coincide with the ones of other Arctic countries – Denmark, Canada, Norway, and the USA" (Skosyrev, 2021).

On the other hand, Western economic sanctions and anti-Russian rhetoric contribute to the Russian-Chinese rapprochement. China, in turn, is concerned about excessive dependence on the Malacca Shipping Route, as well as the Suez and Panama Canals, especially in the field of hydrocarbon supplies by sea, and is interested in diversifying global cargo transportation routes.

The implementation of the Arctic strategy, the joint work on the ISA TR21, and its innovation and industrial belt can become the impetus and basis for establishing a dialogue between India and Pakistan aimed at preventing the harmful effects of climate change, ensuring sustainable development, and guaranteeing inclusive growth in South Asia.

The imbalance and increasing cost of cargo transportation by sea in the Europe-Asia direction associated with the COVID-19 pandemic and measures to combat it increase the competitive advantages of railway routes, including those along the North-South transport corridors. The role of the geopolitical and geo-economic activity of Russia and Central Asian countries in the south is increasing.

According to Anders Samuelsen, the Danish Minister of Foreign Affairs, "the countries of the Arctic region should focus primarily on investments in the 'green' energy. It is the Arctic countries that witness the effect of global climate change and should become an example of respect for nature, shift the focus of development from the extraction and research of traditional energy resources to renewable energy sources, new technologies" (Zmieva, 2019).

Global warming, melting of permafrost, flooding, and waterlogging of the Russian Far North make it urgent to return to the project of diverting part of the flow of Siberian rivers to Central Asia.

This requires external support, the stimulation of the socio-economic development of Afghanistan, its inclusive growth, and its inclusion in the Free Trade Zone of the FTA with the EAEU countries. Thus, after Tajikistan enters the EAEU and the agreements on the creation of an FTA with Afghanistan, Iran, and India are concluded, a free trade regime will apply to all countries located along the ISA TR21.

The countries of the Islamic world are the most suitable partners of Russia and other post-Soviet countries in reindustrializing the economy using the latest technological advances, stimulating inclusive growth, and forming the trade routes of the twenty-first century and their innovative industrial belts in the "North-South" direction, as well as using transport and transit support for implementing energy projects and developing other minerals in the Arctic, increasing the exports of transport services.

References

Alyoshina, A. (2021a, October 29). An order has been given—Containers for the Northern Sea Route. We will do it. Interview with the Dean of the Arctic Faculty of the Admiral S. O. Makarov GUMRF A. Afonin. *RZD-Partner*. https://www.rzd-partner.ru/wate-transport/interview/dan-prikaz-konteynery-na-severnyy-morskoy-put-my-ego-vypolnim/ (in Russian).

Alyoshina, A. (2021b, August 31). The Northern Sea Route: Who will it belong to if the era of global warming comes. Interview with Y. Chizhkov, member of the Presidium of the Arctic Public Academy of Sciences and Head of the North-Western Branch of the International Academy of Transport (IACT). *RZD-Partner*. https://www.rzd-partner.ru/wate-transport/interview/severnyy-morskoy-put-komu-on-budet-prinadlezhat-esli-nastupit-epokha-globalnogo-potepleniya/ (in Russian).

Belov, P. (2021, October 7). 37 billion rubles will be allocated to improve the safety of navigation along the Northern Sea Route until 2026. *RZD-Partner*. https://www.rzd-partner.ru/wate-transport/news/dlya-povysheniya-bezopasnosti-sudokhodstva-po-sevmorputi-do-2026-goda-budet-vydeleno-37-mlrd-rub/ (in Russian).

Chernyshevskaya, Y. (2021, August 5). Rosatom plans to bring containers to the Northern Sea Route. *RZD-Partner*. https://www.rzd-partner.ru/wate-transport/news/rosatom-planiruet-vyvesti-konteynery-na-severnyy-morskoy-put/ (in Russian).

Dobrynin, V. (2018, September 14). Dragon in the Arctic: Why China aspires to the Arctic. *Izvestia*. https://iz.ru/788765/vladimir-dobrynin/drakon-v-arktike-pochemu-kitai-stremitsia-v-zapoliare (in Russian).

Dodin, D. A. (2005). *Sustainable development of the Arctic (problems and prospects)*. Nauka.

Dosenko, V. A., Zvorykina, Y. V., & Zvorykina, E. I. (2017). On the impact of changes in circadian rhythms in the Arctic on the efficiency and safety of transport. *Transport and Logistics in the Arctic. Regular Shipping on the NSR Is the Key to the Accelerated Development of the Far East. Almanac, 3*, 66–71 (in Russian).

Gerdo, V. (2020, January 15). Lavrov: India may become the first non-Arctic state to extract resources in the Arctic. *TASS*. https://tass.ru/ekonomika/7520823 (in Russian).

Gusachenko, N. (2021a, August 30). Until 2024, hydrocarbon raw materials will remain the main cargo on the Northern Sea Route, after that, transportation will grow due to containerized cargo. Interview with the Minister of Transport and Road Management of the Sakhalin region V. Spichenko. *RZD-Partner*. https://www.rzd-partner.ru/wate-transport/interview/do-2024-goda-osnovnym-gruzom-po-sevmorputi-ostanetsya-uglevodorodnoe-syre-posle-perevozki-budut-rast/ (in Russian).

Gusachenko, N. (2021b, September 1). The establishment of the Northern Sea Route as a new international transport corridor is hindered, among other things, by political aspects. Interview with the Head of the Railway Transportation Department of the EU and China regions of the AsstrA Group of companies Vl. Martin. *RZD-Partner*. https://www.rzd-partner.ru/wate-transport/interview/stanovleniyu-sevmorputi-v-kachestve-novogo-mezhdunarodnogo-transportnogo-koridora-prepyatstvuyut-v-t/ (in Russian).

Haidarov, D. R. (2017). LNG transportation in containers—Progress in the energy sector and the rise of the Russian economy. *Transport and Logistics in the Arctic. Regular Shipping on the NSR Is the Key to the Accelerated Development of the Far East. Almanac, 3*, 61–65 (in Russian).

Indzhiev, A. A. (2010). *The battle for the Arctic: Will the North be Russian*? Eksmo (in Russian).

Karabyants, A. (2021, February 11). Another contender for Arctic resources has appeared in the world. *Prime*. https://1prime.ru/energy/20210211/833022883.html (in Russian).

Kommersant. (2021, November 22). India may invest in Vostok Oil and Arctic LNG-2. *Kommersant*. https://www.kommersant.ru/doc/5087834 (in Russian).

Konovalova, Y. A. (2015). Russian-Indian trade and economic cooperation at the present stage. *Science Studies, 7*(2). http://naukovedenie.ru/PDF/157EVN215.pdf (in Russian).

Kudashev, N. (2020, February 13). Russia is interested in cooperation with India in the Arctic. Interview with the Russian Ambassador to India N. Kudashev. *RIA Novosti*. https://ria.ru/20200213/1564588336.html (in Russian).

Mikhailichenko, D. (2020, September 18). The antagonist of the Arctic: The Arctic plans of India. *GoArctic Portal*. https://goarctic.ru/politics/antagonist-priarkticheskogo-arkticheskie-plany-indii/ (in Russian).

Pak, E., & Sarkisov, S. (2014). Ideals and realities of the transport complex of the single economic space of Russia, Kazakhstan and Belarus: Implications for Kazakhstan. *The International Journal of Interdisciplinary Global Studies, 4*, 15–27.

Pak, E. V. (2021). Transport and logistics in the EAEU. In N. A. Piskulova (Ed.), *The economic dimension of Eurasian integration*. Palgrave Macmillan. https://doi.org/10.1007/978-3-030-59886-0_6

RIA. (2021, April 6). Novatek has agreed to supply LNG to China with Arctic LNG - 2 (2021). *RIA Novosti*. https://ria.ru/20210604/postavki-1735576912.html?in=t (in Russian).

RZD-Partner. (2021, October 11). A multipurpose Arctic transport vessel for the Northern Sea Route is being developed in Russia. *RZD-Partner*. https://www.rzd-partner.ru/wate-transport/news/v-rossii-razrabatyvaetsya-mnogotselevoe-arkticheskoe-transportnoe-sudno-dlya-sevmorputi/ (in Russian).

Salygin, V., & Mustafinov, R. (2018). Inclusive model of common electricity markets: The 'Greater Eurasia' concept. *International Trends, 16*(4), 63–73. https://doi.org/10.17994/IT.2018.16.4.55.4 (in Russian).

Sidorov, I. (2018, July 10). China's policy in the Arctic and its impact on the region. *Military-Political Analytics*. https://vpoanalytics.com/2018/07/10/politika-kitaya-v-arktike-i-ee-vliyanie-na-region/ (in Russian).

Skosyrev, V. (2021, May 4). Washington is alarmed by China's creeping expansion in the Arctic. *Novaya Gazeta*. https://www.ng.ru/world/2021-05-04/100_world04052021.html (in Russian).

Sobyanin, A. (2012, July 1). Geographical justification of the Indo-Siberian transport direction. *Conjecture.ru*. http://conjuncture.ru//transeurasia (in Russian).

Strek, A. A., Tsoi, L. G., & Glebko, Y. V. (2016). Development of the Arctic transport fleet at the present stage. *Transport and Logistics in the Arctic. Regular Shipping on the NSR Is the Key to the Accelerated Development of the Far East. Almanac, 2*, 122–127 (in Russian).

Tsvetkov, V. A., Zoidov, K. K., & Medkov, A. A. (2014). *Formation of an evolutionary model of the transport and transit system of Russia in the conditions of integration and globalization*. MEI RAS and Nestor-Istoriya (in Russian).

Tsvetkov, V. A., Ziyadullaev, N. S., Zoidov, K. K., & Medkov, A. A. (2019). *Transit economy: Theory, methodology, practice: Monograph*. Economic Education Publishing House (in Russian).

Yeranosyan, V. (2021, October 29). "Arctic Fever", or the Chinese beyond the Arctic Circle. *Zvezda Weekly*. https://zvezdaweekly.ru/news/20211026125-A3kRj.html (in Russian).

Zanin, V. P. (2017). Arctic ground transport should master magnetolevitation technologies. *Transport and Logistics in the Arctic. Regular Shipping on the NSR is the Key to the Accelerated Development of the Far East. Almanac, 3*, 80–82.

Zbarshchenko, V. S. (2016). Transport potential of the Northern Sea Route (NSR)–North East passage (NEP). *Transport and Logistics in the Arctic.*

Regular Shipping on the NSR Is the Key to the Accelerated Development of the Far East. Almanac, 2, 71–83 (in Russian).

Zmieva, K. (2019). *The Arctic is a territory of confrontation or cooperation.* Russian Arctic. https://russian-arctic.info/info/articles/redaktsionnye-stati/arktika-territoriya-protivostoyaniya-ili-sotrudnichestva/ (in Russian).

Zoidov, K. K., Medkov, A. A., & Zoidov, Z. K. (2017). *Public-private partnership—The basis of innovative development and security in the transit economy.* Economic Education Publishing House (in Russian).

Zoidov, K. K., & Medkov, A. A. (2020a). *The Chinese Belt and Road Initiative are the basis for the evolutionary development and integration of infrastructure and integration projects in the Greater Eurasia region (review of scientific approaches)* (V. A. Tsvetkov, Ed.). MEI RAS (in Russian).

Zoidov, K. K., Medkov, A. A., Kadyrbayev, A. S., Chernyshov, M. M., & Zoidov, Z. K. (2020b). *Increasing the level of economic security of Russia in the conditions of digital transformation of the economy* (V. A. Tsvetkov & K. K. Zoidov, Ed.). MEI RAS (in Russian).

CHAPTER 17

Rethinking Year-Round Navigation via the Northern Sea Route

Vladimir N. Samotuga, Aleksey Y. Smirnov, and Dmitriy S. Khmara

17.1 Introduction

The first attempts to organize year-round navigation along the Northerns Sea Route were made during 1970. In February 1976, the "Lenin" nuclear-powered icebreaker provided assistance to "Pavel Ponomarev" diesel-electric ship during its voyage to Yamal Peninsula. That

V. N. Samotuga · A. Y. Smirnov
St. Petersburg State Marine Technical University (SMTU), Saint Petersburg, Russia
e-mail: samotuga_serduk@mail.ru

A. Y. Smirnov
e-mail: al-sm@rambler.ru

D. S. Khmara (✉)
Department of Organization of scientific work, St. Petersburg State Marine Technical University (SMTU), Saint Petersburg, Russia
e-mail: dskhmara@yandex.ru

V. I. Salygin (ed.), *Energy of the Russian Arctic*,
https://doi.org/10.1007/978-981-19-2817-8_17

started year-round Dudinka-bound navigation in the Northern Sea Route western sector. Basically, the problem of year-round NSR navigation boils down to using the NSR eastern sector during the winter and spring seasons since, at present, the navigation is limited due to complicated ice conditions in that sector.

17.2 Methodology

The authors are using general scientific research methods. The basic method is a comparative analysis that allows for an objective assessment of decisions aimed at ensuring year-round navigation along the Northern Sea Route.

The article also features a system approach that allows for comprehensively assessing the entirety of climatic and economic conditions for cargo shipment in the Russian Arctic.

17.3 Results

17.3.1 General Cargo Shipment Conditions

Global climate warming that coincided with significant national efforts undertaken by Russia to revive and further develop the Northern Sea Route made the problem of Arctic navigation a subject of research in many scientific works. Among foreign research, of greatest interest is an international collective work by Farré et al. (2014) that became an important milestone in analyzing the transit potential of the Northern Sea Route. The authors correctly point out the two factors that determine a growth in Arctic navigation. The first is Arctic ice melting, and the second is changes in the policies of the Artic countries. One should also single out a work by a Polish researcher Tadeusz Pastusiak (2020), where the author presented his own vision of how to optimize transportation routes in the Northern Sea Route area. A research by Gunnarsson and Moe (2021) summarized the Northern Sea Route activities during the past ten years. The authors are correct to point out that intensification of international navigation along the Northern Sea Route depends on the changes in the global energy markets. At the same time, the navigation will be influenced by different factors, such as changes in the Russian legislation, the international geopolitical situation, a sufficiently stable cargo base, climate conditions, and climate policy.

Among the works by Russian researchers, first of all it is worth mentioning Abramov et al. (2019), who shed the light on different aspects of the Northern Sea Route functioning.

Some research works (Ha, 2014) are mostly theoretical, because they do not account for real navigation conditions along the Northern Sea Route during winter seasons.

The existing navigation problems are analyzed in detail in the work by Kholoptsev and Podporin (2019), where the authors note that the most complicated ice situations during winters are observed in the vicinity of the Novosibirsk Islands and are explained by both existing currents and masses of ice wind-blown from the estuaries of the Lena, Khatanga, Anabar, Olenek, and Yana Rivers. During the past decade, the shrinkage of the ice cover in that area has been rather slow.

While in December all the waters around the Novosibirsk Islands are quite navigable for Arc7 class ships, in January the situation is different. In late January the Sannikov Strait is covered by ice over 1.7-m thick, which makes navigation in that area rather risky. At the same time, it is possible to use a high latitude route to bypass the Anjou Islands from the North. In February, independent navigation of Arc7 class ships in that area is out of the question, and in March, ice conditions rule out the navigation of such ships even with icebreaker assistance.

However, the situation should change in future. During an analysis of the Northern Sea Route infrastructure development, one should note that the Russian Federation Government Decree №3120-p, dated December 21, 2019, approved the Plan of the Northern Sea Route Infrastructure Development for the period up to 2035 (Government of the Russian Federation, 2019). The plan provides for a number of important measures aimed at ensuring year-round navigation along the transportation corridor in question.

The plan is to build by 2030 one prototype and four follow-on Project 22220 icebreakers (Arctica class LK-60Ya) as well as two new Leader Project icebreakers (LK-110Ya). The planned measures are proceeding in strict accordance with the schedule. Thus, October 2020 saw the commissioning of Project 22220 Arctica prototype icebreaker. In June 2020, Zvezda shipyard laid down the first Project 10510 Leader icebreaker (christened Rossiya).

Arctica type icebreakers with 60-megawatt shaft power can break three meters of ice, while Leader class ice breakers with 120-megawatt shaft power can break four meters of ice (Samotuga, 2020). Those unrivaled ships can ensure year-round navigation in the most difficult area of the Arctic—in the vicinity of the Novosibirsk Islands (using a high latitude route north of the Anjou Islands).

Analyzing potential cargo traffic along the Northern Sea Route, one should note that a current private company leader in organizing navigation in the NSR eastern sector is PAO Novatek, which is implementing natural gas production projects such as Yamal LNG and Arctic LNG-2. For the former project, that is already in full swing: the company built fifteen Arc7 ice-class tankers. The last of those ships was commissioned in December 2019.

Twenty-one ships are ordered for the Arctic LNG-2 project. Fifteen ships will be built in Russia at a new Zvezda shipyard and six ships will be built by Daewoo Shipbuilding & Marine Engineering Company in South Korea. All those ships are extremely environment-friendly because they will be using LNG as major fuel.

Minimal requirements to Arc7-type ships are that they should be capable of independent navigation in the ice up to 1.4 m thick during winter and spring seasons and should follow an icebreaker in one-year old ice up to two meters thick; however, Novatek's LNG carriers have a higher icebreaking capability.

In cooperation with Novatek, Rosatom is planning year-round cargo shipment along the Northern Sea Route. That is confirmed by concrete practical decisions. Thus, in 2020, the company organized the first extremely early May voyage for Arc7-type Christoph de Margerie LNG carrier along the Northern Sea Route.

The ship started its voyage on May 18, 2020, when the LNG carrier left the port of Sabetta and sailed through the Gulf of Ob, and proceeded to China along the Northern Sea Route. Before that spring, during the most difficult navigation period, LNG had never been transported along that route. LNG was shipped to China along a longer western route around Europe via the Suez Canal and the Strait of Malacca. It takes more than thirty days to sail that route. Shipping Yamal LNG cargoes via the Northern Sea Route is almost thirty percent faster (Rosatom, 2020). However, in practice, the time it takes the carriers to cover that route greatly depends on ice conditions in the eastern NSR sector.

On May 19, 2020, at the Kara Sea ice cover edge, the Yamal icebreaker started providing assistance to the LNG carrier. The total sailing time along the Northern Sea Route was 12.5 days. The ships sailed 2653 nautical miles at an average speed of 8.5 knots (including icebreaker assistance of 2123 nautical miles [80 percent of the route length] at an average speed of 7.9 knots).

The second extremely early icebreaker assisted voyage was from May 24 till June 2, 2020, when Vladimir Voronin LNG carrier sailed the Northern Sea Route led first by Fiftieth Victory Anniversary and then Yamal icebreakers.

Those experimental voyages resulted in developing a number of proposals:

- To create an integrated icebreaker-supported transportation system within NSR waters in order to ensure safe and year-round cargo shipment. Icebreakers should be positioned at certain route points, while LNG carries can sail ice-free route sections independently;
- To create (on the basis of Rosatom sea operations headquarters) an integrated surveillance and monitoring system along the NSR to ensure navigation and to provide timely information for the ships sailing through potentially dangerous route segments; and
- To create a modern Arctic space satellite constellation that would provide real-time monitoring of ice situation.

In early January 2021, Novatek effected late-season icebreaker assisted voyages for two LNG tankers that sailed to Asia via the eastern segment of the Northern Sea Route. Christoph de Margerie LNG carrier sailed from the port of Sabetta on January 5 and Nikolai Yevgenov LNG carrier on January 6. On January 26, Christoph de Margerie LNG carrier reached the East coast of China. That voyage was caused by high LNG prices in East Asia where they exceeded $1200 per 1000 cubic meters due to severe winter conditions. LNG carriers could sail without icebreaker assistance partly because of climate warming: in late October 2020, the Laptev Sea was free of ice for the first time in history.

At the same time, starting from late January-early February 2021 even high ice-class Arc7 LNG carriers still could not sail through the Northern Sea Route eastern segment without icebreaker assistance. That raises the issue of the route's economic efficiency in commercial cargo shipping. It is also worth noting that during winter seasons tanker sailing speeds along the Northern Sea Route are much lower because it is necessary to use icebreakers. While in summer it takes Novatek's LNG carriers six days on the average to sail the entire NSR length, in December it would take them 15–16 days without icebreaker assistance (Olkhovik, 2018). That results in higher cargo shipping costs.

Using positive experience gained in 2020–2021, Rosatom state corporation is planning to organize year-round cargo shipments along the Northern Sea Route, thus creating the Northern Sea Transit Corridor.

The project's implementation deadlines (after 2024) are tied to new icebreaker building schedules and to effecting other measures under the Northern Sea Route Infrastructure Development Plan.

Rosatom experts estimate target cargo shipment volumes at 8–10 million tons per year. We believe that the project can be implemented only under the conditions of further climate warming caused by both anthropogenic and other factors, while the size of target figures seems rather optimistic, especially accounting for crisis phenomena in the world economy due to coronavirus limitations. A lot would depend on economic development pace, which is difficult enough to forecast (Smirnov, 2020).

17.3.2 Navigation Support Infrastructure Services in NSR Waters

At present, the existing navigation support infrastructure generally meets the basic service requirements, thus creating the possibilities for year-round navigation in NSR waters and beyond.

The existing services that support navigation in NSR waters can be tentatively divided into three types:

(1) Communication services;
(2) Navigation and hydrographic support services; and
(3) Search and rescue services.

Besides, shipping development is also supported by additional services such as ship repair and maintenance, fueling, crew recreation, etc. These will be considered in greater detail.

17.3.2.1 Providing Communication Services to Support Navigation Along the Northern Sea Route

At present, the world uses several satellite communication systems that can, on a limited scale, meet existing communication needs in the Arctic up to 70–75 degrees of Northern latitude. In higher latitudes, it is possible to use a foreign Iridium system (USA) based on person-to-person voice communication with low data transmission channel capacity. Use of those systems by clients who are involved in navigation along the NSR will not require budget expenditures.

Today, using the American Iridium system does not meet national security requirements. At present, Russia operates two national satellite communication systems that could potentially provide communications in the Arctic zone. Those are *Gonets* and *Meridian* systems, the latter being a double-purpose system; both systems are, however, morally obsolete. The Gonets system has been used since early 1990s and is designed to transmit brief SMS-type messages. The messages can reach intended recipients with significant time delays.

The Federal Space Program (FSP) provides for "creating a Gonets M1 multi-function person-to-person space communication and data transmission system from low-orbit space satellites" (Roscosmos, 2016). Federal budget expenditures to implement that system within a Federal Targeted Program amount to ₽9.8 billion. Besides, there are GEO satellites in the Russian civilian satellite constellation (Express AM series satellites that belong to Space Communications Federal Enterprise and Yamal series satellites that belong to PAO Gazprom Space Systems).

Those satellites can provide communication coverage for 90 percent of NSR without a guarantee of stable communication with fixed or mobile land, sea, or air recipients which is explained by physical laws of electromagnetic waves propagation at low operating angles (less than five degrees). Real data transmission channel capacities are limited to approximately 1–2 megabit/second.

There is information about a new *Express PB* system (supported by the Russian Ministry of Communications). That system would include three satellites (plus one back-up) at high elliptical orbits (Tundra type, altitude forty thousand kilometers) and land infrastructure. The cost of the new Express PB system in the *Ku frequency* range is estimated at ₽90 billion, including ₽60 billion for the satellite constellation and ₽30 billion for land infrastructure.

A *Polaris system* project (suggested by OAO Gazprom Space Systems) includes three satellites at Tundra type orbits and land infrastructure. The satellites are built according to the HTS principle and would operate in the *Ka-frequency* range. The system is designed to provide different subscribers with internet access at speeds over 10 megabit/second.

The cost of deploying the Polaris system is estimated at about ₽60 billion. There are initiatives to design a medium-altitude HTS system in the Ka-frequency range (the project is suggested by Zond Holding Company supported by Skolkovo Fund). The first stage of the system creation is estimated at ₽15 billion. It envisages servicing the territory of Russia (including the Arctic), and the space segment includes five medium-altitude satellites (designed by OAO ISS, Krasnoyarsk). The second stage consists in providing global coverage. Those projects are not reflected in the Federal Space Program and are not funded from the federal budget.

It is worth noting that the near future holds a high probability of deploying new foreign super informative satellite systems (OneWeb, O3B-D, etc.) that can provide global high-speed communications including those along the NSR.

Because of that, there is a risk that even if the Russian federal government allocates significant budget funding to create any of the national satellite communication systems, civilian users will still prefer foreign satellite communication systems, which would make national system costs unrecoverable even in the long term, considering the developments in B2B and B2G markets related to accomplishing tasks in the Arctic.

17.3.2.2 Navigation Equipment to Support Shipping in the Northern Sea Route Waters

Navigation and hydrographic support for shipping in the NSR waters is affected by the Federal Agency of Maritime and Riverine Transport and its subordinate organization FGUP Hydrographic Enterprise (Rosatom, 2021).

In 2021, Rosmorrechflot optimized available navigation support equipment (NSE) in the NSR waters. The work resulted in formulating proposals according to which the NSR waters should accommodate 1500 NSE items (day and night beacons equipped with radar reflectors and radar transponders). At the time, only 302 items of equipment were operational. They are mainly positioned in the western NSR segment from the Kara straits to Dudinka.

For year-round navigation, it is very important to provide vessels sailing through the Northern Sea Route waters with hydro, weather, and ice situation information.

Maritime Safety Information (MSI) includes navigation and meteorological messages. Many hydrometeorological stations (HMS) are worn and require reconstruction, overhaul, or, frequently, the construction of new facilities.

More economical and stable operations of observation and monitoring systems require power supply modernization, purchasing diesel generators or installing wind generators.

An important modernization component is installing modern observation and monitoring equipment and devices, including automatic and semi-automatic. The above measures would result in obtaining higher quality information about Arctic territories and waters. That information would serve as a basis for better environmental forecasts in the interests of the population, some territories, and vital facilities, would help avoiding environmental threats (increasing maritime operations safety along the NSR), and would assist in the development of the Arctic offshore areas.

The early 1980s saw the beginning of operations of Rosgidromet's Sever automated ice information system for the Arctic. That system is administered by FGBU AANII and uses satellite photography to provide users with detailed ice and weather information along the course of a ship.

In turn, maritime safety information is transmitted by FGUP Hydrographic Enterprise's Information Collection, Procession and Transmission Center (ICPTC) through the INMARSAT International Safety Net (SafetyNet) and onshore NAVTEX service station at the port of Tiksi.

NAVTEX (NAVigational TEleX) is an international automated information system and a part of IMO's Global Maritime Distress and Safety System (GMDSS) in accordance with the International Convention for the Safety of Life at Sea (SOLAS-74/88). At the same time, in the vicinity of the Taymyr Peninsula, there is an NSR segment that is not covered either by a satellite signal or by an onshore station signal.

Thus, reliable MSI transmission requires commissioning another NAVTEX system station at the Andrei Island in the Laptev Sea. The federal budget allocated ₽10 million to build Global Maritime Distress and Safety System facilities along the NSR.

17.3.2.3 Search and Rescue Systems

Russian Government's Decree No. 371 of 23.07.2004 "On Approving the Statute of Maritime and Riverine Transport Federal Agency" charged Rosmorrechflot with responsibility for maritime search and rescue operations. New multi-purpose search and rescue ships are being built for Rosmorrechflot's SAR service. 2014 saw the commissioning of Project P-70202 Baltika multi-purpose search and rescue ship that is operating in the NSR waters. The ship cost 78 million euros to build (Korabli, 2014). However, the NSR's eastern segment has no infrastructure either for ship or life rescue.

In accordance with International Maritime Organization's (IMO) rules, today all passenger ships as well as cargo ships with a dead weight of over 299 tons must be equipped with an Automated Identification System transponder (AIS) that allows real-time transmission of a ship's name, course, and speed. The data can be transmitted either via a satellite or via a network of onshore automated identification stations.

For example, Alaska hosts ninety-five automated identification stations that receive signals from onboard ship transmitters (Marine Traffic, 2016). The stations are spaced at 70-km intervals. Such a density of automated stations enables receiving instant real-time information about all vessels in the waters and sending them immediate warnings if the personnel of the navigation safety monitoring centers believe that the chosen route can end in an accident.

According to the Internet, Russia hosts fifty AISs; however, NSR waters host only two (in Dikson and Anadyr) and both were offline at the moment of the research. Thus, ships transmit their coordinates via satellites, which causes delays in data processing.

Besides, the correct operation of a satellite positioning system requires onshore control and adjustment stations that would transmit corrections to onboard GLONASS/GPS receivers. As of today, NSR waters host six control and adjustment stations at the following locations: Oleniy, Andrey, Stolbovoy and Kamenka Islands, the Sterlegov Cape, and the Indigirka River estuary.

At present, Russian EMERCOM is creating a comprehensive safety system for the Russian Federation's Arctic Zone; that system is to reduce response time and improve search and rescue quality. It will include an Arctic Zone geoinformation system that would highlight potential emergency sources (an electronic map of the region with accompanying data bases), a system to monitor and forecast the emergence of dangerous phenomena, a population warning system, etc. The basis of the monitoring system's information component will be a module that provides a visual picture of the situation in the region, including its waters.

An estimated cost to create a standard EMERCOM's search and rescue center in the Arctic (including an air component) is ₽520 million. Today, four of the ten centers are built.

17.4 Conclusions

At present, year-round navigation along the Northern Sea Route is technically possible in the western sector; however, it is difficult to organize navigation in the eastern sector during the period from late January to mid-May. In the western sector, navigation is commercially attractive and efficient, unlike in the eastern sector. If current climate trends persist, during the next decade it would be possible to forecast longer independent navigation for Arc7 class ships in the eastern sector for the period up to late January. Icebreaker assisted year-round navigation for ships of that class shall be possible after commissioning Leader class icebreakers after 2027. If climate conditions are favorable, Project 22220 icebreakers shall allow even earlier navigation north of the Anjou Islands.

Year-round navigation along the Northern Sea Route can be organized only on the basis of public–private partnership, which would allow the efficient use of those advantages that are characteristic of state-owned companies and of the private sector. In such a partnership, the government mission consists of creating a necessary infrastructure, including an icebreaker fleet, auxiliary vessels, ice reconnaissance, monitoring, and SAR means. The mission for private companies is to ensure high and predictable cargo shipment volumes, which would allow full use of the created infrastructure (ruling out misuse of public funds). Budget policy optimization is an important strategic goal for the government under current economic conditions (Koniagina, 2020). In that respect, cooperation between PAO Novatek and Rosatom State Corporation in organizing super early LNG transportation in May 2020 must be assessed positively. The economic efficiency of year-round navigation depends on the situation in the global commodity markets. Thus, a significant natural gas price growth during winter periods causes the need to transport gas from the Yamal Peninsula to the East. At the same time, lower gas prices make such transportation too expensive and economically unreasonable. Continuous year-round Northern Sea Route functioning requires a wider freight nomenclature, as natural gas only is not enough.

At present, using the Northern Sea Route for cargo transit is not justified economically. The ice situation in the NSR zone remains difficult to forecast, which impedes efficient sea transportation within exact deadlines and prevents exact forecasts of transportation costs. The situation will change with climate warming and ice cover shrinkage in the Arctic Ocean, especially in the vicinity of the Novosibirsk Islands.

A necessary condition for organizing year-round transit shipments is the involvement of major world logistic companies that can provide the necessary volumes of freight shipments on a continuous basis.

For the Northern Sea Route to operate efficiently, it would be necessary to create navigation, ice monitoring, and maritime safety systems. The year 2014 saw the start of an active modernization of existing systems aimed at improving navigation safety on the basis of using modern digital technologies. However, that process is impeded due to a number of technical problems—the most important problem is the lack of a continuously operating high-speed satellite communication system.

A review of how NSR sailing ships are provided with communication, hydro navigational, hydrometeorological, repair/maintenance, and other services demonstrated that, today, the existing infrastructure generally meets the basic requirements in the above services, thus creating the potential for year-round navigation in the NSR waters and beyond. At the same time, there is also great potential in developing onshore and other infrastructures that would ensure a better quality of the services, given that now they lag far behind the international analogues.

References

Abramov, A., Zagorodnikov, M., & Smirnov, A. (2019). Management of innovative projects in the Arctic Zone. Modern tools for sustainable development of territories. special topic. Project Management in the Regions of Russia (MTSDT 2019). *The European Proceedings of Social and Behavioral Sciences*. https://doi.org/10.15405/epsbs.2019.12.05.125

Farré, A. B., et al. (2014). Commercial Arctic shipping through the northeast passage: Routes, resources, governance, technology, and infrastructure. *Polar Geography, 37*(4), 298–324. https://doi.org/10.1080/1088937X.2014.965769

Government of the Russian Federation. (2019). *Northern Sea Route Infrastructure Development Plan for the Period up to 2035*. Retrieved November 27, 2021 from http://static.government.ru/media/files/itR86nOgy9xFEvUVAgmZ3XoeruY8Bf9u.pdf (in Russian).

Gunnarsson, B., & Moe, A. (2021). Ten years of international shipping on the northern sea route: Trends and Challenges. *Arctic Review, 12*, 4–30. https://doi.org/10.23865/arctic.v12.2614

Ha, Y. S. (2014). The northern sea routes and Korea's Trade with Europe: Implications for Korea's Shipping Industry. *International Journal of e-Navigation and Maritime Economy, 1*, 73–84. https://doi.org/10.1016/j.enavi.2014.12.007

Kholoptsev, A. V., & Podporin, S. A. (2019). Prospects for unescorted navigation of Arc7 Ice-Class vessels in the region of the New Siberian Islands during the winter period. *Vestnik Gosudarstvennogo Universiteta Morskogo I Rechnogo Flota Imeni Admirala S.O. Makarova, 11*(5), 867–879. https://doi.org/10.21821/2309-5180-2019-11-5-867-879 (in Russian).

Koniagina, M. N. (2020). Forecast of Budget Revenues From Taxes in the Context of Economy Digitalization (2020). *IOP Conference Series: Materials Science and Engineering, 940*(1). https://doi.org/10.1088/1757-899X/940/1/012040

Korabli. (2014). *The Multifunctional Rescue Vessel "Balti" of the R-70202 Project*. Retrieved November 27, 2021 from http://www.korabli.eu/galleries/oboi/grazhdanskie-suda/baltika (in Russian).

Marine Traffic. (2016). Retrieved November 27, 2021 from https://www.marinetraffic.com

Olkhovik, E. O. (2018). Analysis of speed regime LNG-tankers in the northern sea route in period of winter navigation 2017–18. *Vestnik Gosudarstvennogo Universiteta Morskogo i Rechnogo Flota Imeni Admirala S.O. Makarova, 10*(2), 300–308. https://doi.org/10.21821/2309-5180-2018-10-2-300-308 (in Russian).

Pastusiak, T. (2020). *Voyages on the northern sea route*. Palgrave Macmillan. https://doi.org/10.1007/978-3-030-25490-2

Rosatom. (2020). *The Northern Sea Route Will Become Year-Round*. Retrieved November 27, 2021 from https://strana-rosatom.ru/2020/08/03/

Rosatom. (2021). *FGUP "Hydrographic Enterprise"*. Retrieved November 27, 2021 from http://www.hydro-state.ru/kage.html

Roscosmos. (2016). *The Federal Space Program of Russia for 2016–2025. Approved by the Decree of the Government of the Russian Federation of March 23, 2016, No. 230*. Retrieved November 27, 2021 from https://www.roscosmos.ru/22347/

Samotuga, V. N. (2020). Self-driving forecast and scenarios for the development of the world and domestic shipping market on the northern sea route. *Economy and Management, 7*(177), 721–729. https://doi.org/10.35854/1998-1627-2020-7-721-729

Smirnov, A. (2020). The Northern Sea Route Cargo Transportation Development. *E3S Web Conference, 157*, 05005. https://doi.org/10.1051/e3sconf/202015705005

CHAPTER 18

Revisiting the Prospects of a Multihull Icebreaker for Year-Round Navigation via the Northern Sea Route

Dmitriy S. Khmara and Anna V. Firsova

18.1 Introduction

The development of the Russian Federation Arctic Zone is closely related to the development of the Northern Sea Route (NSR) that allows transporting natural resources and other cargoes through two-to-three-meter-thick ice with appropriate icebreaker assistance.

The list of priority projects—as mentioned in the shipbuilding industry development strategy as those that can increase Russia's competitiveness in cargo shipping—includes liquefied natural gas (LNG) carriers, container carriers, and heavy-tonnage ice-class vessels, as well as building

D. S. Khmara (✉) · A. V. Firsova
St. Petersburg State Marine Technical University,
Saint Petersburg, Russia
e-mail: dskhmara@yandex.ru

A. V. Firsova
e-mail: firsova-a-v@mail.ru

V. I. Salygin (ed.), *Energy of the Russian Arctic*,
https://doi.org/10.1007/978-981-19-2817-8_18

twenty-four icebreakers before 2035 in order to meet the needs of both domestic and foreign markets (Kremlin, 2019).

Today, economic activity has declined, caused by COVID-19 pandemic on the one hand, and climatic factors in different countries which affected LNG shipping via the Northern Sea Route on the other. An example is dispatching two Yamalmax-type carriers to China in early January 2021, when LNG prices in Asia under immediate delivery terms exceeded $1,000 per one thousand cubic meters because of extremely cold temperatures. LNG shipment via the NSR saved two weeks compared to the LNG shipping route via Europe (Eurasia Daily News Agency, 2021).

In particular, the development of the NSR and the Northern Maritime Transit Corridor envisages their use in order to organize a full-scale international transportation and logistical artery that could compete with (or even surpass) such traditionally recognized sea cargo transportation routes as the Suez and Panama Canals.

For example, LNG transportation to Asia via the Panama Canal is complicated because the canal can pass only three LNG carries a day to the Pacific Ocean, while LNG shipments via the NSR even with icebreaker assistance are cheaper and shorter in time (Zagorodnikov, 2018a).

In other words, the Russian Federation is to ensure economically attractive transportation and transshipment of international cargoes and their transit via the NSR.

In order to implement such an ambitious project, it is necessary to build cargo fleet and appropriate port infrastructure as well as specialized vessels and icebreakers to ensure year-round navigation. The costs required to design and build specialized vessels and icebreakers can amount to eighty percent of the total costs related to creating a promising northern transportation and logistics system (Mikhailov, 2020).

Successful cargo shipments via the NSR shall be contributed to by building Project 10,510 Leader nuclear-powered icebreakers that can ensure year-round icebreaker assistance to all vessels sailing in the Russian Federation Arctic zone (Information and analytical Agency "Portnews" 2020).

At present, Rosatom state corporation is cooperating with PAO CDB Iceberg in designing nuclear-powered Arctic container carriers whose width will vary from 43 to 51 meters, which would allow a cargo capacity of 20,000 TEU and would mean economically efficient operations in Arctic conditions (Rosatom, 2020).

However, operating ice-class ships of such a width would cause difficulties in providing icebreaker assistance to them. In heavy conditions, icebreakers can provide assistance to vessels of comparable dimensions. On the one hand, assistance to individual ships is more expensive than to convoys but, on the other hand, icebreaker idling time caused by forming a c'onvoy of similar width ships results in losses for icebreaker fleet owners (Solntsev, 2020).

Today, all operational and under-construction icebreakers have widths smaller than Yamalmax-type carriers or projected tankers and container carriers; thus, there is an important task to design and build icebreakers that would make channels wide enough for high-tonnage vessels and convoys to navigate. Attempts to design such ships (in particular three- and four-hull icebreakers) were already made by FGUP Krylovsky State Scientific Center in collaboration with PAO CDB Iceberg; at that time, however, Yamalmax-type carriers were just established, and considering the fact that those ships could break through two-meter-thick ice during winter and spring seasons, it was decided to build Leader-type icebreakers.

However, taking into account changing geopolitical and socio-economic conditions, it again becomes important to renew activities aimed at designing and building innovation icebreakers, which is precisely the purpose of this research.

18.2 Methodology

An innovation multihull icebreaker is similar to a trimaran: its separate hulls break ice, making a channel wider than existing icebreakers for individual ships or convoys to sail; at the same time, the weight of the hulls (and, consequently, the power consumption) of an innovative icebreaker (IIB) is much lower than that of a traditional icebreaker of the same width, to say nothing of the fact that there are currently no icebreakers with a hull width of fifty-sixty meters (the Leader-type icebreaker under construction is forty-six meters wide at the design waterline (DWL)). Based on the ice conditions allowed for in the research, the authors identified the critical components and necessary capacity of an IIB power plant that can ensure icebreaker-efficient operations under the current economic conditions in Russia (Abramov et al., 2019).

To accomplish the tasks, it is necessary to compare the characteristics of icebreakers that can provide assistance to individual ships or convoys and the characteristics of ships that can transport natural resources (LNG foremostly) and other resources via the NSR.

In assessing technical characteristic and economic efficiency of icebreakers and ice-class vessels, the authors used general research methods and rating the factors under analysis.

A system approach allowed the authors to propose a concept of line hauls via the NSR using IIBs and ice-class vessels smaller than Arc7.

18.3 Results

Table 18.1 demonstrates characteristics and capacities of LNG carries that can sail the NSR. The table shows the data for two icebreaker types (LK-60Ya and IIB) that should be considered as vessels that can support LNG transportation from Yamal fields to Western Europe with transshipment at a Kola Peninsula base.

These characteristics demonstrate that the carriers in question are much larger than the icebreakers that can provide icebreaker assistance. With the operational carriers that service the Varandeya terminal and Prirazlomnaya platform in the Barents Sea (where ice situation is much better that in the Kara Sea), the ships are 34 and 32.2 meters wide, unlike the obsolete fifty-thousand cubic meter capacity carriers with a hull width of thirty-two meters which fully fit in the width of a channel made by an existing icebreaker.

Table 18.1 Characteristics of ice-class LNG carriers and LK-60Ya and IIB icebreakers

LNG carrier capacity, thousand cubic meters	*170*	*90*	*180*	*LK-60Ya*	*IIB*
Maximal length, meters	300	250	299	160 (DWL)	247 (DWL)
Width amidship, meters	50	40	48	33 (DWL)	60 (DWL)
Trunk deck height over the molded base, meters	33	27.5	34	–	–
Draft at summer waterline, meters	11.7	10.7	11.5	8.5 / 10.5	11.5
Draft in ballast, meters	10.6	9.5	10.5	–	–
Cruising speed, knots	18.7	18.3	19.5	20	20
Power plant capacity, kW	54	40	42	60	60

Source compiled by authors on the basis of existing LNG carrier characteristics

Arctic ships that service Norilsk Nickel and operate in the Western Arctic have hulls twenty-three meters wide. Such ships are within existing icebreaker assistance requirements, but at the same time they are obsolete from the viewpoint of economically efficient shipments via the NSR.

Problems of cargo shipments via the NSR multiply with the growth of LNG/container carrier tonnage. The wider the ship, the more effort will be required to widen the channel originally made by an existing icebreaker. Consequently, convoy speed will drop and energy consumption of assisted ships will increase. Turns of large-tonnage ship convoys will result in slower speeds and delays when sailing in drifting ice will be likelier and longer. Similar situations will arise when convoys shall have to overcome ice reefs. Thus, it will take larger ships longer to sail through ice-covered stretches and to make two-way voyages which reduces economic efficiency (Appolonov et al., 2016).

Research assumptions determine basic technical and economic indicators that characterize the efficiency of using the icebreaker fleet. Below, the most important indicators are presented in descending order of priority:

1. *Icebreaker width.* Wider icebreakers will make wider channels that will accommodate ships or convoys (depending on weather conditions) of the same width;
2. *Icebreaking capability.* A higher icebreaking capability enables sailing in the heaviest ice conditions and reduces the probability of jamming both for assisted ships and icebreakers themselves;
3. *Speed in even ice.* This indicator directly affects the economic efficiency of icebreaker assistance, as it shows the speed of sailing in two-meter-thick ice. The higher the indicator, the higher the speed. Higher speeds mean lower voyage costs, both for assisted ships and for icebreakers;
4. *Propulsion power.* High propulsion power affects both icebreaker operations in extreme ice conditions and icebreaker potential for towing;
5. *Displacement.* A larger displacement gives an icebreaker a higher kinetic energy, which means that it can overcome difficult ice reefs (such reefs may be ten-fifteen meters high);

6. *Crew strength.* This indicator directly affects icebreaker assistance cost/income: the lower the manning, the lower the wage, insurance, and provisioning costs. In other words, crew strength directly affects boat day cost of icebreaker operation.

A comparison of the above indicators for different icebreakers is shown in Table 18.2.

A comparison of icebreaker ratings is shown in Table 18.3.

However, assessments of icebreakers and other ships cannot be accepted for consideration without accounting for the priority of vessel parameters in question, including those chosen for the purposes of the research.

Assessment of the parameters of ships that can provide icebreaker assistance was accomplished in accordance with a comparison methodology that includes:

Table 18.2 Comparison of technical and economic indicators for existing and innovation icebreakers

Indicator	*LK-60Ya*	*IIB*	*10,580, "Taymyr"*	*10,521, "Yamal"*	*10,521 M, "50 years of victory"*
Dimensions, meters:					
• length	160.0	247.0	151.8	147.9	147.9
• width	33.0	60.0	29.2	29.9	29.9
• draft	8.5/10.5	11.5	9.0/8.1	11.0	11.0
Icebreaking capacity, meters	3<	3<	1.7	2.0	2.7
Speed in even ice, knots	10	10	< 3	< 3	< 3
Propulsion power at propellers, megawatt	60	60	35	54	54
Displacement, tons	33,500	34,000	19,600	22,625	25,168
Crew strength	75	60	70	79	79

Source Compiled by authors on the basis of the existing icebreaker characteristics

Table 18.3 Comparison of icebreaker ratings (best vessels tinted yellow)

Indicator	Icebreaker ratings:					
Dimensions	LK-60Ya	IIB	10580, "Taymyr"	1052, "Arktika"	10521, "Yamal"	10521 M, "50 years of victory"
• length						
• width						
• draft	5.50	10.00	4.87	4.98	5.00	5.00
Icebreaking capacity						
Speed in even ice	10.00	10.00	6,07	7.14	7.14	9,64
Propulsion power at propellers	10.00	10.00	3.0	3.0	3.0	3.0
Displacement	10.00	10.00	5.83	9.00	9.00	9.00
Crew strength	9.85	10.00	5.76	6.90	6.65	7.40
	8.00	10.00	8.57	7.59	7.59	7.59

Source Calculated and compiled by authors based on Table 18.2 expert assessment

1. Rating indicators;
2. Determining the weight of each indicator in the integrated ship assessment following the Fishburn rule in accordance with Formula 18.1;

$$r_i = \frac{2 \cdot (N - i + 1)}{(N + 1) \cdot N} \qquad (18.1)$$

where:

r_i—weight of i-indicator in the integrated assessment;

N—number of indicators;

i—indicator number according to its importance (rating).

A comparative ship assessment is done according to a ten-point system, under which the ships with the best indicator value are given 10 points and others are rated according to decreasing indicator value;

3. Integrated assessments for each ship are calculated according to Formula 18.2:

$$N = \sum_{i=1}^{N} X_i \cdot r_i \tag{18.2}$$

where:

X_i—assessment of i-indicator;

r_i—weight of i-indicator in the integrated assessment;

N—number of indicators.

Calculation of integrated assessments for the ships in question is shown in Table 18.4.

Table 18.4 Calculation of integrates assessments for the ships in question, (integrated assessments and best ship ratings are tinted yellow)

Indicator	Indicator weight	Icebreaker ratings:					
		LK–60Ya	IIB	10580, "Taymyr"	1052, "Arktika"	10521, "Rossiya"	10521 M, "50 years of victory"
Dimensions	–	–	–	–	–	–	–
• length	–	–	–	–	–	–	–
• width	0.2857	1.57	2.86	1.39	1.42	1.43	1.43
• draft							
Icebreaking capacity	0.2381	2.38	2.38	1.45	1.70	1.70	2.30
Speed in even ice	0.1905	1.91	1.91	0.57	0.57	0.57	0.57
Propulsion power at propellers	0.1429	1.43	1.43	0.83	1.29	1.29	1.29
Displacement	0.0952	0.94	0.95	0.55	0.66	0.63	0.70
Crew strength	0.0476	0.38	0.48	0.41	0.36	0.36	0.36
Integrated assessment (И)	1.00	8.60	10.00	5.20	6.00	5.98	6.65
Ship rating	–	2	1	6	4	5	3

Source Compiled by authors using design data

The resulting assessments show that IIB is an absolute leader because of the width of the ice channel it can break, propulsion unit power, icebreaking capacity, and other factors.

However, such an assessment will not be full without a comparison with the Leader-type icebreaker which has 110-megawatt propulsion power at propellers. The basic characteristics of Leader-type and IIB icebreakers are shown in Table 18.5.

Table 18.6 shows a comparison of ratings for two icebreakers under design.

Calculation of integrated assessments for compared ships is shown in Table 18.7.

A comparison of the IIB and Leader icebreaker (under construction) demonstrates that Leader is much superior due to its higher icebreaking capacity and almost double propulsion power at propellers. However, using a nuclear propulsion plant of such power will require a proportional increase in its operating costs; besides, reliable and safe icebreaker assistance for fifty-meter-wide ships in heavy ice conditions will require two icebreakers. Leader's crew strength is also a disadvantage when compared with an innovative icebreaker.

Today, it is evident that the icebreakers under comparison are not competitors; they have different targets and missions. The Leader-type icebreakers are designed to operate in the heaviest ice conditions; in extreme ice conditions, they can provide icebreaker assistance for ships

Table 18.5 Comparison of characteristics for two icebreakers under design

Indicator	*Values for icebreakers:*	
	Leader	*IIB*
Dimension, meters:		
• length	209.0	247.0
• width	47.7	60.0
• draft	13.0	11.5
Icebreaking capacity, meters	3.76	2.8
Speed in even ice, knots	12	10
Propulsion power at propellers, megawatt	110	60
Displacement, tons	69,550	34,000
Crew strength	127	60

Source Compiled by authors using design data

Table 18.6 Comparison of ratings for two icebreakers under design according to selected indicators (best ship figures are tinted yellow)

Indicator	Ratings for icebreakers:	
	Leader	IIB
Dimensions:		
• length		
• width	7.95	10.00
• draft		
Icebreaking capacity, meters	10.00	7.45
Speed in even ice, knots	10.00	9.33
Propulsion power at propellers, megawatt	10.00	5.45
Displacement, tons	10.00	4.88
Crew strength	5.00	10.00

Source compiled by authors on the basis of expert assessments according to Table 18.5

sailing to practically any destination in the Arctic. In turn, IIBs are designed to operate in less severe ice conditions (desirably where ice is less than 2.5 meters thick) and to provide icebreaker assistance to wider-hull ships. Joint operations of IIBs and Leader-type icebreakers can ensure unimpeded year-round navigation in any direction via all NSR segments.

IIBs' characteristics will allow them to provide icebreaker assistance in the Western segment of the Arctic during winter and spring seasons and in the Eastern segment of the Arctic during summer and fall seasons.

In assessing an IIB lifecycle, the following costs were taken into account:

1. Design—18–24 months long (possibly with partial parallel building preparation that can reduce building time);
2. Preparation for building and building—12–24 months long;
3. IIB operation—40 years with the following cost structure:
- repair and maintenance;
- insurance;
- crew (wages and provisioning);
- administrative and other costs;
- fueling, including disposal of spent nuclear fuel;

Table 18.7 Comparison of integrated assessments for two icebreakers under design, (integrated assessments and best ship ratings are tinted yellow)

Indicator	Indicator weight	Ratings for icebreakers:	
		Leader–110Ya	Leader–alternative
Dimensions:	–	–	–
• length	–	–	–
• width	0.2857	2.27	2.86
• draft			
Icebreaking capacity, meters	0.2381	2.38	1.77
Speed in even ice, knots	0.1905	1.91	1.73
Propulsion power at propellers, megawatt	0.1429	1.43	0.78
Displacement, tons	0.0952	0.95	0.46
Crew strength	0.0476	0.24	0.48
Integrated assessment	1.00	9.17	8.08
Ship rating	–	1	2

Source calculated and compiled by authors based on Table 18.6 using Formula 18.2

- disposal (12 months planned length).

The calculations demonstrated that IIB's **lead ship lifecycle** full cost will be about ₽150 billion, which is only 30 billion less than announced marginal cost **to build** a Leader-type icebreaker.

IIB lifecycle structure is shown in Fig. 18.1.

Calculations show that the average daily costs of icebreaker assistance for fifty-meter-plus-wide ships and convoys provided by IIBs are only 30.5 percent higher compared to LK-Ya icebreaker costs and over 45 percent lower compared to Leader icebreaker costs.

In heavy ice conditions (year-old ice 2–2.8 meter-thick), it is highly probable that two icebreakers will be simultaneously required to make a channel wide enough for ships with hull width of fifty meters and wider.

In such situations, the cost of icebreaker assistance provided by two icebreakers (for example, Leader and LK-60Ya) is 260 percent higher than the cost of icebreaker assistance provided by an innovative icebreaker.

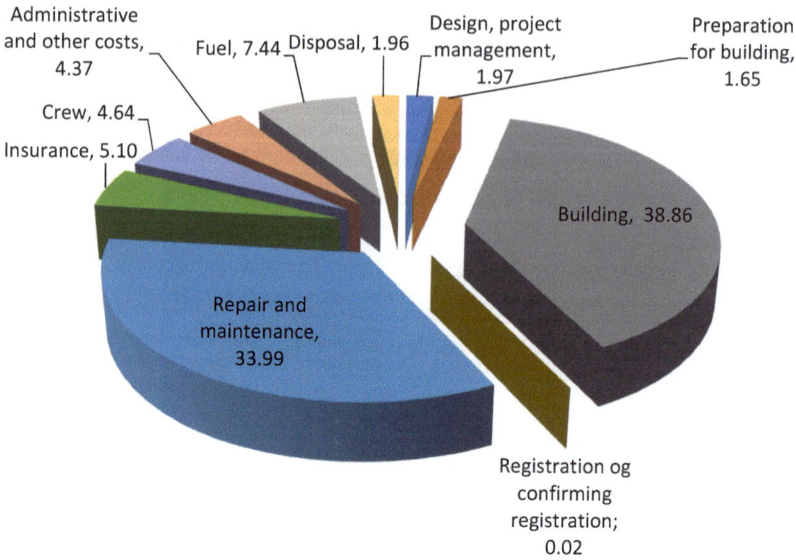

Fig. 18.1 IIB's lead ship lifecycle full cost structure, (percent) (*Source* Calculated by authors)

18.4 Conclusions / Recommendations

Due to the analysis results, the authors can make the following conclusions:

1. IIBs can provide icebreaker assistance to all existing Arctic ice-class ships (that meet the NSR sailing criteria) and to all future ships in even ice up to 2.8 meters thick;
2. IIB lifecycle was calculated;
3. Leader-type icebreaker lifecycle costs are 60 percent higher than IIB lifecycle costs;
4. Compared to existing icebreakers that can provide icebreaker assistance to individual ships and convoys, IIBs are more efficient in terms of:

- making sufficiently wide channels;
- planned icebreaking capacity in relatively heavy ice conditions;

- optimal lifecycle cost compared to Leader-type icebreaker lifecycle cost in ice conditions accepted for the purposes of the research;
- average daily operation costs compared to LK-60Ya costs or Leader and LK-60Ya combination costs when providing icebreaker assistance to 50–60-meter-wide ships sailing via NSR "shortcuts".

5. Analysis of the technical and economic characteristics of the compared icebreakers that can assist cargo shipments via the NSR (IIB and Leader-type) identified the most efficient pattern for their operations, namely:

5.1 During winter seasons, Leader-type icebreakers operate in the Eastern direction, while IIBs operate in the Western direction;

5.2 Vice versa, during the second part of the navigation, IIBs operate in any direction, while Leader-type icebreakers provide assistance north of difficult ice conditions (i.e., closer to the North Pole). Thus, the two icebreakers will ensure year-round navigation;

6. Considering the allowances in and results of the research, the authors suggested a concept for IIB-assisted transit cargo shipments via NSR "shortcuts" by fifty-sixty-meter-wide ships with ice-class lower than Arc7. Such ships are less metal-intensive that Arc7 class and, thus, can sail at higher speeds not only in NSR waters (assisted by icebreakers in 1.5-2- meter-thick ice) but also in "clear" waters. In turn, this will enable much shorter voyage times and increase the economic efficiency of operations for both icebreaker and carrier owners. In time, such attractive conditions will contribute to organizing line hauls via the NSR waters.

Today, icebreaker assistance issues are important not only from the commercial viewpoint, but also from the national security viewpoint (Zagorodnikov, 2018b). It is exactly those issues that will become the subject of future research by the authors.

References

Abramov, A. V., Zagorodnikov, M. A., Smirnov, A. Y., & Khmara, D. S. (2019). Development of icebreaking fleet of Russia. *Actual Problems of Innovative Economics, 6*, 5–15 (in Russian).

Appolonov, E., Abramov, A., Khmara, D., & Zagorodnikov, M. (2016). Evaluation of the design, construction and operation of a promising nuclear icebreaker. *Maritime Herald, 4*(60), 7–11 (In Russian).

Eurasia Daily News Agency. (2021). *The price of gas in Asia soared to the skies: Tankers from Russia rush along the Northern Sea Route*. www.eadaily.com/ru/news/2021/01/12/cena-gaza-v-azii-vzletela-do-nebes-tankery-iz-rossii-speshat-po-sevmorputi

Kremlin. (2019). *Decree of the Government of the Russian Federation No. 2553-r of October 28, 2019*. On the Approval of the Strategy for the Development of the Shipbuilding Industry for the Period up to 2035.

Korabel.ru. (2015). *Krylov State research centre presents a prospective icebreaking device—A multihull icebreaker*. https://www.korabel.ru/news/comments/krilovskiy_gosudarstvenniy_nauchniy_tsentr_predstavlyaet_novoe_perspektivnoe_ledorazrushayushchee_ustroystvo_mnogokorpusniy_ledokol.html (in Russian).

Mikhailov, A. (2020). *The creation of the Northern Sea transit corridor has begun in the Arctic*. https://rg.ru/2020/07/07/reg-szfo/v-arktike-nachalos-sozdanie-severnogo-morskogo-tranzitnogo-koridora.html (in Russian).

Portnews IAA. (2020). *There are no problems with the design of ice-class container ships in Russia*. https://portnews.ru/comments/2866/ (in Russian).

Rosatom. (2020). *What are the advantages of Arctic container ships with a composite hull?* https://strana-rosatom.ru/2020/05/18//чем-хороши-арктические-контейнерово/ (in Russian).

Solntsev, A. (2020). *One transit container ship in the Arctic is not a warrior*. https://www.rzd-partner.ru/wate-transport/comments/odin-tranzitnyy-konteynerovoz-v-arktike-ne-voin/ (in Russian).

Zagorodnikov, M. A. (2018a). Investment support for efficient year-round navigation along the northern maritime transport highway (SMTM). *Horizons of Economics*, *1*(41), 12–19 (in Russian).

Zagorodnikov, M. A. (2018b). Role of icebreaker fleet in the Russian Arctic zone national security. *Economic and Management*, *1*(147), 54–59 (in Russian).

CHAPTER 19

Managing the Ice Situation When Ensuring the Safety of Oil Platforms in the Arctic Zone of Russia

Dmitry V. Nikushchenko and Dmitriy S. Khmara

19.1 Introduction

Ensuring environmental safety during resource development in the Arctic is a priority task for the Russian Government (Garant.ru, 2020). From the existing climatic situation and its future progress scenarios, it is evident that glacier melting, iceberg formation, and drift can cause significant damage to oil and gas producing companies; besides, there is the threat of environmental and reputational damage.

In 2017, PAO NK Rosneft operations faced a threat from an approaching iceberg with a mass of over one million tons. Preventing the collision between the iceberg and the oil platform involved the

D. V. Nikushchenko · D. S. Khmara (✉)
St. Petersburg State Maritime Technical University, Saint Petersburg, Russia
e-mail: dskhmara@yandex.ru

D. V. Nikushchenko
e-mail: nikushchenko@smtu.ru

V. I. Salygin (ed.), *Energy of the Russian Arctic*,
https://doi.org/10.1007/978-981-19-2817-8_19

icebreaker fleet towing the iceberg into a position safe for producing natural resources (Podobedova et al., 2017).

If a collision between a drifting iceberg with an oil platform occurred, it would mean not only risks for oil and gas producers but also a technogenic catastrophe that can impact all activities in the Arctic Zone (AZ), both for the Russian Federation and for world community as a whole.

To avoid such situations, it is evidently necessary to create an ice situation management system that will ensure:

1. Situation monitoring in AZ areas during natural resources development;
2. Taking measures to prevent emergency situations.

When choosing the best option of measures to prevent emergency situations (the collision of icebergs with drilling platforms or other facilities), the authors analyzed published scientific works that suggested both passive methods—for example, towing with synthetic tug lines (Bruneau et al., 1977) or iceberg towing by several ships (Crocker et al., 1998),—and active methods—for example, subjecting icebergs to hot water jets (Chernyavets et al., 2013).

The analysis of emergency prevention options highlighted the following as the most feasible:

1. Iceberg towing with a tug line;
2. Iceberg towing with a net.

In order to develop decisions for ice situation management in ensuring the safety of oil platforms operating in the Arctic Zone of Russian Federation, it is necessary to:

1. Develop an enlarged algorithm for ice situation management;
2. Perform digital and physical modeling of the most frequent types of icebergs that are encountered in the zones of oil platform operations;
3. Perform software-assisted modeling of the optimal option to divert ice formations from oil platforms;
4. Estimate economic efficiency of using an appropriate option to divert ice formations from oil platforms.

19.2 Methodology

To perform digital modeling of water and air flow over and around an iceberg, the authors developed a mathematical model based on Reynolds averaged Navier–Stokes equations and continuities:

$$\frac{\partial \langle u_i \rangle}{\partial x_i} = 0 \tag{19.1}$$

$$\frac{\partial \langle u_i \rangle}{\partial t} + \langle u_j \rangle \frac{\partial \langle u_i \rangle}{\partial x_j} = -\frac{1}{\rho}\frac{\partial \langle p \rangle}{\partial x_i} + \nu \frac{\partial^2 \langle u_i \rangle}{\partial x_j \partial x_j} - \frac{\partial \langle u_i' u_j' \rangle}{\partial x_j} \tag{19.2}$$

To complete the Reynolds equations, the authors used a gradient-diffusion hypothesis:

$$-\langle u_i' u_j' \rangle = -2\nu_t \langle S_{ij} \rangle, \langle S_{ij} \rangle = \frac{1}{2}\left(\frac{\partial \langle u_i \rangle}{\partial x_j} + \frac{\partial \langle u_j \rangle}{\partial x_i} \right) \tag{19.3}$$

$$\nu_t = c_\mu k / \omega \tag{19.4}$$

where ν_t is locally isotropic turbulence viscosity; k is turbulence kinetic energy; ω is turbulence kinetic energy dissipation specific speed; $k - \omega$SST is a turbulence model suggested by Wilcox (2008).

$$\frac{\partial k}{\partial t} + \langle u_j \rangle \frac{\partial k}{\partial x_j} = P_k - \beta^* k\omega + \frac{\partial}{\partial x_j}\left[(\nu + \sigma_k \nu_t) \frac{\partial k}{\partial x_j} \right] \tag{19.5}$$

$$\frac{\partial \omega}{\partial t} + \langle u_j \rangle \frac{\partial \omega}{\partial x_j} = \alpha S^2 - \beta \omega^2 + \frac{\partial}{\partial x_j}\left[(\nu + \sigma_\omega \nu_t) \frac{\partial \omega}{\partial x_j} \right] + 2(1 - F_1)\sigma_{\omega 2} \frac{1}{\omega} \frac{\partial k}{\partial x_i} \frac{\partial \omega}{\partial x_i} \tag{19.6}$$

$$\nu_t = \frac{\alpha_1 k}{\max(\alpha_1 \omega, S F_2)} \tag{19.7}$$

with ratios $\sigma_k = 1.1$, $\sigma_\varepsilon = 1.3$, $c_{\varepsilon_1} = 1.44$, $c_{\varepsilon_2} = 1.92$, $c_\mu = 0.9$.

Equation space approximation was performed using control volume approach.

To integrate Navier–Stokes equations, the authors used the OpenFOAM package (Open-Source Field Operation and Manipulation CFD Tool Box). Today, OpenFOAM is one of the most "complete" and known packages designed to solve equations in partial derivatives using control volume approach.

Special three-dimensional modeling means, based on the cloud of points obtained from underwater hydrological surveillance of a natural iceberg, allowed for building solid-body models that were further used for physical modeling. Modeling was performed using research assets of Saint-Petersburg State Marine Technical University (SMTU University).

Based on the analysis of data obtained from digital and physical modeling, the authors developed mathematical models of hydrodynamic forces and moments at different iceberg towing angles, which were then integrated into potential versions of prototype software that can forecast data related to iceberg drift and iceberg towing to a safe position.

19.3 Results

At present, the following means are used to find icebergs: visual; aerial photography from different aircraft; satellite monitoring; ship radar equipment. To determine an iceberg's exact parameters, it is necessary to obtain data on:

1. The iceberg's tip, usually by aerial photography from a helicopter or an unmanned aerial vehicle (UAV);
2. Any submerged part of the iceberg, using hydro-acoustic survey technology, namely using a multi-beam echo sounder.

In both cases, the obtained data is computer-processed to get a 3D iceberg model.

SMTU University's research and production base allowed the following preliminary work to process available data on iceberg tips and submerged parts:

1. Digital modeling to create a mathematical model of iceberg flow around with the help of 3D modeling, which produced the following results:
   1.1. Developing a mathematical model for a digital model of water and air flow over and around an iceberg based on Reynolds averaged Navier-Stokes equations and continuities;
   1.2. Calculating hydrodynamic forces and moments for the case of iceberg movement at constant speed, for the case of iceberg rotation around a stationary axis running through the center of

masses at a constant angular speed in RANS position and for the case of iceberg rotation around a stationary axis running through the center of masses at a constant angular speed in uRANS position;

2. Physical modeling of towing in SMTU University's experimental tank and wind tunnel, including the following tests:

    2.1. Wind tunnel tests, which determined the hydrodynamic characteristics of iceberg tip and submerged parts by circular blowing at then degree intervals. To imitate air-liquid interface, a flat screen was used. They also determined a self-similarity zone of hydrodynamic characteristics at different iceberg model towing angles. Rotational characteristics of the models were determined with the help of a variable motion plant;
    2.2. The tests in SMTU University's experimental tank, which determined the positional components of the iceberg solid-body model hydrodynamic characteristics. The experiment identified self-similarity zones at different iceberg model towing angles. An analysis of the results demonstrates an insignificant increase in resistance ratio at higher towing speeds, which is due to a bigger wave component of resistance;
    2.3. Other tests, which also determined additional resistance with rise and fall of the waves.

An ice situation management algorithm (a consolidated algorithm is shown in Fig. 19.1) must include:

- Searching for dangerous objects;
- Determining their aero- and hydrodynamic characteristics by technical means;
- Choosing a tug vessel from the database of accessible ships proceeding from the aero- and hydrodynamic characteristics of dangerous objects;
- Computer modeling of the towing process;
- Physical (actual) towing;
- Analyzing and comparing physical towing parameters with modeling parameters with the help of software prototype and making a decision on either accepting or rejecting the towing trajectory.

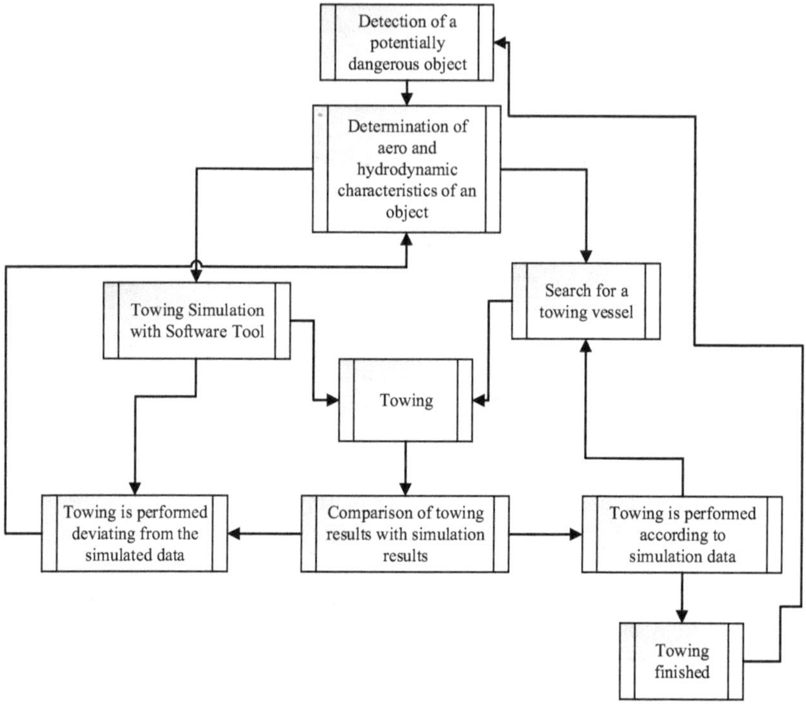

Fig. 19.1 Consolidated control algorithm regarding ice conditions (*Source* compiled by authors)

In the course of iceberg tow modeling, the following works were performed (accounting for a towing system optimal design):

1. An analysis of potential towing system options that took into account:
    - Iceberg shape,
    - Iceberg interaction with the capturing net, and
    - Changes in time of resistance to iceberg movement.
2. The following options of physical impact on the object:
    - Towing by one ship;
    - Towing by two ships;

- Influence by propeller streams; and
- Influence by a fire-fighting system.

The most efficient method of physical impact on the object is using a towing system.

A towing system can be of different, namely:

1. Floating synthetic tow line;
2. Iceberg net.

The main components of a towing system are:

- A two line that can be made of polystele;
- A polyamide line representing a polystele sling with eye splices at the ends;
- A line for lashing and mooring;
- A conductor line;
- A dynamometer;
- Rigging shackles;
- A floating anchor;
- A signaling device; and
- Metal weights to weigh down a line or the lower section of the iceberg net, etc.

Line and net towing technologies are identical (Korshunov et al., 2021). In both cases, tow lines and nets are to the stern of a ship that moves in a safe direction. The towing process consists in the following:

1. A tug vessel approaches a chosen iceberg;
2. A tug vessel circulates around the iceberg paying out a line (net);
3. After the maneuver, the end of the conductor line (net) is caught by a grappling anchor and connected to the stern winch that hauls in the free end of the tow line.

The research developed end element method calculation models designed to analyze towing systems.

The developed models helped in performing calculations of typical iceberg stationary and non-stationary towing at different tow forces.

Calculation of calm water towing results demonstrated that a higher tow force causes a non-linear increase in iceberg speed; at the same time, the asymmetrical form of water flow around the iceberg causes cross oscillations in the towing course. With tight lines and nets, the "ship-iceberg" system behaves as a single whole, causing higher cross direction speeds and making a towing vessel abandon the initial course. Due to that, ship acceleration components are clearly oscillatory, which in turn causes oscillatory forces in the line system branches. Higher tow forces cause higher oscillations with increased amplitudes (Tryaskin et al., 2021).

Iceberg yaw/slew and significant deviation from the initial course cause by the drift result in successive slacks of the towing system and maximal forces work alternately in all branches.

It is established that using nets is reasonable only with towing force not higher than 750 kilonewtons. To prevent iceberg-embracing nets slipping and deforming, there are analyzed options for using a system of weights, snagging rings, and buoys.

Besides, the authors analyzed friction, which influences the reliability of a towing system. Friction results in line wear, which reduces its strength. It is established the dominating factor that influences a towing system's service life is abrasion, which can be internal or external. The authors analyzed three ways of protection against abrasion:

- reducing the roughness of system elements touching the line;
- using line coating that is most resistant to abrasion; and
- using abrasion protection technologies that imply special coatings.

Calculations demonstrated that the most optimal options of towing systems (considering all ships that were or could be used in deviating icebergs from oil platforms) are the following:

- towing force 750 kilonewtons (line and net);
- towing force 1500 kilonewtons (line); and
- towing force 2250 kilonewtons (line).

The research developed a simplified method to determine aero- and hydrodynamic water resistance when towing icebergs in calm waters along a straight line or when movement direction changes.

The key problem in the method development was choosing a coordinates system, a position of its beginning, and the orientation of axes. Traditionally, in determining hydrodynamic forces and moments that impact a ship's hull moving on the water surface, the ship theory uses a fixed orthogonal coordinates system or a semi-fixed coordinates system that are strictly oriented relative to the main diameter plane and the midship plane. Apparently, such notions do not exist for icebergs. Any calculation methods imply a standard use of a coordinates system. The results of calculating the projections of hydrodynamic forces and moments obtained during digital modeling of iceberg towing within the OpenFOAM package caused a non-trivial decision to tie the orthogonal coordinates system's axes orientation to the hydrodynamic moment value and its numerical value derived from the drift angle. Therefore, it was decided to position the beginning of a fixed or semi-fixed coordinates system in the gravity center of the waterline plane, because the waterline contour, the waterline plane, and other parameters characterizing waterline geometry can be determined reliably enough in real conditions. Besides, waterline information is necessary to assess iceberg's stability when towing.

The next important problem that faced the method's developers was a tremendous variety of drifting icebergs' sizes and shapes, which disallows developing a correct enough mathematical model of a towing formation consisting of a towing icebreaker, a towing system, and an iceberg. In this connection, a number of simplifiers were used to estimate the efficiency and safety of such operations. The first and main allowance concerns iceberg shapes. While information on an iceberg's tip can be obtained by aerial assets, the knowledge of an iceberg's submerged part is limited to the fact that its volume is 5.6 times larger than that of the iceberg's tip. At the same time, an iceberg's waterline area and the height of the tip can be estimated, which makes it possible to calculate the volume and mass of the whole iceberg and then calculate its total volume, inertia of the waterline plane, metacentric radius, and initial metacentric height. The three latter values are necessary to estimate a statistical iceberg bank during towing and wave induced oscillation.

To simplify further calculations, it is necessary to approximate the waterline by an analytical expression; for example, circle or ellipse equations. The area of those figures and the eccentricity should be maximally close to the real shape of the towed iceberg waterline.

Software prototypes developed at SMTU University for ice situation management enabled the following:

1. Modeling an iceberg's submerged part, proceeding from available data on the iceberg's tip;
2. Reflecting objects (ships, drilling platforms, and icebergs) and their parameters on the map in real time;
3. Visualizing objects through a camera over/under the water surface (Fig. 19.2);
4. Creating new simulation scenarios by choosing desirable objects from the list of available models of icebergs and tow-vessels, and adding them to the map with selected initial orientation and speed;
5. Allowing a user to set current and sea swell in any number of map points; the data can be used by iceberg models;
6. Modeling iceberg towing regime in real time;
7. Monitoring a free iceberg drift to position it after towing; and
8. Adding to/deleting new objects from the library, editing their characteristics, and uploading towing scenarios.

The research demonstrated that by the beginning of 2023 (taking into account the results concerning iceberg towing scenario models) it will be possible to develop a comprehensive IT-product for ice situation management.

The estimated cost will be only ₽352 million, including: ₽160 million for forming a library of tug ships and different icebergs (the cost includes hiring research vessels and UAVs designed for aerial photography for ice formations); ₽60 million for additional research related to physical and digital modeling of ice formation towing; ₽120 million to develop additional modules for ice situation management software prototype; and ₽12 million for integrating navigation maps and ice situation data. Another ₽260 million will be required to test and calibrate the software prototype in real conditions of ice formations towing.

The funding for the first stage of creating an integrated ice situation management system (considering already obtained results) is insignificant compared to the potential economic results from introducing the system into practice. One could just imagine the damage caused by a collision of a one-million-ton iceberg with an oil platform that could be damaged or

Fig. 19.2 Visualization of objects over and under the water surface (*Source* compiled by authors with the use of software developed at SMTU University)

completely destroyed. In most cases, it will result in suspending geological exploration or hydrocarbon production, which means huge economic losses that could not even be compared with the cost of developing and introducing an ice situation management system. A critical event is a collision of an iceberg with a platform resulting in a technogenic catastrophe.

Estimates show that introducing an ice situation management system will allow: prolonging geological exploration for over two months; ensuring an economic effect of over $2 million from one platform if there is a threat of collision with an iceberg; ensuring a total yearly economic effect not lower than $20 million.

19.4 Conclusions/Recommendations

The article described the digital and physical modeling of iceberg hydrodynamic characteristics during the towing process.

The digital experiment built instant trajectories of liquid particles' movement and vector fields in different planes.

It also defined the iceberg towing directions which have the most minimal moment and the lowest resistance.

The results of digital modeling correlate well with physical modeling, both qualitatively and quantitively.

The developed mathematical model of a towing system differs from existing ones in details of accounting for the specific interaction between an iceberg and a line system presented 3D.

That model's use allowed for analyzing different versions of towing systems and identifying conditions for the emergence of extreme and emergency situations during the process of towing.

The research enabled developing technical projects for three versions of towing systems that cover a wide range of icebergs and developing an original version of a towing system.

The use of simplified methods to determine water hydrodynamic reactions during towing and to determine wind influence over an iceberg's behavior can provide a quick access to the materials required for choosing a safe speed and the direction of towing, as well as for developing recommendations for maneuvering the towing formation. Besides, there is no necessity to use complicated mathematical calculations that require thorough initial data preparation and input in the calculation program (Kornishin et al., 2021).

A promising research direction in that field is using identification to determine iceberg hydrodynamic characteristics in the process of towing. That requires developing special methods.

The developed software enables a wide range of actions in testing iceberg models, towing system configurations, towing simulations, and reflecting the parameters of different object models (including testing in real time and under maximal acceleration); it also provides for expanding the library of objects, etc.

An estimated cost of developing and introducing the first version of an ice situation management system is ₽352 million.

Introducing research results into ice situation management practice in the areas of operational offshore facilities will reduce reaction time in organizing safe iceberg towing. Finally, it will have a positive influence over ice situation management and contribute to a yearly economic effect of not less than $20 million.

References

Bruneau, A., Dempster, R., & Peters, R. (1977). *Iceberg towing for oil rig avoidance, iceberg utilization*. Pergamon Press.

Chernyavets, A., Zhiltsov, N., Zenkov, A., Anosov, V., Fedorov, A., & Chernyavets, V. (2013). *Method of Active Control of Iceberg Danger and Device for Active Control of Iceberg Danger*. Retrieved November 6, 2021 from https://patenton.ru/patent/RU2484209C1 (in Russian).

Crocker, G., Wright, B., Thistle, S., & Bruneau, A. (1998). *An Assessment of Current Iceberg Management Capabilities*. Contract Report for National Research Council Canada, Prepared by C-CORE and B. Wright and Associates Ltd., C-CORE Publications 98-C26.

Garant.ru. (2020). *Decree of the President of the Russian Federation No. 164 of March 5, 2020 "On the Fundamentals of the State Policy of the Russian Federation in the Arctic for the Period up to 2035"*. Retrieved November 6, 2021 from https://www.garant.ru/products/ipo/prime/doc/74710556/ (in Russian).

Kornishin, K. A., Efimov, Y. O., Nikushchenko, D. V., Khmara, D. S., Tryaskin, N. V., Andreev, A. G., Skutin, A. A., & Smirnov, K. G. (2021). Effect of Iceberg Shape on Wind Force Parameters. *Proceedings of the 31-st International Offshore and Polar Engineering Conference,* 967–971. Retrieved November 6, 2021 from https://onepetro.org/ISOPEIOPEC/proceedings-abstract/ISOPE21/All-ISOPE21/ISOPE-I-21-1294/464543

Korshunov, V. A., Mudrik, R. S., Ponomarev, D. A., Rodionov, A. A., Nikushchenko, D. V., Khmara, D. S., Kornishin, K. A., Efimov, Y. O., &

Maksimova, P. V. (2021). Research of the sea state influence on iceberg towing system performance. *Proceedings of the 31-st International Offshore and Polar Engineering Conference*, 940–948. Retrieved November 6, 2021 from https://onepetro.org/ISOPEIOPEC/proceedings-abstract/ISOPE21/All-ISOPE21/ISOPE-I-21-1290/464559

Podobedova, L., & Kazarnovsky, P. (2017). *"Rosneft" Managed to Avoid a Collision of an Oil Platform with an Iceberg and Named it a Titanic-Killer.* Retrieved November 6, 2021 from https://www.rbc.ru/technology_and_media/12/09/2017/59b7a77c9a794750c30e601c

Tryaskin, N. V., Ermolaeva, E. V., Ovchinnikov, K. D., Nikushchenko, D. V., Khmara, D. S., Kornishin, K. A., Efimov, Y. O., Nesterov, A. V., & Skutin, A. A. (2021). Features of tabular iceberg towing. *Proceedings of the 31-st International Offshore and Polar Engineering Conference*, 949–961. Retrieved November 6, 2021 from https://onepetro.org/ISOPEIOPEC/proceedings-abstract/ISOPE21/All-ISOPE21/ISOPE-I-21-1291/464558

Wilcox, D. C. (2008). Formulation of the K-omega turbulence model revisited. *AIAA Journal, 46*(11), 2823–2838.

CHAPTER 20

Specifics of LNG Shipments via the Northern Sea Route

Andrei M. Golubchik and Tautginas Sankauskas

20.1 Introduction

The authors chose liquefied natural gas (LNG) as the object for research as it is one of the newest commodities, with trade and transportation volumes growing at a huge pace. At the same time, LNG is a completely unique commodity because there is no other such product that forms a very complicated system of interrelationships: its technical capabilities mean that unique provisions in chartering LNG tankers must be included, and those unique provisions in their turn affect the provisions included in international sales contracts.

A. M. Golubchik (✉)
Department of Oil and Gas Trading and Logistics, National University of Oil and Gas "Gubkin University", Moscow, Russia
e-mail: pigeon_am@mail.ru

T. Sankauskas
Lithuanian National Association of Forwarders and Logistics (LINEKA), Vilnius, Lithuania
e-mail: tautginas@vilteda.lt

V. I. Salygin (ed.), *Energy of the Russian Arctic*,
https://doi.org/10.1007/978-981-19-2817-8_20

LNG transportation process always generates boiled-off gas that can be dealt with in three ways:

- bleed it to the atmosphere—as was done with the very first LNG tankers merely 25–30 years ago;
- re-liquefy it—such technologies are currently available;
- use it as fuel for ship power plants.

In any case, the situation arises where, because of cargo specifics, a carrier company cannot guarantee compliance with the fundamental principle of transportation—to deliver the cargo to the consignee in the same quality and quantity as it was received from the consignor. Consequently, a ship-chartering contract must include an appropriate provision.

It appears that the LNG transportation process implies absolutely legal and explicable reasons for a reduced quantity of the freight to be delivered from the seller to the buyer. It means that freight's specifics must be reflected in an international sales contract.

Concerning LNG shipments via the NSR, the transportation system can be presented as a total of four components:

- icebreaker fleet;
- specialized LNG tanker fleet;
- transshipment systems and terminals;
- navigation and hydrographic support.

This work will attempt an analysis of all the above components.

20.2 Hydrography, Icebreakers, and Permits to Sail via the NSR

Cargo shipments in the Arctic Ocean seas have always been extremely difficult due to the sparse population and severe climate in the region, the latter of which will probably be most important. Along its entire length, the Northern Sea Route is only free of solid ice for 100–105 days a year, while individual icebergs, large floes, and crushed ice fields are almost always present along the entire NSR length.

At Western Arctic archipelagoes (Svalbard, Franz Josef Islands, Novaya Zemlya, Severnaya Zemlya), icecap areas are much smaller than in Greenland (totaling 92 thousand square kilometers), being not more than 500 m thick, and their iceberg discharge is estimated at about ten cubic kilometers (which is equivalent to nine billion tons). To determine the relative scale of these, and whether it constitutes a lot, reference shall be made to Russian hydrographers A. Vinogradov and B. Tsukerman, who proposed the following assessment methods.

Assuming an "average ordinary" iceberg's mass at 0.1 million tons, it is possible to approximately assess the number of icebergs in the Kara and Barents Seas at ninety thousand, and each iceberg presents a certain danger for ships and offshore installations. Today, unfortunately, neither Russia nor Norway saves specialized services that monitor the conditions of icecaps at the Arctic archipelagoes and control iceberg generation by outlet glaciers. Sporadic information on iceberg monitoring in the waters of the Barents Sea from transiting ships is accumulated in the data base of Arctic and Antarctic Research Institute State Scientific Center (AANII): from 1928 to 2017, it registered over 29,700 "observed" icebergs (Vinogradov & Tsukerman, 2017: 119).

One can see that the iceberg danger is real, which means that any country which is extremely interested in developing the NSR should pay close attention to resolving the issue. Thus, Russia's powerful and modern icebreaker fleet will be a guarantee of safe and successful NSR operations.

Taking into account the importance of the Northern Sea Route for the country, the Decree of the President of the Russian Federation No. 164 of March 5, 2020 approved the Strategy of Development of the Arctic Zone of the Russian Federation and the Provision of National Security for the Period to 2035, where great attention is paid to provisions that concern transport communications in the Arctic region (Decree of the President of the Russian Federation, 2020).

The document provides for priority development of the Arctic transportation system that would ensure year-round connection between major territories of natural resources development in the Arctic Zone of Russia (first of all, between offshore oil and gas producing areas and Arctic Ocean seas' coastal facilities), as well as between domestic and international markets and the centers that consume products from Arctic territories.

Also, it envisages building ten new nuclear-powered and diesel-electric icebreakers, as well as developing port and rescue infrastructure in such locations as Tiksi, Pevek, and Providenia. If there are no effective transport communications, it will be impossible to achieve progress either in Russian Arctic's socio-economic development or in forming a modern defense belt along northern borders of Russia (Vinogradov & Tsukerman, 2017: 118).

As was already mentioned, sailing the NSR is risky and requires special ship and crew preparation. For that reason, to sail even short NSR stretches (to say nothing of its entire length), each ship (including ships flying the Russian flag) must get permission for an ice/non ice class ship to navigate in the water area of the Northern Sea Route. The right to issue such permissions belongs to only one agency: FSUE NSR Administration (Federal State Budgetary Institution the Northern Sea Route Administration).

Permissions may be one-time for a period shorter than a year or for one calendar year. Applying for and getting permissions is free, but pilot and icebreaker services must be paid for according to existing government rates. A payment calculator is available at NSR Administration website, but in the Russian language only (Northern Sea Route Administration, 2021a).

A monopoly to provide icebreaker services belongs to FSUE Atomflot (Atomflot, 2021). At present, continuous NSR navigation is assisted by five operational nuclear-powered icebreakers that belong to Atomflot:

- two Arctica-class icebreakers—Yamal (built in 1992) and 50 Years of Victory (2007);
- two Taymyr-class icebreakers—Taymyr (1989) and Vaigach (1990);
- one Project 22,220 icebreaker—lead multi-purpose nuclear-powered icebreaker christened Arctica (2020).

The period from 2013 to 2017 saw the start of active large-tonnage ship use; at the same time, the number of local and transit voyages via the NSR increased more than tenfold. Just four years ago, speaking at the "Arctic: Present and Future" Seventh International Forum that took place in St Petersburg on December 5, 2017, Nikolay Monko, first deputy director of FSUE NSR Administration, noted that, starting from 2013, they issued between 625 and 718 permissions per year to navigate the NSR for ships that belong to 178 companies (including seventy foreign companies). The result was a continuous growth in cargo shipment volumes which, in 2017, reached 10.2 million tons.

As of October 31, 2021, permissions were issued to 1158 different ships that flew the flags of eighteen countries (Northern Sea Route Administration, 2021b). That is almost one hundred permissions more than during the whole of 2020 which certainly shows the popularity of that transportation route. First of all, the growth was due to more voyages by LNG carriers—in 2021, their voyages amounted to more than fifty percent of the total number. Shipments proceed from the port of Sabetta to the eastern (most frequently) and Western NSR boundary. Transit shipments via the NSR amount to about fifteen percent, voyages being categorized by different service ships, ships that carry cargoes within the framework of Northern Supply, and ships that belong to the Murmansk branch of OAO GMK Norilsk Nickel. It is worth noting that before 2010 the main mission of the NSR was to ensure coastal shipments under Northern Supply programs and to provide transportation support for GMK Norilsk Nickel. During the Soviet period, NSR cargo shipments reached 6.5 million tons. The peak was in 1987, and shipment volumes had already dropped to 1.2 million tons by early the 1990s, and during the most difficult period they collapsed to 30–40 thousand tons per year in the NSR eastern segment. During that period, shipments were by small dead weight ships (12–16 thousand tons) of extremely low ice class that always required icebreaker assistance. In those conditions, there was practically no necessity to take into account any threat from icebergs; consequently, no special measures were taken to neutralize that threat.

20.3 LNG Carriers for the NSR

In October 2012, a Russian 288-m-long "Ob River" LNG carrier made the first commercial voyage via the NSR. It is worth recalling that the above route is forty percent shorter than the traditional way from Western Europe to North Eastern Asia, which enables a radical reduction in negative environmental impacts with simultaneous economic efficiency due to saved time and fuel.

Further NSR and Arctic region development required significant progress in shipbuilding technologies. Thus, 2017 saw successful sea trials of Christophe de Margerie LNG carrier—the first of YAMALMAX-type increased ice class ships ordered by PAO Sovcomflot for Yamal LNG project. In August 2017, the ship made a first commercial voyage from

Kirkenes (Norway) to the port of Pusan (South Korea) and set a record of unassisted individual tanker NSR sailing of six days, twelve hours, and fifteen minutes. Total voyage length was twenty-two days, which is thirty percent shorter than a usual route via the Suez Canal.

In January 2020, the same ship set a new record—it became the first ship in history that sailed the Eastern NSR segment from the port of Sabetta to the Bering Strait without icebreaker assistance (RBC, 2021). Such a voyage became possible because the ship was Artc7 class—the highest ice class for civilian commercial ships. No wonder that in 2017, Christophe de Margerie LNG carrier won a Platts Global Energy Awards prestigious international prize in the nomination "2017 Energy Project of the Year".

The ship is the first of the sixteen LNG carriers of a principally new YAMALMAX Arctic type built specially to service Yamal LNG Project. Their characteristics are: a length of 299.0 m; a width of 50 m; a dead weight of 80.2 thousand tons; a speed of up to 19.5 knots (in ice up to 2.5 m thick from two to seven knots); a draft of 11.8 m; a cargo capacity of 172,630 cubic meters (Polyakova, 2017).

On April 23, 2014, Yamal LNG published shortlists of shipping companies that would service its LNG plant. It includes Sovcomflot, Mitsui, and Canadian Teekay LNG. Each company was to build an LNG carrier fleet: six ships for Sovcomflot, four for Mitsui, and six for Teekay LNG. The cost of one ship is $340 million ($5.5 billion for all sixteen LNG carriers). In August 2015, because of financial difficulties, Sovcomflot ceded the rights for five of the six ships to a Greek DYNAGAS LTD shipping company (Vedomosti, 2015)

As of early November 2021, the Yamal LNG project has been serviced by fifteen out of the sixteen planned ships:

- Vladimir Rusanov, Vladimir Vize, Nikolay Urvantsev (Mitsui LNG);
- Georgiy Ushakov, Rudolf Samoylovich, Eduard Toll, Yakov Gakkel, Vladimir Voronin, Nikolay Yevgenov (Teekay LNG);
- Boris Vilkitsky, Fedor Litke, Georgiy Brusilov, Boris Davydov, Nikolay Zubov (DYNAGAS LTD).

Taking into account the prospects of Yamal LNG project, sixteen initial ships (no matter how modern they are) will not be enough. The project's marketing strategy envisages both spot and contract sales. Specific figures are a commercial secret; it is only known that the main volumes of produced LNG will be sold to project shareholders who have full rights for further disposal.

According to certain information (the reliability of which cannot be guaranteed by the authors), about 90–93 percent of LNG production has been contracted. The authors believe it possible to publish the following approximate data.

Project shareholders (Total and China National Petroleum Company (CNPC)) under long-term contracts will purchase 2.4 and 4 million tons per year accordingly. Gazprom's subsidiary—Gazprom Marketing & Trading Singapore (GM&T)—has a long-term contract for 2.9 million tons of LNG a year, which will be resold to GAIL Indian Company. The Spanish trader Gas Natural Fenosa has contracted 2.5 million tons of LNG a year for twenty-five years, considered tremendous according to existing standards. Novatek's regular trader Novatek Gas & Power arranged for selling one million tons of LNG to French Engie, 0.9 million tons of LNG to British Shell International Trading Middle East, and 0.5 million tons of LNG to Gunvor trader.

All long-term LNG selling contracts are tied to specific ports or regions. Approximately, 86 percent of produced LNG shall be shipped to the Asia–Pacific region (Vedomosti, 2017).

The transportation and technical system to transport LNG from the Arctic port of Sabetta can be regarded as a complicated logistical system. The difficulty in creating the system and ensuring its efficient functioning is attributed to the following:

- permafrost in LNG production areas;
- building pipelines to the gas liquefaction plant (loading points);
- low temperatures;
- difficult ice and hydrometeorological conditions at port terminal locations;
- difficult ice and hydrometeorological conditions at locations from which LNG will be shipped by LNG carries assisted by icebreakers;
- the extra sensitivity of the Russian Arctic zone to environment pollution.

Therefore, because they begin as early as building a ship, the specifics must be sorted. Just fifteen years ago, all LNG carriers were built to service a specific gas project. This means that every ship was built to the specifications of a certain charterer. Therefore, if a ship is built for continuous long-term use in a certain LNG project, the charterer could have full control over the building process. On a potential charterer's part, to use that right they were either to send their own supervising team to the shipyard or to hire a respectable surveyor company (Golubchik & Ahtyrskij, 2018: 31).

Sea transportation is an integral part of the production and sales chain that underlies the global LNG industry. Objectively, it is a necessary link between gas production and consumption. Of course, of equal importance are liquefaction and regasification facilities; all the above components have their own ranges of technological tasks to accomplish. For this reason, no element can be excluded without violating the viability of the entire system whose one and only purpose is to deliver the final product to consumers. During the lifecycle of any LNG project, transportation plays a vital role, but sea transportation is characterized by huge volumes and wide geographic coverage, and in case of transportation via the NSR, difficult climatic conditions are of utmost importance.

In practice, there is no significant difference between chartering an LNG carrier and a conventional oil tanker. Formally, both ships are designed to transport liquid cargo. However, LNG characteristics such as extremely low temperatures necessary for storage and transportation and safe transportation technologies themselves are absolutely different. Consequently, it becomes evident that an LNG shipping contract should reflect the specifics of LNG transportation.

Crew qualification requirements are important for every charter (with the possible exception of bareboat charters) and can include minimal service periods for key ship officers and sufficient competences for other crew members. Many remember that ten to fifteen years ago, a skyrocketing international LNG market caused a serious lack of qualified personnel to operate LNG carriers, especially senior officers who were to have necessary experience of service aboard such ships.

A usual requirement to senior officers was at least six months of service in the same or similar position aboard an LNG carrier during the past five years. The problem acquired special importance during 2008–2010, when 30–50 ships were launched per year, which exceeded ten percent of the LNG tanker fleet at that time. Resolving that problem required great effort from ship owners.

In the near future, a similar situation is unlikely, despite the fact that approximately thirty "standard" size tankers of 160–170 thousand cubic meters capacity are launched every year.

LNG carriers designed for year-round operations at the NSR are required to have ice class not lower than Arc5, while all sixteen ships for Yamal LNG project will have Arc7 class.

Besides, standard charter forms lack limitations for ship operations in any regions or geographic locations. When signing freight contracts and determining navigation areas, one should take into account not only ship characteristics but also the requirements of specific ports and terminals. Special attention will be required from ship owners, because they shall be responsible for guaranteeing that ships meet the requirements.

In its turn, a shipment contract can provide a charterer with an opportunity to demand transferring LNG to another ship at sea—a ship-to-ship operation (STS). Such a provision can be limited by conditions required for crew rescue or expanded to include a commercially reasonable reloading. At the same time, any transshipment can be affected only in strict compliance with industry rules, standards, and guidelines.

Before 2005, STS operations were extremely rare, because there were no clear-cut and understandable rules for such operations and, essentially, there was no concept to ensure safety. Nevertheless, such operations were greatly demanded by charterers, with the demand actively growing with every passing year. High demand was explained by introducing Floating Storage and Regasification Units (FSRU).

In 2005, the International Society of Gas Tanker and Terminal Operators (SIGTTO) developed safety instructions for STS operations (known as "general rules"). From that time on, STS operations became routine, but all freight contracts contain a requirement to coordinate such operations with a charterer.

As far as the authors know, an STS permit was initially integrated in the texts LNG carrier charter contracts for ships that service Yamal LNG Project. Moreover, one of the fifteen operational ships is always in ballast, scheduled as a "standby" vessel ready to sail any time to rendezvous a ship in a critical situation where emergency transshipment is required.

20.4 Boiled-Off Gas: Accounting and Disposal Specifics

Boiled-off gas is the gas that appears as a result of liquefied gas natural evaporation in the tanks of an LNG carrier. A standard daily norm of losses at a modern LNG carrier is 0.15 percent of the transported cargo mass.

At the beginning of the article, the authors looked at three methods to dispose of boiled-off gas. Bleeding into the atmosphere was ruled out at once, because it greatly damages the environment and is absolutely ilogical from the viewpoint of economy.

Modern LNG carries have a technical capability to capture up to 95 percent of boiled-off gas, to re-liquefy it, and to return it to the tanks. Such a technology was enthusiastically welcomed by active green energy adepts, but it is highly energy-intensive; all onboard power is produced by diesel generators that run either on diesel fuel or marine fuel oil (MFO). Also, that is a very expensive and environmentally unsound method of power generation.

There remains a third method—using boiled-off gas as fuel for ship power plants. For that, practically all modern LNG carriers designed to service Yamal LNH Project are equipped with advanced three-fuel diesel-electric (TFDE) power plants that can run on light and heavy oil products and on boiled-off gas. TFDEs have a number of advantages compared to other power plants, which makes such hybrid power plants very attractive both commercially and otherwise (Shcherbanin, 2019). The advantages are as follows:

1. TFDE-equipped ships do not carry a diesel engine as the main propulsion plant. They have two relatively small electric engines. Thus, less space is occupied and more can be used for cargo transportation;
2. There is almost "zero" maintenance for electric engines compared to diesel or steam engines;
3. The absence of the main diesel engines makes a ship's lower deck "almost empty". Thus, there is great flexibility in where to place the machinery;
4. There is also the full absence of mechanisms to drain used oil from diesel engines. Such mechanisms are just not required, so there are no great volumes of oil used for an electric generator connected to

a standard diesel. For that reason, a ship can be built without an oil sump and bilge water collectors;

5. There is no need to have a special compressor and piping to start the main diesel engine;
6. There is no need for additional exhaust purification systems (scrubbers), as aboard ships with main diesel engines, and, consequently less maintenance is required alongside lower maintenance costs;
7. A significantly lower vibration and noise level is present, as compared to diesel engines;
8. Electric engines can operate at even 2–3 rpm, which creates additional serious advantages during navigation and those advantages are vitally important during commercial ship operations;
9. Unlike in a diesel engine, there are no limits for the number of starts;
10. Response time to commands is very short as compared to a diesel engine; consequently, the engine can start faster;
11. There is no need for complicated electric or electronic circuits and transformers for the engines and speed control equipment; this in turn simplifies TFDE use;
12. Piping for the sea water necessary for a cooling system becomes much smaller;
13. Regular fueling is not required, because ship power plants run on gas which, if necessary, can be extracted from the ship's tanks;
14. Problems with electric engines start are very small compared to problems in starting and reversing a diesel engine;
15. Power shaft electric brakes are more efficient than mechanical brakes installed in diesel and steam engines;
16. Total ship operational costs are much lower compared to ships with diesel engines.

Thus, the authors have all reasons to assert that using TFDEs aboard LNG carriers is the best option in all respects.

However, gas boiling-off in the transportation process can cause significant material losses and disputes between ship owners and charterers. The 0.15 percent that constitute a standard evaporation norm and that are free for use by ship engines are by far not enough for normal ship propulsion; depending on the speed, a ship will have to consume two-four times more per day. The need to boil-off additional gas volumes arises.

Without doubt, one of the most important financial conditions is an agreement on boiled-off gas quantity used as engine fuel free of charge for the ship owner. The LNGVOY charter pro-form, developed in 2016 jointly by Baltic and International Maritime Council (BIMCO) and International Group of Liquefied Natural Gas Importers (GIIGNL), reads that parties should agree on a certain maximal volume of gas evaporation. However, such a condition can make charterers over-increase ship technical and operational requirements in order to conserve greater cargo volumes, since even in the absence of agreed limits for LNG evaporation a ship under normal operational conditions cannot consume more gas than required for its engines.

Voyage Charter-Party (VCP), another charter's pro-form developed by GIIGNL and used since 2012, fixes the allowable volume of gas boiling-off from dropping last outward sea pilot till sending a notice of readiness at the destination port. However, difficulties emerge in proper cargo quantity measurements for a number of technical reasons.

For reference, both indicated points are almost always at high seas; waves contribute to continuous liquid movement in the tanks and disallow exact liquid level measurements. The port of Sabetta is at the very exit from the Gulf of Ob to the Kara Sea. That sea is very rarely calm, with a 2–3 magnitude always being there.

The importance of knowing the exact liquid level can be illustrated by a simple example: for an LNG carrier of 170,000 cubic meter capacity, a liquid level shift of just one centimeter causes a 36–42 cubic meter difference in measurements; a liquid level shift of 5–6 cm will cause a measurement error close to a daily gas evaporation norm calculated at 0.15 percent of the total gas volume. It is worth noting that the allowable accuracy of equipment to measure LNG level in the tanks is 7.5 mm, which means 20 cubic meters. A 5–6 cm liquid level shift will occur at force 3–4 sea; such force and higher is observed at the exit from the Gulf of Ob 200 days a year.

To reduce the influence of sloshing cargo, special LNG level measurement techniques are used, such as not less than 5–7 level measurements made at 30–60 s intervals. Despite that, this condition is unattractive for a ship owner, because it is very difficult to correlate measurement results with more accurately calculated actual LNG volumes at the moments of ship loading and unloading.

Since VCP charters are oriented exactly at the spot market, the provisions that regulate the order and conditions for preparing tanks for loading have been significantly changed. They envisage three reservoir conditions: cooled and ready for loading; warm and under gas vapor; warm and inverted. Upon arrival at the port of Sabetta, a ship must be ready for immediate loading; all operations to cool the tanks to the required temperature have to be completed and only after that the tanks are filled with gas.

Besides, if required, the time needed for ship preparation is not usually counted as stalling or demurrage time. Special attention should be paid to Article 18, devoted to the time of a ship's arrival to the port of unloading. A ship owner's risk is an exact calculation of ship's speed, the quantity of boiled-off gas, its temperature, pressure in the tanks, and the cost of fuel required for reaching set targets.

Thus, the charter's article under consideration significantly influences the process of agreement on freight rate in specific charters, because delivery deadlines are directly linked to the ship owner's expenses for bunker fuel. The provisions of that article are usually directly linked to the main gas sales contract and the buyer's requirements concerning delivery deadlines that are obligations of a gas seller who frequently is the ship charterer. At the same time, Article 18(a) provides only for a ship owner's obligation to guarantee a necessary agreed speed without a direct obligation to ensure the ship's arrival at set deadlines. The article also describes the ship owner's action if naturally evaporated gas is not enough for achieving necessary speed, which means the necessity to agree with the charterer on the additional forced evaporation.

The difference with other forms can also be seen in determining LNG price, which is important for the charter, be it residual LNG in the tanks after unloading or consumed en route over the agreed norm of boiled-off gas consumption. The LNGVOY form fixes LNG price in the first part, on the face side of the charter. Most probably, it is the same as LNG price under the main LNG sale contract. At the same time, to keep LNG contract price a commercial secret, the ShellLNGTime1 form for the purposes of the charter determines LNG price as that equal to the price of ship's engine fuel.

Due to the fact that LNG tankers consume their cargo as fuel and that gas evaporates naturally, parties must indicate a gas price calculation method to compensate the charterer if a ship is decommissioned with cargo in the tanks.

20.5 Specifics of LNG Sales Contracts: Why DES and DEQ?

The logic of this subject does not envisage an analysis of LNG trade commercial side, but it cannot be disregarded that LNG shipments by modern carriers have their own specifics, related to the fact that before actual voyage termination nobody knows exact cargo volumes remaining in the tanks. Moreover, nobody can ever reliably calculate actual cargo volumes in the tanks after unloading. For that reason, it is important to touch upon contractual work issues.

The above factors indicate that all LNG sales contracts contain certain unique provisions as compared to other commodity sales contracts. The most popular delivery terms are either delivery from a ship at a port of destination (*DES* Incoterms 2000) or delivery from a ship berthed at the port of destination (*DEQ* Incoterms 2000). LNG is always delivered by the Seller. Despite the fact that the 2020 version has been in use for a while, Incoterms 2000 is actually used.

The logic for building a contract chain is as follows: major traders are related to Yamal LNG shareholders through long-term contracts (minimum three-year-long, normally five–seven years long) and purchase LNG from Sabetta plant. The traders charter LNG carriers and organize deliveries to their clients exactly under DES or DEQ terms. The practice demonstrated that those terms are the best for LNG trade specifics (Kholopov & Sokolova, 2021).

Buyers receive the cargo at a port of destination. The exact price is determined for each LNG ton received by a regasification plant at the temperature of –163°C and at the pressure close to atmospheric.

Payment terms most frequently include an advance up to 50 percent of estimated delivery price. Full payment is made after cargo handover and signing a delivery and acceptance certificate. Payments are made by bank transfers or confirmed letters of credit. The latter method is relatively rare because the cargo is already physically handed over to the buyer.

Using concrete delivery terms depends on whose pumps are used for unloading. If unloading is done by a ship's pumps, DEQ terms are in force; if unloading is done by shore facilities, it is DES terms that are in force.

20.6 Conclusions

Despite the fact that LNG is the newest mass commodity, today it is the most popular commodity that enjoys ever-growing demand. LNG transportation technology that uses part of the cargo to ensure ship movement forms special terms included in the contracts of charter parties who, in turn, form specific terms in LNG sales contracts.

Russia is one of the major players in the LNG market, while the Yamal LNG project is the key and most powerful Russian asset. However, that asset is located at the Yamal Peninsula on the Arctic Ocean coast very far from major LNG consumers.

The routes for shipping LNG from the port of Sabetta run via the eastern or western NSR segment, which necessitates special technical requirements for LNG carriers and higher qualification requirements for the crew. At present, Yamal LNG Project is serviced by fifteen out of the sixteen planned special YAMALMAX LNG carriers that belong to four major specialized LNG shipping companies: Sovcomflot, Mitsui LNG, Teekay LNG, and DYNAGAS Ltd. All tankers have one of the highest ice classes Arc7, while others authorized for year-round navigation have an ice class not lower than Arc5.

Besides, the very fact of the sixteen specialized ships' availability simply confirms that the majority of new LNG carriers are built to service a specific long-term project, and only 20 percent of the ships are built for operations in the spot market (Feeney, 2012).

Building ships with such a high ice class is explained by extremely complicated conditions for ship operations: many voyage segments run via the NSR, which is free of ice only 100–105 days a year. Most of the year, ships sail through crushed or solid ice fields under real threat from icebergs. Besides, the coast along the NSR is very sparsely populated; ports are very rare and small and tend to have obsolete loading equipment, being practically without facilities for repair or maintenance. One possible exception is the port of Dikson, which hosts a major part of rescue forces for the entire NSR.

To document the shipments, companies use standard VCP and LNGVOY charter pro-forms that have long been successfully used. Those charters mostly meet the parties' requirements that concern control of and accounting for both naturally boiled-off gas and artificially evaporated gas used to provide ship propulsion.

The authors are convinced that shipments via the NSR will continue growing, not only LNG but also all other shipments. Of course, climate conditions in the region will not change. At the same time, technological and organizational conditions (first of all related to navigation safety) will undoubtedly improve, which will eventually render the NSR not an extreme route, but merely a complicated one—complicated exclusively because of its climatic conditions.

References

Atomflot. (2021). *Russia's Icebreakers in Operation*. http://www.rosatomflot.ru/flot/atomnye-ledokoly/deystvuyushie/ (in Russian). Accessed October 22, 2021.

Feeney, A. (2012). Does the LNG chartering market need a new form? *UK P&I Club*. https://www.ukpandi.com/knowledge-publications/article/does-the-lng-chartering-market-need-a-new-form-5594/

Golubchik, A. M., & Ahtyrskij, A. A. (2018). LNG-specific freight operations and charter-proformas. *Herald of Transport, 10*, 29–34 (in Russian).

Kholopov, K. V., & Sokolova, O. V. (2021). Synergy of LNG exports and prospects of the northern sea route. *Herald of Transport, 1*, 12–17 (in Russian).

Kremlin. (2020). *Decree of the President of the Russian Federation No. 164 of March 5, 2020. (2020). "On the Fundamentals of the State Policy of the Russian Federation in the Arctic for the Period up to 2035"*. http://www.kremlin.ru/acts/bank/45255 (in Russian). Accessed October 19, 2021.

Northern Sea Route Administration. (2021a). *Icebreaker Assistance Value Calculating*. http://www.nsra.ru/ru/ledokolnaya_i_ledovaya_lotsmanskaya_provodka/raschet_stoimosti_ledokolnoy_provodki_v_akvatorii_smp.html (in Russian). Accessed November 1, 2021.

Northern Sea Route Administration. (2021b). *Permissions for Navigation in the Water Area of the Northern Sea Route*. http://www.nsra.ru/en/rassmotrenie_zayavleniy/razresheniya.html (in Russian). Accessed November 1, 2021.

Polyakova, A. (2017). *The First Two Supertankers Departed from the Loading Port Sabetta after Fuelling by LNG at "Yamal LNG" Enterprise*. https://www.yamal.kp.ru/daily/26765.5/3800639/ (in Russian). Accessed October 25, 2021.

RBC. (2021). *LNG-tanker for the First Time to Sail via the NSR in its East Direction without an Icebreaker Assistance*. https://www.rbc.ru/business/18/01/2021/600598679a7947431777917a (in Russian). Accessed October 23, 2021.

Shcherbanin, Y. A. (Eds.). (2019). *Freight and documentary operations at oil and LNG fleet*. RUSAJNS (in Russian).

Vedomosti. (2015). *Dynagas to ship LNG from "Yamal LNG" enterprise*. https://www.vedomosti.ru/business/articles/2015a/08/11/604277-grecheskaya-dynagas-zaimetsya-perevozkoi-gaza-yamal-spg (in Russian). Accessed October 21, 2021.

Vedomosti. (2017). *First shipment done from "Yamal LNG" enterprise*. https://www.vedomosti.ru/business/articles/2017/12/08/744652-yamal-spg (in Russian). Accessed October 21, 2021.

Vinogradov, A. N., & Tsukerman, V. A. (2017). Icebergs in the western arctic and ways to reducing risks when shipping and drilling off-shore. *The North and the Market: Economic Space, 5*, 115–128 (in Russian).

PART IV

Greener Perspective for Energy of the Russian Arctic

CHAPTER 21

Renewable Energy in the Russian Arctic: Energy Transition and Opportunities in the Context of Post-pandemic Realities

Valentina Ignatyeva and Sébastien Gadal

21.1 Introduction

The COVID-19 pandemic has not only caused unprecedented changes to society and the economy around the world, but also has inadvertently demonstrated the Earth's incredible resilience and its ability to clean up and recover after years of accumulated emissions. The pandemic has provided us with a peek into what a low-carbon future could hold in store if we act now and go full speed down the decarbonization path.

V. Ignatyeva (✉) · S. Gadal
Aix-Marseille University, Aix-en-Provence, France
e-mail: valentina.ignatyeva@etu.univ-amu.fr

S. Gadal
e-mail: sebastien.gadal@univ-amu.fr; szh.gadal@s-vfu.ru

S. Gadal
North-Eastern Federal University, Yakutsk, Russia

V. I. Salygin (ed.), *Energy of the Russian Arctic*,
https://doi.org/10.1007/978-981-19-2817-8_21

As this has been happening, the recent collapse of oil prices in 2020 has also shown the fragility of the fossil fuel sector in these volatile and changing times and has underlined the importance of economic and fuel diversification. Considering the overwhelming role that fossil fuels and fossil fuel-driven industries play in Russia's economy—and particularly their prevalence in the Russian Arctic region—this should provide a wake-up call for the country: it needs to reconsider its energy strategy, and understand the role that decarbonization and renewable energies need to play in it.

The long-term sustainability of economic development in the Arctic regions will be influenced by global environmental and climate regulation, as well as, of course, the ongoing transformation of energy markets. As renewable energy sources are already completely intertwined with the concept of sustainable development and decarbonization, they can play a significant role in ensuring the sustainable development of the Russian Arctic. A more robust incorporation of renewable energy systems (RES) would not only provide a sustainable route for the economic development of the region, but would also see health benefits for local populations and regional environments.

Moreover, more and more of Russia's economic partners are going down the path of carbon neutrality and increasing their goals to reduce emissions and switch over to greener energies; therefore, for Russia to stay relevant and competitive in the long-term, it should consider embracing some of these global trends and begin incorporating appropriate policy changes. Now is the perfect time to adopt the necessary incentives and prioritize resilient investments to begin changing the country's economic structure in order to help drive Russia toward overcoming future economic and climate challenges and turn some of these potential challenges into growth opportunities.

21.2 Pandemic Lessons and Changing Trends

The outbreak and exponential spread of the COVID-19 coronavirus (SARS-CoV-2) thoroughly shook the foundation of every society across the globe, claiming over five million lives (as of 15/11/21). After nearly two years of intermittent quarantines, border shutdowns, travel restrictions, civil unrests, economic downturns, and continued volatility, the pandemic is still far from over.

This global health crisis has provided a jarring realization of how unprepared we are as a society for shocks of this magnitude, and forced many of us to reevaluate our priorities and how we operate. Countries were required to implement emergency measure that redefined how societies function for months on end, with some countries enforcing country-wide lockdowns, prompting people to adapt to new realities of working from home, thus reshaping the energy demand curve.

These draconian measures, however, also led to an unprecedented decrease in global emissions not seen in years. One study published in the journal Nature Climate Change showed that at the peak of quarantine (April/May 2020) there was a 17 percent decline globally, taking emissions down to around the level of 2006 (Nature Climate Change, 2020). For weeks, the media showed the impact of confinement on nature and how cities permanently encapsulated in smoke were seeing clear blue skies for the first time in years; stars spread across the night sky in big cities as both air and light pollution cleared away; the canals of Venice saw fish and swans for the first time in decades; and wild animals ventured and roamed into nearby cities and communities across the globe. People were fascinated how quickly the silence from human activity had led to positive reverberation across global environments.

While the quarantines have had a short-term effect on pollution levels, the overall concentrations of greenhouse gases continue to grow in the atmosphere, showing just how vital it is to act now to avoid an irreversible environmental catastrophe in the future. Of course, as the initial lockdowns began to ease and people ventured out and resumed economic activity, the levels of pollutions quickly shot back up, negating all the positive effects that were incurred during the initial quarantines. Nonetheless, energy related CO_2 emissions fell by 7 percent in 2020 (IEA, 2020).

While the pandemic is still far from over and the far-reaching effects of it will continue to shape how we operate, countries are adopting recovery packages to salvage their economies in the months and years to come. Many are therefore looking—and indeed pushing—to use this opportunity to not just restore the economies to the way they were, but to do it better by incorporating green solutions and decarbonization measures. Even some international oil giants understand that green recovery is key to long-term survivability not just of their companies but more importantly of the planet.

21.3 Ringing the Arctic "Warming" Bell

At the same time, while economies and societies are still trying to find their way through the mess of the pandemic and market fluctuations, there are even more worrying things happening with the climate.

For years, there has been a rise in climatic anomalies, which have been getting more intense in the recent years. Over the past 25 years, the number of dangerous meteorological events has more than doubled. Intense heat waves, violent storms, tornadoes, floods, and wildfires have dominated the news across the globe. The Russian Arctic has been making headlines in recent years as the effects of global warming are undeniably and forcefully resonating through the area.

While temperatures are steadily increasing across the globe, in the Russian Arctic they have been increasing on average 2.5 times faster than in other parts of the world. In June 2020, the city of Verhoyansk (in Russia), typically known as the coldest city on Earth, hit an unprecedented 38 degrees Celsius—the highest temperature ever recorded above the Arctic Circle. In recent years, the Russian Arctic was on average 4 degrees warmer, both on land in the water, than previous years. In 2020, the average temperatures in north-central Siberia were about 8 degrees Celsius above average (Litvinova & Borenstein, 2020). According to the United States' National Oceanic and Atmospheric Administration (NOAA) and NASA, 2020 nearly tied with 2016 as the hottest year on record; now, 2021 is also well on its way to crown this grim pedestal, with July of this year being the world's hottest month ever recorded (NOAA, 2021).

These persistent high temperatures have also been the cause of devastating wildfires that have been ravaging the Siberian and Arctic landscapes more and more frequently over recent years. 2021 squeezed out 2012 as the top recorded year in the twenty-first century for forest fires in Russia, reaching a total of 18.16 million hectares, with over 4 million hectares in Yakutia alone (Bonch-Osmolovskaya, 2021).

According to meteorologists, this continuous warming will continue to narrow the difference in temperature and pressure between the Arctic and lower latitudes, therefore slowing the Arctic jet stream, leading to weather systems that bring extreme heat or rain getting stuck over places for much longer periods (Litvinova & Borenstein, 2020). Considering the white nights of the North in the summer, where the sun never fully sets at night, the prolonged exposure to its rays paired with the rising temperatures means that the ground will not have sufficient time to cool off. This will perpetuate even more wildfires in the future.

In addition, it will also give rise to another major problem, whose ramifications would be detrimental not only for the Arctic but also for the entire global community. The high temperatures and increasingly severe and prolonged wildfires are leading to the thawing of permafrost. This can have a destructive effect on local infrastructure, as most of the building and pipeline foundation is built on permafrost. According to the Russian Minister of Natural Resources, this thawing is the underlying reason for 23 percent of all technical system failure in the region, and up to 29 percent of oil and gas production facilities can no longer be operated (TASS, 2021). A recent study showed that the cost of overall infrastructure and structural damages caused by climate change and the melting permafrost could easily surpass $100 billion (Streletskiy et al., 2019). This cost, however, does not consider the potentially irreparable damages to the environment that could ensue from this structural undermining. This was the case in 2020, where subsidence of the foundation piles (due to permafrost thawing) at the Norilsk-Taimyr energy plant led to largest oil spill on record for the Russian Arctic.

Moreover, the thawing of permafrost has also been calculated to lead to methane release, which is a strong greenhouse gas estimated to contain more than twice the amount of carbon as already exists in the atmosphere (Tamsin, 2020). There are fears that the melting permafrost might also have some other hidden surprises in store as it thaws out, potentially releasing unknown or dormant viruses back into the air, in which case the current pandemic world order might become a dangerous new normal.

21.4 Adapting to New Realities

Seeing how the current pandemic destabilized society globally, there has been a growing understanding that the shocks from a changing climate will be more intense, prolonged, and—if we wait too long—irreversible. Therefore, as countries are adopting pandemic recovery packages, many have taken this opportunity to accelerate the energy transition by incorporating more aggressive decarbonization strategies to ensure a more sustainable future.

Energy transition is being defined by decarbonization, energy efficiency, and the robust integration of non-fossil energy sources. Renewable energy sources have become practically synonymous with this trend and have been growing at exponential rates. According to Bloomberg NEF, the world committed a record $501.3 billion to decarbonization in 2020, beating the previous year by 9 percent; global renewable energy capacity additions beat all previous records despite the economic disruption caused by the COVID-19 pandemic (Saul & Mathis, 2021). The International Renewable Energy Agency (IRENA) estimates that more than 260 GW of renewable energy capacity were added last year, exceeding the growth in 2019 by almost 50 percent (IRENA, 2021).

As one of the largest oil and gas producers in the world, and whose primary exports are fossil fuels (accounting for over 50 percent of export commodity trade), Russia's economy is particularly exposed to these changing global energy trends.

Russia's two major export markets—the EU and China—have actively emphasized their goals to become carbon neutral by 2050 and 2060 respectively, and both regions are looking increasingly toward renewable energies as the solution to reach this goal, as well as to increase their energy security and reduce import dependence.

The EU has pledged to allocate €550 billion to green projects over the next seven years, which is the largest single climate pledge ever made (Abnett, 2020). Furthermore, under the new Green Deal, the EU is planning on restricting funding for natural gas projects under the revised green finance rules because of the risk they pose to the collective climate goals, with the addition of the proposed Carbon Border Adjustment Mechanism (CBAM), which would impose additional taxes on imports to preventing carbon leakage. Under the pandemic recovery program, the European Commission is considering fast-tracking the adoption of this mechanism and passing the necessary legislation in 2021 in order to introduce it in 2023.

Global institutions are also cutting their financing for fossil fuel projects. In 2019, the European Investment Bank (EIB) declared that it would end financing fossil fuel energy projects from the end of 2021 (Cohen, 2020). In 2020, other private investment banks (including Deutsche Bank, Morgan Stanley, Citi Bank) followed suit, announcing their strategies to reduce exposure to the fossil fuel sector (Cohen, 2020).

As the world steps more solidly on the decarbonization path and toward a more sustainable energy structure, Russia's overreliance on fossil fuel production and revenue can expose its economy, which is why diversification and sustainability should be the key drivers in shaping the country's strategy in the years to come.

21.5 RES in Russia—A Path to Sustainability and New Opportunities

With significant hydrocarbon reserves, renewable energy generation has not been a top priority for the Russian government. Nonetheless, some effort is being made to keep up with the global trends and to allocate some importance to renewables in future energy strategies, mostly with aims to improve technological knowhow in this rapidly evolving sector and to reduce the carbon footprint.

While currently renewables (excluding large hydro and nuclear) make up less than 2 percent of Russia's total energy mix (Tikhonov, 2020), new generation capacity has been growing every year, with an end goal to reach 12.5 percent by 2050 (Maksimov, 2021). This share of renewable energy refers to predominantly wind, solar, biomass, and small hydro, as large hydro has been widely present in Russia's energy mix for decades, accounting for roughly 20 percent of total generation in 2020 (Russian Power System Operator, 2021). At the same time, natural gas, which currently accounts for more than 50 percent of the energy mix, is set to remain as the primary energy source in the years to come (Finmarket, 2021).

The Russian Government published several decrees to develop mechanisms to promote renewable energy generation for integration into the grid, as well as for generation in isolated regions and territories, aiming to resolve various environmental and social issues at the same time. Among these, they have introduced an incentive for long-term power supply contracts for RES, where the state guarantees a return of investments with a yield of about 12 percent on the invested capital over 15 years (Decree №499). Furthermore, they have local content requirements, meaning that these installations must use materials that are either partly produced or assembled in Russia (Order №3788). This measure was introduced to stimulate economic activity and created jobs in the renewables sector.

At the same time, a lot of freedom is given to the regional governments to establish their own mechanisms (like the Nenets Autonomous Region's program for cogeneration plants [mostly wind and diesel]), which is why active collaboration with local administrations is crucial to successfully promote regional RES development in the Russian Arctic.

When it comes to strategies, the Arctic regions have emerged as a central focus point for Russia. In October 2020, the Government adopted the official Strategy for the Development of the Russian Arctic Regions to

2035, as well as numerous supporting legislations aimed at providing tax and financial incentives for socio-economic development in the regions. Still, this goal of aggressive development comes at a time of heightened environmental fragility of the Arctic and growing global trends toward decarbonization and curbing climate change. It is thus paramount that this development does not come at an unacceptable cost.

The long-term sustainability of the economic development of the Arctic regions will be influenced by the global environmental and climate regulation, as well as the ongoing transformation of energy markets. As more and more of Russia's economic partners are working toward carbon neutrality and increasing their goals to reduce emissions through improved energy efficiency measures and integration of green energy sources, it is important that Russia considers embracing these global trends beyond setting carbon neutrality targets by predominantly relying on the carbon offsetting provided by the rich forest coverage it is endowed with, especially as in the context of global warming, millions of hectares of these forests are being decimated by fires, from which it will take years to recover and to return to the previous level of carbon absorption. Moreover, these fires are further exacerbating the situation by contributing to overall carbon emission levels: in 2021 the amount of CO_2 released from the forest fires in the Arctic reached almost 400 megatons, double what was released the year prior (Tanayev, 2021).

It is here that RES could play an integral role for Russia, not only in curbing the environmental costs of economic development, but in achieving the outlined goals of improved energy security, living conditions, connectivity, and infrastructural integration.

A lack of energy security and isolation has been the fundamental roadblocks to regional development and growth, and one of the main contributors to poor living conditions in the region. The isolation of many communities and economic hubs presents major logistical issues, as the lack of proper infrastructure significantly complicates the delivery and increases the cost of goods, in particular energy. Since the majority of the Russian Arctic is not connected to the main electricity grid, there is an overreliance on imported coal and diesel to run regional generators. Limited supplies of fuels have led to intermittent power and heat availability for local populations, with some areas only having access to a few hours of electricity a day. This "Northern Delivery," which can often only be carried out in the summer months due to accessibility issues, can see up to 8 million liters of diesel fuel and 20–25 million tons of coal (IRENA, 2017) being brought into these Arctic regions. Moreover, the federal and regional budgets allocate hundreds of millions of rubles each year to cover the cost of these deliveries, and the share of the transport component in these costs can account for over 70 percent (Berdin et al., 2017: 3). This high cost of the "Northern Delivery" has led to extremely high tariffs for electricity in some Arctic regions, which can be 4–5 times higher than the average electricity price in the country. Curbing high energy costs could provide additional impetus to justifying the profitability of RES in addition to providing energy security.

One of the major advantages of renewable energy systems is that, by decreasing the overall cost of their energy, they would not only be able to provide these isolated regions with secure and continuous power supply, but would also significantly reduce the financial burden both on federal and regional budgets, as well as on the individual consumers. Moreover, these RES solutions would also help in reducing the anthropogenic effect on the local environment, improving living conditions for the local population, and significantly reducing exposure to health risks.

In addition to isolation, many issues have led to a lower-than-average life expectancy in the region, as compared to other parts of the country: the harsh weather conditions of the Arctic, a lack of economic opportunities, intermittent energy/heat supply, and low access to proper health care. This situation is worsened by the fact that most regional power and heat generation is derived from either diesel generators or fossil fuel-based (mostly coal) plants which strongly contribute to high levels of air pollutions in these communities. This is further exacerbated by the fact that these are predominantly used throughout the nearly endless cold weather months, where the slow-moving and dense cold air ends up trapping these particles in place, causing them to accumulate and subsequently inhaled in higher concentrations by the local population.

Numerous studies have shown that sustained exposure to high levels of particulate matter (PM) can have significant adverse effects on respiratory and cardiovascular systems, strongly contributing to higher mortality rates in affected areas. Some studies have also shown that this exposure can create underlying conditions that "exacerbate the severity of COVID-19 symptoms and worsen the prognosis of this disease" (Wu et al., 2020). As global warming thaws permafrost layers, as weather patterns and environments change, and as the global population continues to be increasingly mobile, the spread for new diseases, revival of dormant ones, and the overall likelihood of future health crises become more probable. It is thus paramount that everything is done to improve living conditions for communities across the globe to ensure that environmental conditions are not a factor in mortality rates.

To this extent, renewable energy systems could help address multiple elements of these underlying problems, by helping alleviate (if not altogether removing) the problem of air pollution in these communities.

When it comes to isolation in these regions, it is not only present in terms of road and energy infrastructure but in terms of telecom networks as well. The lack of sustained and secure energy systems has not created the necessary foundation to allow for the establishment of proper digital connectivity throughout large parts of these regions. While on one hand, this could present a potential barrier for successful integration of renewable energy systems, especially considering how those systems are reliant on digital platforms for remote operations and grid management, on the other hand, they could actually serve as the necessary impetus to help drive the development of better regional connectivity.

As the world is increasingly becoming more reliant on digital solutions across various economic sectors, including energy management, it is impossible to consider sustainable development of the Russian Arctic Zones without touching on the digitalization of energy systems and the need for more robust development of network coverage in the area.

The onset of the pandemic has completely rewritten the working culture across industries. Over the course of the last two years, hundreds of thousands of companies across the globe were forced—either by government-implemented restrictions or by own choice—to go remote, requiring their teams to reorganize and reenvisage how to operate in this new context. Since then, most have adapted to living and working under these new conditions, to the point that most companies have now accepted either a hybrid or even maintaining a fully remote operation going forward. This drastic change to the work culture has increased reliance on technology, internet, telecom services, and of course electricity.

This transition to remote and "online" operation has not been limited to just the professional sectors. Many schools and universities have had to go online for parts of the years, communication with friends and family has become increasingly virtual, and various entertainment platforms which have provided respite have also expanded in the virtual space. As this digital reality provides the basis for more basic needs, the demand for electricity will continue to soar, as will the need for data centers, which constitute a "backbone of Internet infrastructure" (Middleton & Rønning, 2020).

This growing electricity and processing demand could turn RES and the Arctic regions into perfect partners going forward, as renewable energy installation can not only help meet the local demand for basic power and heat in a sustainable and reliable way, but could also ensure a more rapid transition of these regions into the progressive communities envisaged by the Arctic development strategies. Moreover, data centers require significant amounts of energy to not only operate but to cool their processors; therefore, there is also tremendous business potential in integrating more of these centers throughout the Russia Arctic. They can thus serve an integral role in developing new and reliable energy systems across these urban centers, while also meeting the growing demand of the digital economy.

Furthermore, with the aspirations to have robust development of the Arctic regions, renewable energy should be strongly considered not just as mechanisms for minimizing environmental impact and sustainable development, but also as strong drivers for job creation, as RES project can be "as much as three-times more demanding of capital and jobs compared to traditional energy developments" (Cohen, 2020). The growing need for data centers with increased volumes of digital traffic (as well as their maintenance) will create additional opportunities and requirements for human capital in the regions.

Finally, renewable energy systems also provide another market opportunity for the region. With the exponential growth of RES across the globe, as well as the ongoing digitalization of economies, there is a growing demand for rare earth minerals that are necessary for RES installation, battery storage, defense, and tech. Having the fourth largest reserves of rare earth minerals in the world at around 12 million tons, Russia is severely underdeveloped, accounting for only 2 percent of global production (MiningWorld Russia, 2020). While the Russian Government has adopted new legislation to help increase Russia's share of global rare earths output to 10 percent by 2030, there is still ample room for growth and development in this sector, as well as a need to reduce its own import dependence for these elements. Moreover, as demand for these elements is on the rise and as other importing countries (such as the US and EU) are actively trying to ramp up their RES capacity and decrease their import reliance for these elements from China, Russia could become an integral player in diversifying this trade dynamic.

21.6 Conclusions

In conclusion, as a new reality begins, where attention on climate change is heightened and economic recovery strategies are aimed at preventing new irreparable global shocks, Russia, as it focuses on revamping its northern territories, has a perfect opportunity to develop with this future in mind. Renewable energy systems have an important role to play in this energy transition and in helping develop these isolated areas into growing and modern urban centers. Renewable energies could also provide new economic opportunities for Russia that could facilitate and accelerate the desired development of its Arctic regions. However, global trends of decarbonization and digitalization will require a powerful and symbiotic relationship between green energy, sustainability, and economic development, and it is important that the right structure is set in place to ensure that this relationship is at the core of Russia's Arctic development.

References

Abnett, K. (2020). EU to restrict support for gas projects, citing climate goals-draft. *Reuters*. https://www.reuters.com/article/eu-energy-gas/eu-to-restrict-support-for-gas-projects-citing-climate-goals-draft-idUSL4N2IP2NL

Ajadi, T., Cuming, V., Boyle, R., Strahan, D., Kimmel, M., Logan, M., & McCrone, A. (2020). *Global trends in renewable energy investment 2020*. Frankfurt School-UNEP Centre/BNEF 2020. https://www.fs-unep-centre.org/wp-content/uploads/2020/06/GTR_2020.pdf

Berdin, V., Kokorin A., Yulkin, G., & Yulkin, M. (2017). *Renewable energy sources in isolated communities in the Russian Arctic*. World Wildlife Fund (WWF). https://wwf.ru/en/resources/publications/booklets/renewable-energy-in-off-grid-settlements-in-the-russian-arctic/

Bonch-Osmolovskaya, K. (2021). 2021 became a record year for the area of forest fires in Russia since the beginning of the 21st century. *Novaya Gazeta*. https://novayagazeta.ru/articles/2021/09/21/2021-god-stal-rekordnym-po-ploshchadi-lesnykh-pozharov-v-rossii-s-nachala-xxi-veka-news. (in Russian).

Cohen, A. (2020). Goldman Sachs: Renewable power will become the largest area of spending in the energy industry in 2021. *Forbes*. https://www.forbes.com/sites/arielcohen/2020/11/17/goldman-sachs-renewable-power-will-become-the-largest-area-of-spending-in-the-energy-industry-in-2021/?sh=5350eb4663eb

Decree No. 499 of the Government of the Russian Federation from May 28, 2013 (as amended on July 12, 2021). *On the mechanism for stimulating the use of renewable energy sources in the wholesale electricity and capacity market*. http://www.consultant.ru/document/cons_doc_LAW_146916/ (in Russian).

Finmarket. (2021). Gas consumption in Russia can be 490–550 bcm by 2035—General scheme project. *Information Agency "Finmarket"*. http://www.finmarket.ru/news/5399751

IEA. (2020). World energy outlook 2020. *IEA*. https://www.iea.org/reports/world-energy-outlook-2020

IRENA. (2017). *REmap 2030 renewable energy prospects for Russian federation* (Working paper). IRENA. https://www.irena.org/-/media/Files/IRENA/Agency/Publication/2017/Apr/IRENA_REmap_Russia_paper_2017.pdf

IRENA. (2021). *World adds record new renewable energy capacity in 2020*. IRENA. Press Release. https://www.irena.org/newsroom/pressreleases/2021/Apr/World-Adds-Record-New-Renewable-Energy-Capacity-in-2020

Litvinova, D., & Borenstein, S. (2020). The Arctic is on fire: Siberian heat wave alarms scientists. *ABC News*. https://abcnews.go.com/International/wireStory/arctic-fire-siberian-heat-wave-alarms-scientists-71421935

Maksimov, A. (2021). The installed capacity of renewable energy sources in Russia can reach 97.4 GW by 2050. *Ministry of Energy of the Russian Federation "News"*. https://minenergo.gov.ru/node/21980

Middleton, A., & Rønning, B. (2020). Data centers as critical infrastructure in the Arctic. *High North News*. https://www.highnorthnews.com/en/data-centers-critical-infrastructure-arctic

MiningWorld Russia. (2020). How is Russia developing rare Earth metals? *MiningWorld Russia*. https://miningworld.ru/Articles/how-is-russia-developing-rare-earth-metals

Nature Climate Change. (2020). Unexpected times. *Nature Climate Change, 10*(479). https://doi.org/10.1038/s41558-020-0817-x

NOAA. (2021). *It's official: July was Earth's hottest month on record*. National Oceanic and Atmospheric Administration. https://www.noaa.gov/news/its-official-july-2021-was-earths-hottest-month-on-record

Order No. 3788 of The Ministry of Industry and Trade of The Russian Federation from September 24, 2018 "On Approval of the Procedure for Determining the Degree of Localization in Relation to a Generating Facility Operating on the Basis of the Use of Renewable Energy Sources". https://docs.cntd.ru/document/551494053 (in Russian).

RIA Novosti. (2021). Smoke from forest fires in Yakutia has reached Arctic stations. *Russian Information Agency "Novosti"*. https://ria.ru/20210803/pozhar-1744191539.html (in Russian).

Russian Power System Operator (2021). *Unified energy system of Russia*. https://www.so-ups.ru/functioning/ees/ups2021/ (in Russian).

Saul, J., & Mathis, W. (2021). Spending on global energy transition hits record $500 billion. *Bloomberg*. https://www.bloomberg.com/news/articles/2021-01-19/spending-on-global-energy-transition-hits-record-500-billion

Streletskiy, D. A., Suter, L. J., Shiklomanov, N. I., Porfiriev, B. N., & Eliseev, D. O. (2019). Assessment of climate change impacts on buildings, structures, and infrastructure in the Russian regions on permafrost. *Environmental Research Letters, 14*(2), 025003.

Tamsin, E. (2020). The Arctic heatwave: Here's what we know. *The Guardian*. www.theguardian.com/commentisfree/2020/jun/25/arctic-heatwave-38c-siberia-science

Tanayev, N. (2021). Expert named the causes of growing wildfires in the Arctic. *TASS*. https://tass.ru/obschestvo/11897687 (in Russian).

TASS. (2021). *The Ministry of Natural Resources estimates the damage from the degradation of permafrost in the Arctic at 5 trillion rubles*. https://tass.ru/ekonomika/11490969?utm_source=yxnews&utm_medium=desktop (in Russian).

Tikhonov, S. (2020). Aleksandr Novak: The share of RES in Russia's energy balance will reach 4–5%. *Rossiiskaya Gazeta*. https://rg.ru/2020/09/04/aleksandr-novak-dolia-vie-v-energobalanse-rossii-dostignet-4-5.html (in Russian).

Wu, X., Nethery, R. C., Sabath, M. B., Braun, D., & Dominici, F. (2020). Air pollution and COVID-19 mortality in the United States: Strengths and limitations of an ecological regression analysis. *Science Advances, 6*(45). https://www.science.org/doi/10.1126/sciadv.abd4049

CHAPTER 22

Climatic Risks of Hydrocarbon Production in the Arctic Zone of the Russian Federation

Olga A. Derendyaeva

22.1 Introduction

The Arctic is the region most vulnerable to various climatic risks (Hassol & Corell, 2006). Their assessment is a paramount task for taking the necessary measures to adapt to climate change. At the state level, this process is a system of measures of a political, legislative, regulatory, economic, and social nature, implemented by the federal executive authorities and executive authorities of the constituent entities of the Russian Federation.

At the regional level, the executive authorities of the constituent entities of the Russian Federation develop and implement regional adaptation plans, which include action plans for the municipal authorities in the area under consideration. In the Arctic region, it is extremely important to develop an effective adaptation plan, especially in the field of gas and oil production in this region.

O. A. Derendyaeva (✉)
All-Russian Research Institute for Civil Defense and Emergencies, Balashikha, Russia
e-mail: derendyaeva94@mail.ru

V. I. Salygin (ed.), *Energy of the Russian Arctic*,
https://doi.org/10.1007/978-981-19-2817-8_22

22.2 Methodology

In this article, a comprehensive assessment of the climatic risks of hydrocarbon production in the Arctic Zone of the Russian Federation was made. To solve this problem, systemic and institutional approaches were used, with the allocation of three main levels:

1. Assessment of climatic risks in the field of hydrocarbon production in the Arctic Zone of the Russian Federation, based on the identification of the main sources of risks.
2. Study of the institutional system of oil and gas production in the Arctic Zone of the Russian Federation.
3. Analysis of financing projects for hydrocarbon production in the Arctic Zone of the Russian Federation.

As sources of information for assessing climate risks, the following were used: reports; reviews; bulletins of the Federal Service for Hydrometeorology and Environmental Monitoring and its organizations, published on an initiative basis or posted in the public domain in accordance with the Decree of the Government of the Russian Federation of November 15, 1997, No. 1425 "On information services in the field of hydrometeorology and monitoring of environmental pollution" (Collected Legislation of the Russian Federation, 1997, No. 47, Art. 5410; 2008, No. 13, Art. 1314) (Roshydromet) and the order of the Government of the Russian Federation dated 10 July, 2013, No. 1187-r "On Lists of Information on the Activities of State Bodies, Local Self-Government Bodies Placed on the Internet in the Form of Open Data" (Collected Legislation of the Russian Federation, 2013, No. 30, Art. 4128; 2018, No. 15, article 2174) (Order of the Government of the Russian Federation dated July 10, 2013, No. 1187); safety passports of the territory of a constituent entity of the Russian Federation and/or a municipal formation, prepared in accordance with the order of the Ministry of Emergency Situations of Russia dated October 25, 2004, No. 484 "On approval of a standard safety passport for the territories of constituent entities of the Russian Federation and municipalities" (registered with the Ministry of Justice of Russia on November 23, 2004, No. 6144) (Order of the Ministry of Emergency Situations of Russia dated October 25, 2004, No. 484); and the Passport of climatic safety of the territory of the constituent entity of the Russian Federation (Order of May 19, 2021, No. 16-r).

The works of foreign economists and representatives of institutional theory were used as a theoretical basis: particularly Douglas North, American economist and Nobel Prize laureate. One of the most significant works was North's *Institutions, Institutional Change and Economic Performance* (North, 1990). Along with Ronald Coase and Oliver Williamson, North helped found the International Society for the New Institutional Economics (ISNIE) (Menard & Shirley, 2014), which held its first meeting in St. Louis in 1997. His research included property rights, transaction costs, the institutional basis of markets, and economic organization in history, as well as economic development in developing countries.

The choice of an institutional approach seems to be the most correct, as it reflects the structural and social problems caused by climatic risks in the territory of the Arctic. It is necessary to pay attention to the fact that most of the costs caused by the risks are indirect, affecting the transaction costs.

In addition, the strengthening of the environmental component of this problem leads to paying more attention to the social aspects of the development of the hydrocarbon economy of the Arctic. The indigenous population of this region is under the closer attention of the global community. Its well-being largely depends on the effectiveness of economic decisions, including the development of hydrocarbon policies. Many companies are increasingly paying so-called reputational costs, whose risks are significantly increasing with the development of projects in the Arctic region, especially in the field of traditional energy.

To write this work, data on the energy resources of the Arctic region were used, taken from the book *Governance in the Arctic: Volume 2. Energy, Living Marine Resources and Shipping*. The Polar North is known to be home to large gas and oil reserves and its position holds significant trading and military advantages, yet the maritime boundaries of the region remain ill-defined. In the twenty-first century, the Arctic is undergoing profound change. As the sea ice melts as a result of accelerating climate change, global governance has become vital.

In this first of three volumes, the latest research and analysis from the Nansen Institute (the world's leading Arctic research body) are brought together. *Arctic Governance: Law and Politics* investigates the legal and political order of the Polar North, focusing on governance structures and the Law of the Sea. Skillfully delineating policy in the region and analyzing the consequences of treaty agreements, *Arctic Governance*'s uncovering of a rather orderly "Arctic race" will become an indispensable contribution to contemporary International Relations concerning the Polar North (Rottem & Soltvedt, 2019). One of its authors is Svein Vigeland Rottem, a senior fellow at the Nansen Institute in Norway, whose main research interests are in Arctic and global environmental governance, Arctic natural resources, Arctic maritime security, science and policy influence, and the Arctic Council. Further key information came from the book *Energy of the Arctic* by Morgunova and Tsunevsky (2012), edited by Vitaly V. Bushuev (General Director of the Energy Strategy Institute, Doctor of Technical Sciences). In this paper, the economic, geopolitical, and socio-solid aspects of the development of the Arctic and the minerality of the rye base of the Arctic Zone of the Russian Federation were analyzed, the prospects for mastering the resources of hydrocarbon raw materials of the Arctic shelf were shown, and the role of the Northern Sea Route with their development was evaluated, including the expected global and regional warming. Based on the results of research on climate change, the main risks in the development of hydrocarbon resources in the Arctic Zone of the Russian Federation are considered. According to materials of Russian and foreign researchers, the geopolitical aspects of the development of the Arctic and international relations in the region are considered. Much attention is paid to a qualitatively new approach to the perception of the Arctic: the emphasis on social, cultural, historical, and complex energy significance of the North is placed. (Morgunova & Tsunevsky, 2012).

In the field of research on climate change in the Arctic, one particular scientist should be distinguished: Edward K. Carmack. He is a Distinguished Research Fellow at the Department of Fisheries and Oceans of Sydney. In 2007, Carmack led a research project called "Canada's Three Oceans" (C3O) to study changes in the Arctic Ocean due to global warming, including ice retreat, species extinction, hypoxia, and acidification. He studied the distribution of fresh water anomalies in the northern oceans by analyzing the criteria that were used to calculate fresh water reserves, and then comparing the distribution of fresh water in the subarctic Pacific, subarctic Atlantic, and Arctic oceans (Carmack et al., 2008).

This article also used the works of Russian researchers of the Arctic—members of the RAS Scientific Council for the Study of the Arctic and Antarctic, among them: A.D. Gvishiani (Academician of the RAS, V.M. Kotlyakov, B.N. Porfirev, A.M. Fedonkin, and V.A. Bernikovsky). For example, the monograph *Socio-economic Development of the Russian Arctic in the Context of Global Climate Change*, edited by RAS Academician Boris N. Porfiriev, presents the results of a study of the impact of climate change on the socio-economic development of the Arctic, as well as on ensuring the national security of the Russian Federation (Porfiriev, 2017).

22.3 Results

Climatic risks in the field of hydrocarbon production in the Arctic Zone of the Russian Federation can be assessed by the following sources of risk:

1. atmosphere: very strong wind (including hurricane, squall, tornado), frost, abnormally high/low temperatures, abnormal atmospheric precipitation, and others;
2. hydrosphere: flooding, ice melting;
3. cryosphere and lithosphere: avalanches, landslides, snow flows, and others.

Each source of risk can vary in terms of intensity, prevalence, and duration of exposure. Thus, all risks can be divided into several levels:

- extremely dangerous risks;
- very dangerous risks;
- dangerous and moderately dangerous risks.

Methods for assessing losses due to various natural and climatic risks suggest acknowledging only direct economic damage from the impact of climatic risks, which can be assessed in monetary and other terms, including damage to the infrastructure of hydrocarbon production in the Arctic, problems with their production, or transportation, all arising from the impact of climatic factors.

Other risks may include possible damage to the health and life of the workers involved in the production of hydrocarbons in the Arctic. In this case, all kinds of risk assessment are subject to uncertainty, and, therefore, the use of risk criteria may be inappropriate. One possible approach is provided by the UK Health and Safety Executive Office (HSE) to determine the permissibility of the risk and its concept of minimum possible level (Alarp), based on a wide range of risks and not on specific criteria.

In the HSE software document *Reducing Risks, Protecting People: HSE's decision-making process* (2001) presents the structure of risks permissibility—an inverted triangle, increasing from a widely acceptable area through the permissible region to an unacceptable region (HSE, 2001). This wide structure is used in the HSE permitting manual—which is the decision-making guidance on the minimum possible reasonable level (Alarp) when monitoring the hazards of major accidents (COMAH)—and provides for the following wide risk ranges:

- The upper area in which Alarp has not been demonstrated and the risk is unacceptable;
- The average area where risk is permissible if Alarp is demonstrated by arguments based on relevant best practices, additional risk reduction methods, and extremely disproportionate costs for further risk reduction;
- The lower area where the risk is generally acceptable and does not require further decline, since appropriate advanced practice is applied.

Despite the fact that wide risk ranges seem compatible with the standard of work of the Law on Safety and Hygiene, interpretation does not include aspects of continuous improvement. This means that in the low-risk range, some risks can remain left unreduced, even if a further reduction in risk can be feasible in practice.

The risk matrix classifies the risk of individual large incidents based on the judgment of the assessment group on the order of the probability and consequences of the incident (HSE, 2001). Typical risk matrices for hazardous industrial objects have a size of 3×3 to 5×5. As a rule, the probability is on the Y axis and the consequences are on the X axis of the matrix. It is recommended that the frequency scale (or likelihood) can be one order of magnitude for each row or column.

The risk increases diagonally in the matrix, and the ranges of wide risk levels can be installed perpendicular to the direction of its increase. It can be seen that these bands generally belong to risk bands, and therefore can be used to designate areas where the risk is unacceptable or where the risk is permissible, subject to the adoption of all feasible measures and continuous improvement. Wide risk ranges can also be associated with the need to take urgent measures.

According to the features of the relief in the Arctic, one is particularly distinguished: the shelf with a sharpness of the mainland and the surrounding outskirts of the mainland. The relief of the Russian Arctic Zone is plain in its advantages: there are some mountainous reliefs especially on the islands; the central part is the Arctic Ocean, the region of deep-water and underwater ridges, of which the Lomonosov Ridge is most significant; most of the Arctic (about 13 million kilometers squared) occupies the Northern Arctic Ocean, and almost half of the area of the Northern Ocean is the shelf; the shelf from the northern shores of European Russia has an exceptionally large depth.

The severe climate causes the low temperature of the oceanic waters. In the field of distribution of drifting ice throughout the year, the temperature of the surface layer of water (thickness of 100–200 meters) is close to 2 . In the areas of purifying from ice in summer, water is heated by several degrees above zero, but a large amount of heat is brought to the Arctic of the ocean flows. The climate of almost the entire Atlantic area of the Arctic is influenced by the warm northlander flow. The impact of the warm water of the Pacific Ocean is much weaker than the smaller inflows.

The significant climate fluctuations were marked in the Arctic. Since 1920, the air temperature began to rise, compared to the end of the nineteenth century. The air temperature of some winter months increased by 5–7 degrees Celsius, as a result of which the Arctic ice has become less strong and the glacier area was reduced (Morgunova & Tsunevsky, 2012).

The Arctic is divided into 5 sectors, the foundation of which is the northern borders of Russia, the USA, Canada, Denmark, and Norway, and the sidelines of which are in the meridians. All lands and islands located within each sector are part of the territory of adjacent states. The system of sectors has developed on the basis of a long-term factual delimitation of the rights and interests of the respective states, recognizing them with priority in the study and mastering of various areas of the Arctic.

Climate change can negatively affect the ice structure, which becomes more vulnerable to storms and wind storms. All this can lead to the acceleration of the movement of the ice sheet (NSIDC, 2021). This poses a potential risk for the oil and gas industry, as increased ice movement can interrupt drilling (Dell & Pasteris, 2010). In addition, greater movement of icebergs can damage rigs and vessels. However, many oil companies are starting to produce more advanced equipment that allows oil rigs and ships to be less vulnerable to ice movement (Ryerson, 2009; Salygin & Lobov, 2021).

Icebergs can become a serious threat to infrastructure facilities. Other climatic events can include extreme winter temperatures and huge sea waves. To combat these natural phenomena, it is necessary to introduce new technologies that can allow earlier detection of icebergs (for example, ice radars), as well as more advanced vessels that can operate in more unpredictable ice conditions (Allen & Kassner, 2001).

Global warming increases the risk of natural events such as hurricanes, which are more common in warmer climates. Recent research suggests that the number of polar storms and hurricanes is likely to increase in the coming years (Kolstad & Bracegirdle, 2008), which will potentially worsen conditions for the oil industry. The increased frequency of storms can pose a serious risk to the industry as it can threaten drilling, production, and transportation.

All of the above can increase the average cost of hydrocarbon production in the Arctic. Operating costs and ten-year lead times require low enough break-even production to justify starting an Arctic oil project.

For example, there is no recoverable oil in the Chukchi Sea at a price of $30 per barrel (BOEM, 2016). Oil production from the Chukchi Sea at $60 per barrel will cost $72.9 billion and could generate net benefits totaling $6.2 billion (Dvorak, 2017).

In foreign practice, banks have recently made economic decisions to stop financing new Arctic oil and gas projects, recognizing the high level of environmental and economic risks. With lower oil prices and increased attention to the environment, Arctic oil is no longer among the "top ten oil projects to finance," according to David Livingston, senior analyst at Eurasia Group. In addition, Livingston calls these decisions a "Lagging Indicator" of the decline in the importance of Arctic oil (Geman & Harder, 2020). Figure 22.1 shows the dynamics of prices for Brent crude oil in 2010–2020.

However, there are more than financial reasons for making operations in the Arctic difficult (Osipov et al., 2021c). The growing attention to the ecological well-being of the Arctic, as well as the increase in "green" energy projects, are increasing public attention to hydrocarbon production in the region. Foreign banks, when making decisions, take into account both reputational risk and social and environmental considerations. Banks are striving to update their environmental policies and move to more sustainable practices (Brown, 2018).

Nevertheless, the use of new technologies can contribute to the development of hydrocarbon production in the Arctic (Piskulova & Pak,

Fig. 22.1 Dynamics of prices for Brent crude oil in 2010–2020, $/barrel (*Source* compiled by the author based on Brent Oil Futures Historical Data [2021]. http://investing.com/https://www.investing.com/commodities/brent-oil-historical-data; BP Statistical Review of World Energy 2020 [2020]. *BP*. www.bp.com/content/dam/bp/business-sites/en/global/corporate/pdfs/energy-economics/statistical-review/bp-stats-review-2020-fullreport.pdf)

2017). PJSC Gazprom Neft launched a project to develop the northern section of the Novoportovskoye field and dispatched the first shipment to ChemChina. It plans to establish a joint venture with Shell to explore and develop hydrocarbons in the Gydan Peninsula. PJSC Novatek has completed the shipment of the first ice-class LNG tanker Arc7 to Japan from the Yamal LNG project. In 2020, PJSC Gazprom Neft launched several new projects at the largest gas condensate fields in the Arctic, including Kharasavey, Bovanenkovo, and Urengoyskoye.

In the Arctic Zone of the Russian Federation, with the support of the state, six large investment projects will be implemented. This includes the creation of a mining and metallurgical plant in the Murmansk region, the development of the Syadasay coal deposit on the Taimyr, and the opening of the enterprise for the extraction and processing of lead–zinc ores on the New Earth archipelago (South Island). The list also features a specialized deep-water terminal for the transshipment of mineral fertilizers in the port of Murmansk, a plant for breeding salmon and trout (Murmansk region), and the project of technical re-equipment of the port of Vitino and the Belomorsk oil refurbs.

It is expected that the total investment in these projects will exceed ₽200 billion. Companies that implement them will be able to count on an irrevocable subsidy from the state in the amount of up to 20 percent of their investments. Funds can be spent on creating a new infrastructure for industrial facilities.

State support is available to companies whose initiatives correspond to a number of conditions: the cost is at least ₽300 million; there will be coverage of subsidies in the form of tax revenues for up to 10 years; workplaces will be created due to the launch of new industries or upgrade existing ones.

The decision to increase the investment attractiveness of the Arctic will have a positive effect on the socio-economic situation and the welfare of the inhabitants of the region (Government of the Russian Federation, 2021).

The Arctic is not only a resource-rich region with colossal oil and gas reserves, as well as other valuable minerals; the Arctic was largely—and remains—a place of innovative development for the economy of Russia and its fuel and energy complex.

The scale and direction of the development of the Arctic region can be assessed by the main targets submitted in the updated *Strategy for the Development of the Arctic Zone of the Russian Federation and Ensuring National Security for the Period up to 2035* (hereinafter referred to as the Strategy). Thus, in the Arctic Zone of the Russian Federation, there are plans to create conditions for the migration growth of the population and a sevenfold increase in the number of jobs. The volume of oil production in the Arctic should, by 2035, be 26 percent of all Russian oil, and gas 79 percent. The production volumes of liquefied natural gas (LNG) may increase more than tenfold. This growth, based on the target strategy indicators, should be accompanied by a sevenfold increase in freight on the Northern Sea Route (NSR).

Break-even estimates for projects vary greatly. Onshore projects adjacent to existing infrastructure have an exit point close to the Russian average of $20 per barrel, but for more complex projects in more remote areas, this figure can reach more than $100 per barrel, at least if the government helps (S&P Global Platts, 2020).

The Russian Arctic is a region where hydrocarbon production is constantly increasing, which leads to an increase in harmful emissions. Figure 22.2 shows the reserves of the main Arctic gas fields in Russia.

In the period from 2010 to 2019, oil production increased from 45.4 to 87.7 million tons, and the volume of APG produced increased from 9.1 to 27.4 billion cubic meters. The Arctic Contaminants Action Program (ACAP) has conducted a study on the flaring of associated petroleum gas

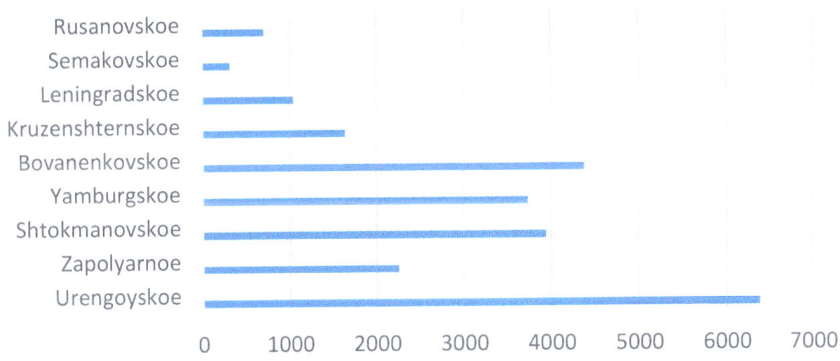

Fig. 22.2 Volumes of reserves of the main Arctic gas fields in Russia in 2016–2017, billion cubic meters (based on McGee [2020] (*Source* compiled by the author https://www.thearcticinstitute.org/mapping-russia-arctic-hydrocarbon-development-scheme/)

(APG) in the Russian Arctic. The report shows that significant economic and environmental benefits can be achieved if the best available technologies (BAT) and best environmental practices (BEP) are applied (Akishin et al., 2020).

In addition, increasing the efficiency of this sector of the Arctic economy is possible through the implementation of digitalization projects that extend to the energy sector, the development of "Smart Greed" (Osipov et al., 2021a), and the development of institutions for financing responsible production so as to maintain environmental security of the region (Osipov et al., 2021b). The use of BAT and NEP could reduce emissions of short-lived climate pollutants (SLCPs) from the current 25 million tons of carbon dioxide to 7 million tons equivalent by 2025 in the Russian Arctic region (McGee, 2020).

22.4 Conclusions

Climate risks are increasing in the Arctic Zone of the Russian Federation due to climate change and the emergence of new sources of risk (for example, hurricanes, which were previously characteristic of warmer latitudes). In addition, an increase in temperature in the Arctic, the rate of which is faster than the global average, leads to instability of the ice cover, which creates additional risks for infrastructure facilities for hydrocarbon production.

All this leads to increased risks and increased costs for the institutional base of hydrocarbon production in the Arctic. There is a need for the innovative development of technologies for the extraction of hydrocarbons, which increases their cost as well as operating costs. Also, many technologies are experimental, which leads to additional risks if they are ineffective. In addition to infrastructure, there is an increase in indirect costs associated with the potential reputational risk of companies; this is due to the fact that at the international level, the course toward renewable energy sources, alongside the social and environmental well-being of the regions, is being strengthened. The Arctic is one of the most vulnerable areas. Decarbonization of this zone is one of the key requirements of many organizations that are fighting to implement the international environmental agenda. In this situation, many banks are striving to work out a more advanced development strategy, which could potentially reduce their activity in supporting projects for the production of hydrocarbons in the Arctic.

However, despite all the above risks, many Russian companies continue to actively develop projects for the production of hydrocarbons in the Arctic Zone of the Russian Federation. In recent years, Gazprom Neft PJSC and Novatek PJSC have launched a number of new projects. The prospects of this direction, primarily, are due to the large area of the Arctic Zone of the Russian Federation: there are both fields with a fairly low production cost, and with quite costly elements in terms of developing the territory. In addition, the development of the best available technologies and the best environmental practices allows companies to keep pace with the development of green technologies in the hydrocarbon sector, which may be applied in other industries in the future.

The study of climatic risks in the development of the economy of the Arctic region is impossible without the development of a methodological basis for assessing the impact of climate vulnerability on the cost of capital and the access of companies operating in the Arctic to financial sources. This line of research is very promising and requires additional study.

References

Akishin, D., Timonin, I., Haugland, T., & Isakova, I. (2020). *Significant economic and environmental gains can be achieved by applying best available technology in the oil sector in the Arctic*. Arctic Council. https://arctic-council.org/news/best-available-technology-in-the-oil-sector-in-the-arctic/

Allen, S., & Kassner, C. (2001). *Global analysis of the Terra Nova Riser system*. Paper presented at the Offshore Technology Conference, paper number: OTC-13021-MS. https://doi.org/10.4043/13021-MS

BOEM. (2016). *Assessment of undiscovered oil and gas resources of the nation's outer continental shelf*. Bureau of Ocean Energy Management. https://www.boem.gov/sites/default/files/oil-and-gas-energyprogram/Resource-Evaluation/Resource-Assessment/BOEM-National-Assessment-Fact-Sheet.pdf

BP. (2020). *BP statistical review of world energy 2020*. https://www.bp.com/content/dam/bp/business-sites/en/global/corporate/pdfs/energy-economics/statistical-review/bp-stats-review-2020-fullreport.pdf

Brown, G. (2018). *Gwich'in nation joined by institutional investors in call to defend Arctic refuge from drilling*. Sierra Club. https://www.sierraclub.org/press-releases/2018/05/gwich-nation-joined-institutional-investors-call-defend-arctic-refuge

Carmack, E. C., et al. (2008). *Freshwater storage in the northern ocean and the special role of the Beaufort Gyre: Arctic-Subarctic ocean fluxes* (pp. 145–169). Springer.

Dell, J., & Pasteris, P. (2010). Adaptation in the oil and gas industry to projected impacts of climate change. In *Proceedings of the SPE international conference on health, safety and environment in oil and gas exploration and production* (pp. 12–14).

Dvorak, M. (2017). *An economic assessment of oil development in the Alaskan Arctic: Masters of marine affairs*. University of Washington.

Geman, B., & Harder, A. (2020). Why Big banks are breaking up with some fossil fuels. *Axios*. https://www.axios.com/banks-fossil-fuel-pledges-jpmorgan-021496ee-cd86-4cee-bbf8-fac3eb886c34.html

Government of the Russian Federation. (2004). *Order of the Ministry of Emergency Situations of Russia dated October 25, 2004, No. 484 "On approval of the standard safety data sheet for the territories of the constituent entities of the Russian Federation and municipalities" (Registered with the Ministry of Justice of Russia on November 23, 2004, No. 6144)*. Russian Emergency Situations Ministry. https://www.mchs.gov.ru/dokumenty/normativnye-pravovye-akty-mchs-rossii/763

Government of the Russian Federation. (2013). *Order of the Government of the Russian Federation of July 10, 2013 No. 1187-r "On Lists of Information on the Activities of State Bodies, Local Self-Government Bodies Placed on the Internet in the Form of Open Data"*. Judicial and regulatory acts of the Russian Federation. https://sudact.ru/law/rasporiazhenie-pravitelstva-rf-ot-10072013-n-1187-r/

Government of the Russian Federation. (2020). *Strategy for the development of the Arctic zone of the Russian Federation and ensure national security for the period up to 2035*. Official Internet Portal of Legal Information. http://publication.pravo.gov.ru/Document/View/0001202010260033

Government of the Russian Federation. (2021). *Approved list of investment projects with state support in the Arctic zone*. http://government.ru/news/41438/

Hassol, S., & Corell, R. (2006). Arctic climate impact assessment. *Avoiding Dangerous Climate Change, 21*, 205–213.

HSE. (2001). *Reducing risks, protecting people: HSE's decision-making process*. HSE Books.

HSE. (2021). *ALARP—As Low as reasonably practicable*. https://www.hse.gov.uk/comah/alarp.html

Investing.com. (2021). *Brent oil futures historical data*. https://www.investing.com/commodities/brent-oil-historical-data

Kolstad, E., & Bracegirdle, T. (2008). Marine cold-air outbreaks in the future: An assessment of IPCC AR4 model results for the northern hemisphere. *Climate Dynamics, 30*(7–8), 871–885.

McGee, R., (2020). *Mapping Russia's Arctic hydrocarbon development scheme*. The Arctic Institute. https://www.thearcticinstitute.org/mapping-russia-arctic-hydrocarbon-development-scheme/

Menard, C., & Shirley, M. (2014). The contribution of Douglass North to new institutional economics. In S. Galiani (Ed.), *Institutions, property rights, and economic growth: The legacy of Douglass North*. Cambridge University Press.

Ministry of Natural Resources of Russia. (2021). *Order of May 19, 2021, No. 16-r on the approval of a model passport of climatic safety of the territory of a constituent entity of the Russian Federation*. Judicial and regulatory acts of the Russian Federation. https://sudact.ru/law/rasporiazhenie-minprirody-rossii-ot-19052021-n-16-r/

Morgunova, M., & Tsunevsky, A. (2012). *Energy of the Arctic*. EC Energy (in Russian).

North, D. (1990). *Institutions, institutional change and economic performance*. Cambridge University Press.

NSIDC. (2021). *National snow and ice data center CIRES*. University of Colorado. http://nsidc.org/

Osipov, V., Bulgarov, M., Krivtsov, O., & Vagin, S. (2021a). Financial provision of energy security of regional economy in transition to industry 4.0. In E. Zavyalova & E. Popkova (Eds.), *Industry 4.0*. Palgrave Macmillan. https://doi.org/10.1007/978-3-030-75405-1_34

Osipov, V., Deberdeeva, N., Nadyseva, D., & Berzon, N. (2021b). Algorithm for institutionalization and financing of responsible production and consumption in Russian regions in Industry 4.0 in order to maintain environmental safety. In E. Zavyalova & E. Popkova (Eds.), *Industry 4.0*. Palgrave Macmillan. https://doi.org/10.1007/978-3-030-75405-1_37

Osipov, V. S., Bulgarov, M. A., Krivtsov, O. F., & Vagin, S. G. (2021c). Financial provision of energy security of regional economy in transition to industry 4.0. In E. B. Zavyalova & E. G. Popkova (Eds.), *Industry 4.0*. Palgrave Macmillan. https://doi.org/10.1007/978-3-030-75405-1_34

Piskulova, N. A., & Pak, E. V. (2017). Prospects of Russian engagement in green economy. *International Trends, 15*(4), 40–58. https://doi.org/10.17994/IT.2017.15.4.51.3 (inRussian)

Porfiriev, B. (Ed.). (2017). *Socio-economic development of the Russian Arctic in the context of global climate change*. Scientific Consultant (in Russian).

Roshydromet. (1997). *Decree of the Government of the Russian Federation of November 15, 1997, No. 1425 "On information services in the field of hydrometeorology and monitoring of environmental pollution"*. https://www.meteorf.gov.ru/services/gosuslugi/28/doclist28/142/11983/

Rottem, S., & Soltvedt, I. (2019). *Arctic governance: Volume 2 energy, living marine resources and shipping*. Series Arctic Governance. I.B. Tauris

Ryerson, C. C. (2009). *Assessment of superstructure ice protection as applied to offshore oil operations safety*. US Army Engineering Research and Development Center. https://digitalcommons.unl.edu/usarmyresearch/49/

Salygin, V. I., & Lobov, D. S. (2021). Defining major oil and gas companies' development strategies in the era of energy transition. *MGIMO Review of International Relations, 14*(5), 149–166. https://doi.org/10.24833/2071-8160-2021-5-80-149-166

S&P Global Platts. (2020). *Insight from Moscow: Russian Arctic oil and gas development continues despite climate concerns*. https://www.spglobal.com/platts/en/market-insights/blogs/oil/090320-insight-from-moscow-russian-arctic-oil-and-gas-development-continues-despite-climate-concerns

CHAPTER 23

Territorialisation, Urbanisation, and Economic Development in the Russian Arctic: Energy Issues

Sébastien Gadal, Moisei Zakharov, and Jūratė Kamičaitytė

23.1 Introduction

The development of the Arctic, and in particular of the Arctic Zone of the Russian Federation, is intrinsically correlated to energy, its implementation, and efficiency. The exploitation of natural resources, the

S. Gadal (✉) · M. Zakharov
Aix-Marseille University, Aix-en-Provence, France
e-mail: sebastien.gadal@univ-amu.fr; szh.gadal@s-vfu.ru

M. Zakharov
e-mail: moisei.zakharov@etu.univ-amu.fr

M. Zakharov
North-Eastern Federal University, Yakutsk, Russia

V. I. Salygin (ed.), *Energy of the Russian Arctic*,
https://doi.org/10.1007/978-981-19-2817-8_23

implementation of mining and port cities, and especially those territories undergoing metropolisation (such as Yakutsk) or urban areas (such as Murmansk) are dependent on energy capacity. This is based on a multiplicity of energy production methods: fuel, gas, and coal-fired power plants, nuclear power, wind power farms, or solar power plants (as in Batagaï in the Republic of Sakha Yakutia). These types of energy production are complementary and depend on the forms of territorialisation: micro coal-fired power plants in villages, gas-fired power plants, or even nuclear power plants for Arctic cities, etc.

Since the 2010s, the complementarity between renewable energy and coal or gas-fired power plants appears as a new energy pattern for urban centres in the making or the process of deployment in some territories. The mono-energy model of the Soviet period is gradually being replaced by a mixed energy production model combining renewable energy and fuel, coal, gas, or nuclear power plants (Overland, 2010). This energy mix is due to the economic needs and issues related to the exploitation of natural resources, as well as a strategy with the need to territorialise the Siberian Arctic regions and the search for autonomous energy production systems (Péné-Annette et al., 2017). Renewable energy is an option under testing and implementation. However, there are many disparities between the European and Asian parts of the Russian Arctic: these differences are due to geographic inequalities in terms of territorial structures, to urbanisation, and to demographics, which occur due to the very exploitation of the Arctic Zone of the Russian Federation.

Energy plants are the fundamental infrastructure for human establishment (bases, ports, mines, urban centres) in the Arctic Zone. This raises not only the question of their integration into the territory in terms of location, energy efficiency for Arctic urban centres such as Tiksi, or mobile and modular city projects, but also the spatial modelling of their inclusion in the territory: optimal location, integration into the energy grid, responses to energy needs, environmental and landscape visual impacts, etc. The choice of spatial configuration and their territorial integration is often forgotten, yet is a fundamental aspect in the Arctic environment. It is crucial for the implementation of mobile urban blocks and

J. Kamičaitytė
Kaunas University of Technology, Kaunas, Lithuania
e-mail: jurate.kamicaityte@ktu.lt

the rehabilitation of Arctic urban centres, which will relate economic and territorial development policies with the future Northern Sea Route and the exploitation of natural resources.

23.2 The Exploitation of Natural Resources: The Main Driver of the Arctic Territorial Development

The question of the development of the Russian Arctic, with the processes of territorialisation and urbanisation as geographical corollaries of the dynamics of anthropisation, is closely intertwined to and dependent on energy issues: availability, power, technological choices linked to the energy transition, and the needs of urban and industrial centres and natural resource extraction areas. This was, and is, the driving force behind the processes of territorialisation and urbanisation in the Arctic and sub-Arctic. Without energy and the exploitation of fossil resources, there is no territorial and economic development and urbanisation dynamics (Anisimov and Kokorev, 2016). It is a key element in the development and territorialisation of Arctic areas.

Because of the economic and political stakes for the Russian Federation, the energy issue is at the core of development strategies (Fedorova et al., 2021). The exploitation of natural resources—minerals, oil, gas, etc.—of the Russian Arctic regions is fundamental for the federal state, in an economy that moreover depends (directly and indirectly) on the public and para-public sectors for 85 per cent. The Yamal Peninsula provides 9.3 per cent of the consolidated revenues dedicated to the federal budget in 2021. The significant role of the Arctic territories and the exploitation of natural resources is increasingly marked and will become more pronounced, particularly with the exploitation of new—admittedly still confidential—deposits in the Far East, such as in Yakutia (Fig. 23.1).

The intensification of the exploitation of natural resources on the Russian Arctic continental shelf—minerals, fossil fuels such as coal, gas, and oil, and fish in the Arctic Sea—is certainly driven by extraction opportunities related to the impact of global warming, as is often emphasised, but it is essentially linked to the need to compensate for the loss of yields from European and Siberian oil and gas fields, while at the same time increasing revenues for the federal-state budget (Statistica, 2020). The latter's dependence on the exploitation of natural resources is growing

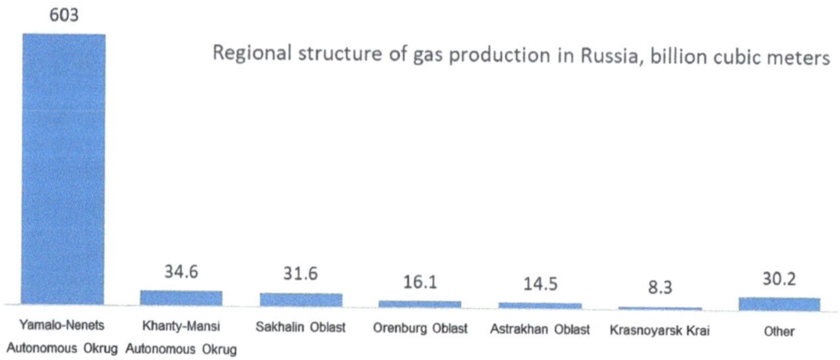

Fig. 23.1 Regional structure of the gas production in Russia in 2019, billion cubic metres (*Source* adapted by the authors from Analytical Center under the government of the Russian Federation [2020]. *Fuel and Energy Complex of Russia*. URL: https://ac.gov.ru/uploads/2Publications/TEK_annual/TEK.2019.pdf [in Russian])

to represent 60 per cent of the Russian GDP (Knight et al., 2021). This is even more important as the Russian economy is directly or indirectly linked to the public sector. The development of new deposits in the Arctic and circumpolar regions is essential for the economic survival and financing of federal-state policy (Fig. 23.2).

The increase in the exploitation of natural resources in the Russian Arctic space is related to the growing international demand for raw materials and energy in the traditional markets of the European Union—especially Germany—and the emerging markets of China and Turkey. For example, the Yamal Peninsula contributed 9.3 per cent of the Russian state-federal budget in 2021. 20 per cent of the gas consumed in France came from the Yamal Peninsula in 2020. Germany's energy transition is based on the exploitation of Russian gas, most of which comes from the Arctic and the Yamal Peninsula.

If the exploitation of mineral and fossil resources is highlighted, this must not be done at the expense of the equally important exploitation of fisheries resources and their role in food security. More than 25 per cent of the fish consumed in North America comes from the waters of the icy Arctic Ocean.

The importance of the processes of the re-territorialisation of the Arctic and sub-Arctic spaces is not only linked to the exploitation of

Fig. 23.2 Contribution of the region in the Russian GDP in 2019 (*Source* Adapted by the authors from Analytical Center under the government of the Russian Federation [2020]. *Fuel and Energy Complex of Russia*. https://ac.gov.ru/uploads/2Publications/TEK_annual/TEK.2019.pdf [in Russian])

natural resources, as was the case during the various periods of the Soviet regime (Potravny, 2021). If they were one of the pillars of the policy of territorial control subject to the exploitation of natural resources during the USSR, the 1990s were characterised by massive anthropic and urban abandonment, particularly with the closure of mines, transport infrastructures, etc. Tiksi, a major port centre, designed to accommodate 50,000 inhabitants, now in 2021 has 4745 inhabitants, where it had 11,649 inhabitants in 1989. However, some territories, such as Yakutsk, have been experiencing a continuous metropolisation process since the 1990s, with more than 330,615 inhabitants in 2021, compared to 193,000 people in 1991. The processes of territorialisation of the Arctic space are not new.

23.3 Urbanisation, Hydrocarbon Exploitation Areas, and the Emergence of Territorial Structures

Spatial analysis of changes in territorial structures—based on the evolution of urban structures and areas of exploitation of fossil and mineral resources between 1992 (blue), 2005 (green), and 2013 (red) identified from DMSP (Defence Meteorological Satellite Program) (Sun et al., 2020) satellite images taken in the visible near-infrared (VNIR)—highlights an intensification of the exploitation of the Russian European Arctic space and western Siberia (Yamal Peninsula) that has been ongoing since the break-up of the USSR (Fig. 23.3).

The European part of the Russian Arctic Zone, which is highly structured (white) with its communications networks linking urban centres and natural resource exploitation built during the Soviet period, is characterised by transformation (pink) and peri-urbanisation dynamics between 2005 and 2013, notably with the exploitation of new deposits (green and red spots). The Yamal Peninsula is experiencing unprecedented economic development with the exploitation of gas reserves (green and red spots). This massive territorial and economic dynamic is accompanied by the

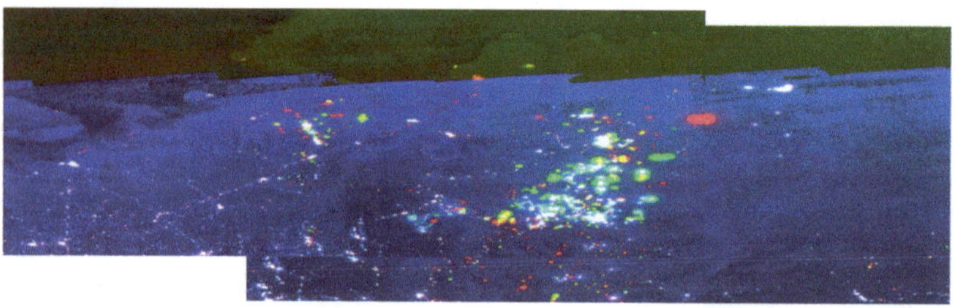

Fig. 23.3 Urban, oil and gas extraction area change in the European and West Siberia Arctic zones (1992–2013) (*Source* Gadal, S., November 2017–September 2021, NOAA-USAF DMSP F10, F15, and F18 satellites, VNIR OPL-PS sensor, mean annual spectral measurement, Urban and exploitation areas land change modelling by Ensemble' algorithm with linear model-data fusion (Li and Zhou, 2017; Ma et al., 2020). Blue: 1992 [no change]; Green: 2005 [emergence of urban/natural resources spots]; Red: 2013 [emergence of urban/natural resources spots]; Yellow: between 2005 and 2013 [change recognition]; White: between 1992 and 2013 [no change)]

emergence of new urban centres or the extension of those built during the USSR, such as Nadym located in the southeast of the peninsula. The region further to the east between the Ob and Yenisey rivers is characterised by an intense hydrocarbon exploitation activity associated with the accompanying urbanisation dynamics (Fig. 23.4).

The Arctic and sub-Arctic regions of the Siberian Far East are characterised by urbanisation and metropolisation dynamics unique to Yakutia, which does not depend on the dynamics of urban pioneer fronts generated by the exploitation of natural resources. The Yakutsk region, a sub-Arctic territory characterised by an Arctic climate, has experienced sustained urban growth since the 1990s, with the emergence of a vast suburban territory that stretches along the Lena River for more than 100 km from the north to the south of the metropolis (between Pokrovsk in the south and Namtsy in the north), and in the east in central Yakutia with a demographic growth affecting villages and urban centres (spots in red). These urbanisation dynamics based on the demographic growth of villages also concerns the regions of Tommot, Aldan, and Nerioungri in

Fig. 23.4 Urban, mineral, oil and gas extraction area change in the Eastern Siberia Arctic zones (1992–2013) (*Source* Gadal, S., November 2017–September 2021, NOAA-USAF DMSP F10, F15, and F18 satellites, VNIR OPL-PS sensor, mean annual spectral measurement, Urban and exploitation areas land change modelling by Ensemble' algorithm with linear model-data fusion (Jeswani et al., 2019). Blue: 1992 [no change]; Green: 2005 [emergence of urban/natural resources spots]; Red: 2013 [emergence of urban/natural resources spots]; Yellow: between 2005 and 2013 [change recognition]; White: between 1992 and 2013 [no change])

the south of the Republic of Sakha Yakutia. The urban centres (in blue) located in the Oyimyakonsky district have disappeared or experienced a sharp decrease in population with the cessation of mining activities. Only Oyimyakon (a cold city linked to tourism) has not been affected by the urban decline (Fig. 23.5).

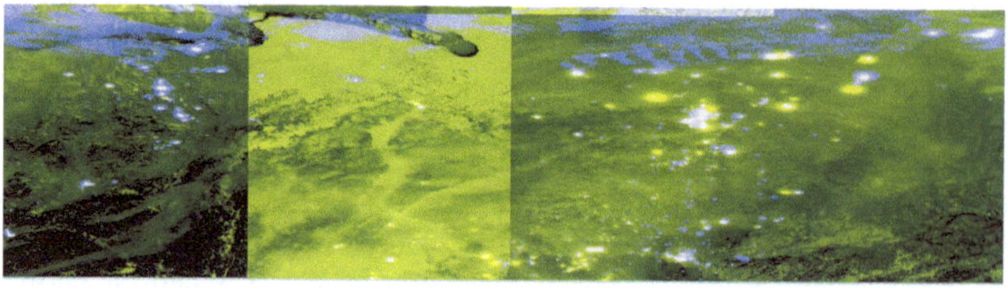

Fig. 23.5 Urban, oil and gas extraction area change in the European and West Siberia Arctic zones (2015–2021) (*Source* Gadal, S., Zakharov. M., 23 December 2021, NOAA-Suomi National Polar-orbiting Partnership, VIIRS sensor, Urban and exploitation areas land change modelling by Ensemble algorithm with histogram equalisation model-data fusion. Blue: 2015 [decline]; Red: 2013 [emergence of urban/natural resources spots]; Yellow: between 2015 and 2021 [change recognition]; White: between 2015 and 2021 [no change])

Fig. 23.6 Novy Urengoy and the areas of gas exploitation captured at night (*Source* photos taken by Gadal, S., 02 December 2021, Novy Urengoy, Yamal, Russia)

The use of VIIRS (Visible Infrared Imaging Radiometer Suite) data from the NOAA-Suomi NPP (Suomi, 2021) sensor highlights the dynamics of territorial transformations significantly different from the period 1992–2013. In the Yamal Peninsula, the commissioning of new gas fields can be observed (yellow spots), several gas extraction sites have not changed over the last 6 years, and several urban centres have declined (blue spots). The European part of the Russian Arctic is characterised by stagnation, even a decline in the intensity of hydrocarbon production between 2015 and 2021 (Figs. 23.6 and 23.7).

The territorial dynamics observed during the last 6 years in central Yakutia are significantly different from the period 1992–2013. While the urban region of Yakutsk is experiencing a significant process of growth and intensification of urbanisation, as well as certain urban centres in central Yakutia, there are processes of decline in certain villages and urban centres (blue spots).

Fig. 23.7 Urban, mineral, oil and gas extraction area change in the Eastern Siberia Arctic zones (2015–2021) (*Source* Gadal, S., Zakharov, M., 23 December 2021, NOAA-Suomi National Polar-orbiting Partnership, VIIRS sensor, Urban and exploitation areas land change modelling by Ensemble' algorithm with histogram equalisation model-data fusion (Kimijima et al., 2021). Blue: 2015 [decline]; Red: 2013 [emergence of urban/natural resources spots]; Yellow: between 2015 and 2021 [change recognition]; White: between 2015 and 2021 [no change])

23.4 Energy Security and Economic Development

The question of energy is integral to the economic development of the Russian Arctic territories, as well as for regional and global sustainability (Ilinova et al., 2020). Energy security is a major issue in the Russian Arctic, due to the strong regional isolation factor, which has been the fundamental roadblock to stimulating regional growth (Ghani et al., 2019). Most of the Russian Arctic is either connected to small independent power grids or is served by decentralised generation systems using mostly petroleum-derived products.

The reliance of many of these communities on external deliveries for their energy supply limits the number of hours that the population has access to electricity: in some cases down to only several hours a day. Wind energy—and renewables in general—can help provide the isolated settlements with uninterrupted power (increasing electricity/heat consumption up to 24 hours a day) and significantly reduce the cost of delivering hydrocarbon fuel (by roughly 30 per cent). At the same time, the low carbon footprint of these plants will help significantly reduce the human impact on the environment and improve the living conditions for the local population, both in terms of quality of life and health in extreme conditions (Fig. 23.8).

Access to reliable energy sources is imperative for revitalising Arctic settlements (including three new mobile military modular cities) and

Fig. 23.8 Carbon power plant, Verkhoyansk, Republic of Sakha Yakutia (*Source* photo taken by Gadal, S., Zakharov, M., 17 July 2017, Verkhoyansk, Republic of Sakha Yakutia, Russia)

Fig. 23.9 Solar power plant of Batagaï, Republic of Sakha Yakutia (*Source* photo taken by Gadal, S., Zakharov, M., 16 July 2017, Batagaï, Republic of Sakha Yakutia, Russia)

supporting regional economic activity (gas, oil, mines, fisheries, tourism, etc.). Furthermore, considering the vital role that the development of the Northern Sea Route will play not only for Russia but for the entire global community, the economic development and the energy security of these regions become essential factors in turning these Arctic waters into the new major trade artery (Gadal and Ignatyeva, 2020).

Alternative green energy—such as wind and solar—has already been implemented in several regions of the Russian Arctic—for example, Batagaï in the Republic of Sakha Yakutia—to test the operating capacity of these systems, as well as to see if these solutions can help alleviate some of the carbon and fuel dependence of the regions. The identification of the benefits and challenges of such installations in the region and can help build a roadmap for more successful and effective integration of such systems throughout the region (Fig. 23.9).

23.5 The Paradigm of Autonomous Cities

The political, technological, and societal paradigm of the energy transition driven by "green" energies has a triple interest: economic, in terms of reducing the cost of infrastructure aimed at bringing electricity, gas, coal, and fuel oil to Arctic towns and urban centres; environmental, in terms of the impact on the health of the population (coal and fuel oil power stations); and through the implementation of autonomous production units (Suter et al., 2019). The establishment of autonomous energy units based on green energy, and the implementation of nuclear micro-power stations, make it possible to meet the growing energy needs related to the renovation of port infrastructures (and other such infrastructures) in the Arctic, and to meet the energy deficits of certain urban centres, as well as to support the development of mining, gas, and oil activities, particularly in territories not covered by any energy grid. The economies of scale in terms of the costs of implementing new areas of natural resource exploitation in the Arctic and sub-Arctic regions are theoretically particularly interesting. The "green" energies promoted by the energy transition are one of the vectors of this. The autonomous structure of the "green" energy grids makes it possible to open new areas for the production and extraction of natural resources.

While several Arctic villages and small towns have developed autonomous energy grids since the 1960s based mainly on the use of coal and/or oil-fired power stations, others are connected to the gas network. Coal and oil-fired power plants tend to be replaced by dual-energy plants if they cannot be connected to the gas grid. Their addition to the gas network is linked to the presence of gas pipelines. The construction of the gas pipeline between the Yamal Peninsula and the People's Republic of China has made it possible to connect several villages to the pipeline (Meynkhard, 2020). Batagaï's energy grid combines a coal-fired power plant with a solar panel farm providing more than 30 per cent of the urban centre's electricity needs; it is now the largest solar panel farm in the Arctic.

If the implementation of new green and autonomous energy grids makes it possible to substitute totally or partially oil and coal-powered power plants and to support the energy needs for the exploitation of natural resources, then the development of infrastructures and urban centres (for which 11 micro nuclear power plants on barges [or ships] are one of the flagship elements) would become possible. The question of their implementation is a technical issue, which sometimes becomes critical due to the extreme conditions (temperatures, wind, etc.) and land movements linked to the permafrost dynamics and its thawing in certain places. This is an upcoming problem faced by energy power plants such as the nuclear plant of Bilibino localised at the Chukotka Autonomous Okrug in Eastern Siberia. Accelerated melting of permafrost is an emerging problem for both power plants and mining, gas, and oil extraction areas, as well as buildings and infrastructure (Streletskiy et al.,

2015). Although the consequences regarding the destruction of buildings, infrastructure, etc. are known, the magnitude of the consequences and impacts are not well known yet.

23.6 The Example of the Wind Farm Experiment in Tiksi

Starting from 2017, the construction of new modern wind farms is ongoing in the Russian Federation (in the Ulyanovsk Region, the Republic of Adygea, the Republic of Sakha, in the Stavropol Territory, in the Rostov and Murmansk Regions). In November 2018, 3 Japanese wind turbines with a capacity of 0.3 megawatts each were put into operation in the Arctic city of Tiksi, an important transport hub for the Northern Sea Route. The successful operation of 3 wind turbines together with a diesel generator in the winter season of 2019 showed the efficiency of electricity generation in difficult weather conditions and the feasibility of further development of this area. In 2020, in the Bulunsky district, the wind farm of alternative energy sources will be replenished with three new Japanese wind generators.

The wind turbines provide great savings to Sakhaenergo. The founders of the wind farm—the companies Yakutskenergo and RusHydro—have become pioneers in attracting investments to the remote Arctic region, particularly the village of Tiksi, in which wind loads allow the power plants to operate.

The organisation of the urban Tiksi settlement is spatially dispersed and divided into three micro districts: Tiksi-1, Tiksi-2, and Tiksi-3. In total, about 4600 people live in the village. Tiksi is one of the strategically important transport hubs of the Russian North: there is a seaport, the main hub of the infrastructure of the Northern Sea Route, and an airport of federal significance. The transport and economic significance of the territory makes it necessary to develop innovative approaches to rational nature management and the issue of energy supply in the Arctic.

Despite the equipment's high stability to climate change, to ensure the safe operation of wind turbines and to increase the efficiency of their work, a crucial factor is the availability of a monitoring and forecasting system for meteorological conditions in the Tiksi wind farm (Strijhak et al., 2021). In addition, the issues of reliable operation of wind turbines and assessing the impact of a wind farm operating in difficult climatic conditions on the local microclimate have not been studied. There are currently no scientific publications on this wind farm or this topic in general.

While the diverse landscape of the Russian Arctic provides many opportunities for the integration of various renewable energy systems, wind energy has emerged as one of the primary candidates for implementation in the Arctic (Tabachkova et al., 2020). The constant and intense winds along the northern coast of Russia, on average sustaining 5–7 metres per second, provide good conditions for wind power generation.

The first regional implementation of wind turbines in Tiksi (designed by the Japanese company NEDO and RusHydro under the supervision of Sakha Energy) and the proposed future extension of the park (to 16 turbines) could be analysed as the example of green renewable energy utilisation in the extreme Arctic climate. Such an analysis would allow a better understanding of the potential for the renewable energy system integration in the Russian Arctic, and the role it can play in helping to improve energy access and security while reducing the environmental impact of energy and economic related activity (Ivanov et al., 2019). At the same time, the study should also consider the existing barriers for the widespread deployment of such systems, including the technical challenges, such as the need for adaptation of materials to the harsh local climate, the lack of sufficient infrastructure, and the absence of specialised human capital in the prospective regions.

The physical modelling of the energy efficiency of the wind power farms, their microclimatic and environmental impacts, and the benefits and challenges of the implantation of this type of renewable energy system would allow a better understanding of the possibilities and limitations of such systems implementation. The model for energy efficiency of wind power systems should consider not only effective electricity generation but also the local environmental impacts. It should integrate different geographic data from the metrological station, microclimatic exposure and evolutions, high-resolution (centimetres) terrain relief models generated from drones, digital elevation model (DEM) (2 × 2 metres covering large areas), high-resolution mapping of the land covers and land use, landscape permafrost, and their change dynamics at different seasons modelled by high and medium spatial satellite remote sensing Sentinel 2A (Fig. 23.10).

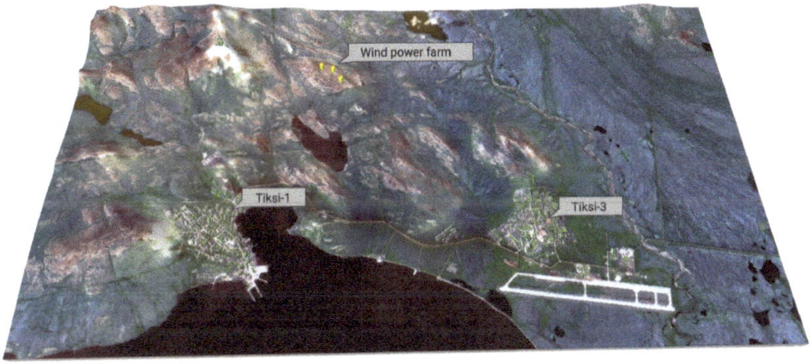

Fig. 23.10 3D map of the Tiksi area with the location of the 3 wind turbines (*Sources* compiled by Zakharov, M., 23 December 2021, Arctic Digital Elevation Model, Colour composite from Sentinel 2A satellite MSI data)

23.7 Conclusions

The question of energy is integral to the economic development of the Russian Arctic territories, as well as for regional and global sustainability. With its abundant supply of hydrocarbons, Russia has never placed a strong emphasis on renewable energy development; however, some recent efforts have been made to keep on top of global trends and to start integrating the renewable energy system into national energy strategies and to develop the necessary mechanisms to stimulate this energy transition. The rapidly growing strategic importance of the Arctic territories for the Russian government provides the perfect opportunity to study how the integration of these renewable energy systems can help address several economic, social, and environmental challenges that exist in the region.

Energy security is a major issue in the Russian Arctic, due to the strong regional isolation factor, which has been the fundamental roadblock to stimulating regional growth. Most of the Russian Arctic is either connected to small independent power grids or is served by decentralised generation systems using mostly petroleum-derived products. The reliance of many of these communities on external deliveries for their energy supply limits the number of hours that the population has access to electricity, in some cases down to only several hours a day.

Access to reliable energy sources is imperative for revitalising Arctic settlements (including three new mobile military cities planned by the

Russian Army) and supporting regional economic activity (gas, oil, mines, fisheries, tourism, etc.). Furthermore, considering the vital role that the development of the Northern Sea Route will play—not only for Russia but also for the entire global community—economic development and energy security of these regions become essential factors in turning these Arctic waters into the new major trade artery.

Wind energy—and renewables in general—can help provide the isolated settlements with uninterrupted power (increasing electricity/heat consumption up to 24 hours a day) and significantly reduce the cost of delivering hydrocarbon fuel (by roughly 30 per cent). At the same time, the low carbon footprint of these plants will help significantly reduce the human impact on the environment and improve the living conditions for the local population, both in terms of quality of life and health.

The implementation of green energy has an internal contradiction, even a controversial dimension: on the one hand, the green energy grids are the vectors of the acceleration of the processes of exploitation of natural resources in the Russian Arctic and of territorialisation; on the other hand, they provide a key element contributing to human security: the comfort of life with access to gas and electricity, as well as the reduction of the health and environmental impacts of oil and coal power plants. Perhaps this could be considered as the "ecology policy as a support of the economic development as well as the protection of Arctic natural resources and territorialisation with the implementation of the green energy power plans". The development of the green energy policy contributes one form of support to the Arctic territorial development and the exploitation of the natural resources, as well as providing for human security (health, living environment). The green energy power plans deployment is contributing to the global warming policy agenda; this is possible because of the autonomous energy grids of the green energies for the human settlements, mines, etc. The generalisation of the autonomous energy grids accelerates the emergence of new urban and rural settlement models characterised by self-development and self-urban dynamics, especially in the Siberian Arctic Zone. The distances, the extreme climate and environment, and the absence of transportations infrastructures (excepting the airports) connecting the "arctic urban islands" (including mobile settlement bases) are one key-face of the Arctic geography. Autonomous energy grids provided by the energy transition reinforce this territorial organisation.

References

Analytical Center under the government of the Russian Federation. (2020). *Fuel and energy complex of Russia*. https://ac.gov.ru/uploads/2Publications/TEK_annual/TEK.2019.pdf (in Russian).

Anisimov, O. A., & Kokorev, V. (2016). Cities of the Russian North in the context of climate change. In R. W. Orttung (Ed.), *Sustaining Russia's Arctic cities: Resource politics, migration and climate change* (pp. 141–174). Berghahn Books.

Fedorova, V., Kadzhaeva, E., & Vovkodav, K. (2021). Transformation of the energy sector in the Arctic in the context of sustainable development of the region. *IOP Conference Series Materials Science and Engineering, 1201*(1), 012069.

Gadal, S., & Ignatyeva, V. (2020). Alternative energy sources as part of sustainable development in the Arctic regions. In *Public administration and development in Russia: National goals. Russia's Arctic regions. The role of national projects in shaping sustainable development*, 1 (pp. 732–742). Scientific Library Publishing House (in Russian).

Ghani, R., Kangash, A., Shakeel Virk, M., Maryandyshev, P., & Mustafa, M. (2019). Wind energy at remote islands in the Arctic region: A case study of Solovetsky Islands. *Journal of Renewable and Sustainable Energy, 11*, 053304.

Ilinova, A., Solovyova, V., & Yudin, S. (2020). Scenario-based forecasting of Russian Arctic energy shelf-development. *Energy Reports, 6*, 1349–1355.

Ivanov, A. V., Strijhak, S. V., & Zakharov, M. (2019). Modeling weather conditions in the port area and coastal zone of Tiksi Bay. In *Proceedings of the Institute for System Programming of the RAS (Proceedings of ISP RAS), 31*(6), 163–176 (in Russian).

Jeswani, R., Kulshrestha, A., Gupta, P. K., & Srivastav, S. K. (2019). Evaluation of the consistency of DMSP-OLS and SNPP-VIIRS night-time light datasets. *Journal of Geomatics, 13*(1), 98–105.

Kimijima, S., Sakakibara, M., & Nagai, M. (2021). Detection of artisanal and small-scale gold mining activities and their transformation using earth observation, nighttime light, and precipitation data. *International Journal of Environmental Research and Public Health, 18*(20), 10954.

Knight, D., Emelyanova, O., Rostovtseva, I., Freije-Rodriguez, S., Kolchin, V., Nagle, P., Wheeler, C. M., Ruch, F. U., Levitanskaya, K., Wuester L. J., Aviomoh, H., Mbowe, A., & Prokhorova, A. (2021). *Russian economic report 46*. World Bank. https://documents1.worldbank.org/curated/en/099050011302118976/pdf/P17756206d40310aa0a5e109d6fa60bc55a.pdf

Li, X., & Zhou, Y. (2017). Urban mapping using DMSP/OLS stable night-time light: A review. *International Journal of Remote Sensing, 38*(21), 6030–6046.

Ma, J., Guo, J., Ahmad, S., Li, Z., & Hong, J. (2020). Constructing a new inter-calibration method for DMSP-OLS and NPP-VIIRS nighttime light. *Remote Sensing, 12*(6), 937.

Meynkhard, A. (2020). Priorities of Russian energy policy in Russian-Chinese relations. *International Journal of Energy Economics and Policy, 10*(1), 65–71.

Overland, I. (2010). Russia's Arctic energy policy. *International Journal, 65*(4), 865–878.

Péné-Annette, A., Gadal, S., & Kamicaityte, J. (2017). Pioneering cities of mining: Comparison of the Eastern Venezuela and Eastern Siberia. In *Digital cities and spatial justice* (pp. 135–145).

Potravny, I. (2021). Cities and towns of the Russian Arctic: Approaches to sustainable development. *Systems Analysis and Modelling of Economic and Environmental Systems, 338*(2), 237–242 (in Russian).

Statistica. (2020). *The Russian oil industry*. https://www.statista.com/study/65069/oil-industry-in-russia/

Streletskiy, D., Shiklomanov, N., Kokorev, V., & Anisimov, O. (2015). Arctic cities, permafrost and changing climatic conditions. In *68th Canadian geotechnical conference—GEOQuébec 2015, Quebec, Canada*.

Strijhak, S., Gergel, V., Ivanov, A., & Gadal, S. (2021). On the problem of choosing the optimal parameters for the wind farm in the Arctic town of Tiksi. In D. Balandin, K. Barkalov, V. Gergel, & I. Meyerov (Eds.), *Mathematical modeling and supercomputer technologies*. Springer.

Sun, C., Liu, Y., Zhao, S., & Jin, S. (2020). Estimating offshore oil production Using DMSP-OLS annual composites. *ISPRS Journal of Photogrammetry and Remote Sensing, 165*, 152–171.

Suomi, N. P. P. (2011). *Building a bridge to a new era of Earth observations*. NASA. https://www.nasa.gov/pdf/596329main_NPP_Brochure_ForWeb.pdf

Suomi NPP VIIRS Sensor. (2021). *VIIRS-NRT*. NASA. https://earthdata.nasa.gov/earth-observation-data/near-real-time/download-nrt-data/viirs-nrt

Suter, L., Streletskiy, D., & Shiklomanov, N. (2019). Assessment of the cost of climate change impacts on critical infrastructure in the circumpolar Arctic. *Polar Geography, 42*(12), 267–286.

Tabachkova, X., Prosekov, S., & Sokolinskaya, N. (2020). Energy system structure in the Russian Arctic: Coal production forecast. *International Journal of Energy Economics and Policy, 10*(3), 476–481.

CHAPTER 24

Nuclear and Hydrogen Prospects for the Russian Arctic

Angelina A. Kolomeytseva, Matthias P. Finger, and Andrey K. Krivorotov

24.1 Introduction

For many people, the notion of "Russian Arctic energy" is almost entirely associated with developing the enormous oil and gas potential of Northern Siberia and of the continental shelf. These megaprojects play

A. A. Kolomeytseva (✉)
Department of International Economic Relations and Foreign Economic Affairs, MGIMO University, Moscow, Russia
e-mail: angelinakolomeytseva@yandex.ru

M. P. Finger
Istanbul Technical University, Istanbul, Turkey
e-mail: finger@itu.edu.tr

Northern Arctic Federal University (NArFU), Arkhangelsk, Russia

A. K. Krivorotov
Department of Innovation Management, MGIMO University, Moscow, Russia
e-mail: krivorotov@starlink.ru

V. I. Salygin (ed.), *Energy of the Russian Arctic*,
https://doi.org/10.1007/978-981-19-2817-8_24

a crucial role in providing Russia's domestic energy supply and export capacities, and also serve as a tool for establishing and maintaining its geopolitical presence.

The actual energy picture, however, is much more varied. Both energy production and consumption in the Russian Arctic feature a broad mix of solutions, ranging from the traditional coal and fuel oils to unique installations like the tidal power plant in Kislaya Guba (Murmansk District). This creates numerous opportunities for experiments and innovative high-tech solutions.

This chapter aims to describe the prospects of two niche sectors: the nuclear and hydrogen energy. While the first has existed for decades and the second is just starting to take shape, they have both gained a strong momentum in the past few years within the framework of Russia's clean energy transition. Moreover, they may produce important synergies by applying nuclear energy to hydrogen production.

24.2 Literature Review

The past couple of years has brought about new or updated official papers on Russian energy policies, including notably in the nuclear and hydrogen sectors (Government of the Russian Federation, 2020a, 2020b, 2021a, 2021b), triggering broad media coverage. Meanwhile, a theoretical framing is still in the making, with the industry prospects addressed from various viewpoints.

The global energy transition toward decarbonization plays an increasing role in Russia's energy strategy development. The country has vast renewable sources of energy, major plans for nuclear and hydrogen industry, and development routes for national carbon legislation. Therefore, several authors study the Russian hydrogen prospects and related challenges within the broader context of transforming the national energy sector (Henderson & Mitrova, 2020; Zhiznin et al., 2020). Other publications seek to evaluate the international prospects of the Russian hydrogen industry as such. Mitrova et al. (2019) identify hydrogen as a promising, cleaner alternative to the traditional hydrocarbon exports, to reinforce national positions in the global energy markets. Westphal and Zabanova (2021) present a broad overview of the Russian potential, plans, and actors in the industry, reviewing the development against the European (notably German) trends. Prospects for natural gas-based hydrogen exports—in particular, the benefits and costs balance of hydrogen cooperation between the Russian Federation and the EU—are also examined by Konoplyanik (2020). Some authors, while stressing the broad industry prospects, also address the related technological, economic, and business challenges (Burykin, 2021; Kopteva et al., 2021; Novak, 2021).

The nuclear power industry is not predicted to increase its share in the global energy balance, as a nuclear renaissance at the global level is unlikely to take place. Russia is the only country to integrate civilian nuclear energy into the national Arctic strategy, which may explain the virtual absence of international analytical publications on the issue. However, nuclear technologies have played and will continue to play a key role in the Russian Arctic development. In connection with this, small modular reactors may become attractive tools to support nuclear energy development. The future of nuclear energy is analyzed by Boyko (2021).

24.2.1 Nuclear Energy and Hydrogen in the New Climate Agenda

In the course of the past few years, the Russian engagement in low-carbon transition has grown significantly, due to a set of interrelated trends.

Traditionally, Russia was less involved in the global climate debate, due to realistic, nation-specific reasons. The radical economic reforms of the 1990s resulted in a painful restructuring and a closedown of much Soviet heavy industries, thus leaving the country with large unused emission quotas as compared to 1990 (the benchmark year for the Kyoto Protocol). The economic growth of the last two decades has been largely driven by industries with lower carbon footprints, like services, banking, retail trade, or housing construction. Altogether, Russian greenhouse gas emissions have nearly halved, from 3.1 billion tons CO_2 equivalent in 1990 down to 1.6 billion tons nowadays, which is unmatched among developed economies. Besides, the vast Russian forests (first of all, the Siberian taiga) and swamps possess the world's second biggest CO_2 absorption capacity, next to the Amazon rainforests only.

The climate focus started growing gradually stronger since Russia joined the Paris Agreement in September 2019 and the EU started making its carbon tax real. In 2020–2021, President Vladimir Putin issued several instructions and public statements on further reduction of the greenhouse gas emissions, aiming to make Russia climate neutral by 2060 at the latest—eventually much earlier, subject to successful international cooperation.

The Russian authorities maintain that nuclear power generation, with its low emissions of greenhouse gases, is a major and reliable tool for energy transition. "Low-carbon energy develops at a good rate in Russia," said Putin at the G20 Summit on October 31, 2021, "the share of energy generated from nearly carbon-free sources—and these, as we know, include nuclear, hydro, wind, solar power stations—exceeds 40 percent nowadays. Together with natural gas, the hydrocarbon fuel with the least carbon footprint, this share makes 86 percent. This is among the best mixes worldwide." He underscored that nuclear energy has a four times lower carbon footprint as compared to solar (President of the Russian Federation, 2021). In this respect, Russia has teamed up with France, which gets over 70 percent of its power from nuclear stations and rejects the idea of shutting these down.

Russia's long-term low-carbon strategy through 2050, adopted by the Cabinet on October 29, 2021, has an equally strong focus on both nuclear and hydrogen technologies. The document reiterates the aim of creating a large national hydrogen cluster and promotes a massive introduction of hydrogen in metal and chemical industries to improve their carbon profiles (Government of the Russian Federation, 2021b).

Climate issues also take an important place among the priorities of the Russian Chairmanship in the Arctic Council (2021–2023). While the Russian approach to protecting the Arctic environment is very comprehensive (including issues like ecosystems management, oil spill preparedness, or cooperation within hydrometeorology), the country also undertakes to host a special event on climate change and permafrost melting, as well as to conduct an efficient Chairmanship of the Expert Group on Black Carbon and Methane (Information and Analytical Centre, 2021). Within the framework of the Sustainable Development Working Group, Russian entities have participated in a US-sponsored project on hydrogen energy and launched another one, AHEAD, themselves.

In the area of energy planning, the country has enhanced its efforts on structural changes, especially in exports. Since 2019, the Russian oil companies Rosneft and Lukoil (both with large Arctic upstream assets), as well as government officials, indicated for the first time that they may encounter problems in increasing the crude exports in the longer run. The revised 2020 edition of the Russian Energy Strategy anticipates oil exports (in physical volumes) to fall by between 3.3 and 6.5 percent in

2035 as compared to 2018. Meanwhile, the exports of natural gas (especially liquefied) shall increase dramatically (Government of the Russian Federation, 2020a). This official forecast reflects the need to rethink the nation's role in the future energy supply, as well as to enhance more technologically advanced exports, better tailored to future market needs.

Hydrogen suits this way of thinking perfectly. Gazprom, the national gas company, sees it as an opportunity to enhance its strong position in the European market and since 2018 has proposed a large-scale decarbonization program to the EU, to rely on hydrogen supplies from Russia (Konoplyanik, 2020).

The ongoing massive entry of ESG principles into the Russian business practices, also in the North, has reinforced these trends. Numerous Arctic companies, often operating in extractive industries, were confronted in 2020 and especially in 2021 with harsh ESG requirements on behalf of banks, investors, and rating agencies. In December 2021, CEOs of some 30 leading Russian companies from various branches launched a national ESG Alliance set up by the Sber, Russia's biggest bank and business ecosystem. The key mission of the alliance is to facilitate the members' efficient low-carbon transition. Russian corporate managers—even in the most challenging industries like metals and transport—pledge that ESG goals are doable, but acknowledge that much effort may be needed, especially within the "E": minimizing the environmental—including notably the climatic—footprint (RBC, 2021).

24.3 Nuclear Energy in the Arctic: Russia's Large-Scale Plans

Despite the well-known and widely acknowledged concerns, nuclear energy could become an effective tool to curb global warming and contribute to the successful implementation of the fourth energy transition (Boyko, 2021). Russia is moving steadily forward with plans for an expanded role in the nuclear industry, including the development of new reactor technology. The Russian government actively supports the development of the nuclear industry. Thus, it was decided to allocate ₽80 billion ($1 billion) for the development of the "New Nuclear Energy" program for 2021–2024. The main purpose of the program is the construction of small modular reactors (SMR). The International Atomic Energy Agency (IAEA) defines "small" units as those with a capacity of up to 300 megawatts.

The crucial importance of SMR is related to the fact that there are small countries whose energy systems do not require large reactors (LMR) with the usual capacity of 1000 megawatts, but which still have difficulties with ensuring stability of their energy systems. Besides, there are countries in which the construction of large power plants is expensive. The cost of a modern large nuclear power unit is very high—up to $5–6 billion. In addition, there are large countries that do not provide electricity to their whole territory, but which need energy on remote islands, or on territories where mineral deposits are discovered, and it is unprofitable to build stationary generation facilities.

Thus, SMRs are attractive for use in remote regions with undeveloped power grid infrastructure where more powerful LMRs make no economic sense. Small nuclear power units have evident advantages: they can provide energy generation and energy self-sufficiency in remote regions. These factors are of crucial importance for Arctic countries. A subcategory of very small reactors (vSMRs) is proposed for units under about 15 megawatts, especially for remote communities.

Many remote communities in Russia, Alaska, Canada, and Greenland are not connected via transmission lines or pipelines to one another or to more populous southern regions of their respective countries. Instead, they rely heavily on imported fuels—primarily diesel—for local heating and power generation via microgrids. These diesel fuel shipments may also pose threats to energy security, as the shipments are susceptible to interruptions from weather and seasonal changes in sea ice.

At the beginning of the twenty-first century, countries with a developed nuclear industry began considering SMRs as key elements of the smart energy concept. In 2020, the IAEA published an update of its SMR book "Advances in Small Modular Reactor Technology Developments," with contributions from developers covering 70 designs. Nevertheless, Russia is the only country within the Arctic Council to use SMRs in the Arctic today. In addition to the two existing nuclear power plants in the Arctic regions of Russia—the *Kola* (440 megawatts) in Murmansk District and the recently decommissioned *Bilibino* (12 megawatts) in Chukotka—the country launched the world's first floating nuclear power station (FNPS): named *Akademik Lomonosov*, it has two KLT-40S reactors (70 megawatts). The generated electricity is sufficient to supply more than 100,000 Russian households. It is berthed at the Russian port of Pevek in Chukotka Autonomous Okrug. The reactors KLT-40 have successfully proven themselves since many years of reliable operation on

Russian nuclear icebreaker vessels. In December 2019, a floating power unit generated the first electricity to the isolated electricity network of Chaun-Bilibino hub in Chukotka. The magazine *Power* recognized this as one of the six key events of the year in the global nuclear energy industry.

In August 2021, Rusatom Overseas was licensed to construct the world's first onshore power station with a RITM-200H reactor (55 megawatts) in the Ust-Yansk Region of Yakutia for the power generation of the Kuchus gold mine and the surrounding area. The project is cost-effective. Today it is problematic to deliver fuel to the north of Yakutia because of harsh Arctic conditions. The project shall be completed by 2028. According to preliminary forecasts, up to 700 new jobs will be created during the construction of the SMR, and up to 300 during its operation. SMRs make a significant contribution to the socio-economic development of remote regions. Their creation leads to an increase in the employment, well-being, and quality of life of the local population, as well as reducing greenhouse gas emissions.

Earlier, Rosatom launched the construction of five small floating nuclear power stations for Chukotka. The company plans to electrify the Baim ore zone with the help of four modernized floating power units with a RITM-200 reactor (400 megawatts) and a backup FPU to ensure replacement during overload and repair. In the future, the reserve small floating nuclear power unit can be used to provide energy to the *Akademik Lomonosov*'s consumers. An interesting fact is that the construction of the first two floating nuclear power units will be handled by the Chinese yard, as the Baltic Shipyard is now overloaded with the construction of four LK-60 nuclear icebreakers. It is planned to put the first two power units into operation by 2027, the third by 2028, and the fourth by 2031. In the conditions of energy transition and political restrictions consistently imposed on the extraction and use of fossil fuels, nuclear energy serves as an environmentally friendly solution that meets the most stringent requirements in the field of decarbonization.

Rosatom has had plans for the development of nuclear energy in the Arctic Zone for a long time: at least since 2018, when the company became a single infrastructure operator for the development of the Northern Sea Route and, in addition, the owner of a nuclear icebreaker fleet. In terms of the Arctic maritime development, Russia is ahead of the whole planet today, well in line with its active Arctic strategy at large. The nuclear-powered civilian fleet currently includes five nuclear icebreakers, complemented by container ship *Sevmorput*, the only operational cargo ship with a nuclear power unit in the world.

In the coming years, Rosatomflot shall get four universal nuclear icebreakers of Project 22220. These nuclear icebreakers are designed for independent year-round wiring of large-tonnage vessels in the western Arctic. The two-stage design of nuclear-powered vessels with RITM-200 allows them to be used both at seas and in the mouths of Arctic rivers (Rosatom, 2021a). The icebreakers of type 22220 are the largest and most powerful in the world. The length of the vessels is 173.3 m, the width is 34 m, and the water displacement is 33.5 thousand tons. These icebreakers can conduct caravans of ships in Arctic conditions, breaking through ice up to 3 m thick. They will ensure the transportation of vessels with hydrocarbons from the fields of the Yamal and Gydan Peninsulas to the markets of the Asia–Pacific region.

In addition, Rosatomflot shall include three nuclear icebreakers of Project 10510 by 2030. The unique nuclear icebreakers will have exceptional technical characteristics equipped with RITM-400 (twice as powerful as Project 22220) that will allow year-round wiring in the eastern Arctic. The nuclear icebreaker *Rossiya* is planned to start operation by 2027 (Table 24.1).

Table 24.1 Nuclear icebreaker fleet of the Russian Federation (the existing and developing projects)

Type	*Name*	*Power (megawatts)*	*Construction year*
Taymyr	Taymyr	36.8	1989
	Vaygach	36.8	1990
Arktika	Yamal	55	1992
	50 Let Pobedy	55	2007
Nuclear container ship	Sevmorput	29.4	2016
Project 22220	Arktika	60	2020
	Sibir	60	2021
	Ural	60	2022
	Yakutia	60	2024
	Chukotka	60	2026
Project 10510 "Leader"	Rossiya	120	2027
	N/a	120	2028
	N/a	120	2029

Source Compiled by the authors based on Rosatomflot data. URL: http://www.rosatomflot.ru

In 2019, President Putin declared that by 2035 the country's Arctic fleet would include 13 icebreakers, nine of them nuclear-driven. Taking into account that three icebreakers—*Taymyr*, *Vaygach*, and *Yamal*—will soon be decommissioned at maturity—in 2026, 2027, and 2028 respectively—Russia should undoubtedly foster the development of its nuclear icebreaker fleet in the nearest future.

No other country in the world has such a far-reaching program for the development of its Arctic coast as Russia has. On the one hand, the use of nuclear energy for the development of shipping is an obvious option for the Russian Federation. The use of nuclear energy onboard ships provides advantages of independence, portability, energy intensity, and sustainability. For example, one load of nuclear fuel in the RITM-200 is equivalent to 540,000 tons of Arctic diesel fuel (Go Arctic, 2020). Rosatom is launching a large number of new icebreakers that will ensure the nation's positions in the Arctic, facilitating both the export deliveries, transit, and domestic supplies to remote coastal settlements along the Northern Sea Route. On the other hand, the use of small modular reactors installed on the latest icebreakers can be effective on container ships and oil tankers and can become another promising direction of Russian nuclear energy development in the Arctic. It should be added that in recent years no serious accidents occurred on the nuclear icebreaker fleet, and the experience of Fukushima was taken into account when developing the RITM-200. However, international concerns about the incremental risks of the nuclear energy pertain. Several European nations plan to phase out their national power plants, and ships with nuclear power units onboard are not allowed to enter many foreign ports. These attitudes may represent a challenge to the Russian Arctic development plans in the longer run.

24.3.1 Hydrogen: An Emerging Industry with Long-Term Perspective

Hydrogen energy is, on the other hand, at its initial stage in Russia, well in line with the global development, although the relevant research dates back decades ago. Soviet engineers worked on hydrogen applications since the 1930s and made several breakthrough efforts. These included notably the world's first-ever cryogenic aircraft: the Tupolev 155, which fueled one of its three engines with liquid hydrogen and made a series of successful test flights in the 1980s. A dedicated interest organization, the National Hydrogen Energy Association, was established to promote the industry in 2003, and recent reports by several Russian think tanks urged for building up the national hydrogen production (Mitrova et al., 2019). However, it was not until the past couple of years that hydrogen has started to develop as an actual, government-backed business.

In 2020, Russia adopted two key policy documents outlining its hydrogen development plans. The above new edition of the national Energy Strategy through 2035, while still premised on hydrocarbons, also set the goal of making Russia one of the world's leading producers and exporters of hydrogen. The export targets in the Strategy stand at 0.2 million tons by 2024 and 2 million tons by 2035. By comparison, the present global hydrogen consumption estimate is 75 million tons per year. It is noteworthy that, despite the declared goal of increasing the domestic use of hydrogen, the quantified goals were set for exports only (Westphal & Zabanova, 2021).

The second paper, the Roadmap for Hydrogen Development through 2024, adopted in October 2020, emphasizes Russia's competitive advantages in hydrogen, such as technological know-how and R&D, the existing resource base, significant spare capacity in the power generation system, a developed transport infrastructure, and geographic proximity to major consumers. It lists several key focuses: strategic planning and monitoring; incentives and public support for hydrogen energy; establishing production capacities; launching pilot projects in 2023–2024; research, and high-tech solutions therein; enhancing the relevant national legislative base and standards; human resources development; and international cooperation. The Government allocates over ₽9 billion ($123 million) to support hydrogen energy in 2022–2024 (Government of the Russian Federation, 2020b; Novak, 2021).

As a follow-up, the Russian Government adopted a detailed national long-term concept for phased-in hydrogen energy development on August 5, 2021. The first stage (2021–2024) envisages the creation of specialized clusters and the implementation of pilot projects for the production and export of hydrogen, CCUS, as well as introducing hydrogen energy in the domestic market. At stages 2 (2025–2035) and 3 (2036–2050), the country shall establish and gradually build up large export-oriented production facilities and switch to the serial use of hydrogen technologies in various sectors of the economy, from petrochemicals to housing and utility services. Three large hydrogen clusters shall emerge, one of them in the Arctic. The Concept aims at exporting 2 million tons in 2035 and 15 million tons in 2050, but also sets "optimistic goals" of 12 million tons and 50 million tons respectively (Government of the Russian Federation, 2021b).

The Roadmap essentially relies on state-owned corporations (primarily Rosatom and Gazprom) and on St. Petersburg Mining University as key actors within applied research and launching pilot projects. Indeed, Rosatom, together with leading Russian and international players, is actively involved in the development of hydrogen-fueled transport projects in the Russian Far East (Rosatom, 2020b) and in developing clean technologies for water electrolysis using nuclear energy (so-called yellow hydrogen with very low-carbon emissions). Russian experts highlight the symbiosis of nuclear and hydrogen energy as an important contribution to creating a new, sustainable global energy system (Zhiznin et al., 2020).

Gazprom, on its part, actively evaluates the opportunities of extracting hydrogen from natural gas, both for exports and domestic use. In 2021, Gazprom set up a dedicated subsidiary, *Gazprom Vodorod*, and later signed an agreement with the Russian Government on developing hydrogen energy and decarbonization of industry and transport based on natural gas. Following an in-house meeting on climate agenda on November 30, 2021, President Putin instructed the cabinet and Gazprom to evaluate the opportunities of exporting hydrogen to Europe by pipelines in a hydrogen/methane mix (Voropaeva, 2021).

Meanwhile, other entities have also recently shown major interest in the industry, creating a more diverse and vibrant set of potential actors, which test a variety of technological and logistical options. These include, among others:

- NOVATEK, the biggest Russian independent gas producer, which plans to start producing ammonia and hydrogen in 2027 close to its Yamal LNG plant in Sabetta;
- *Hydrogen Technological Valley* consortium, launched in November 2020 by a number of universities, heavy industry, and transportation companies from various parts of Russia,
- *H2 Clean Energy*, a project company established up in May 2021 by the CEO of Polyus, the biggest Russian producer of gold and platinum (the latter is an important catalyst in hydrogen production); the company is mulling constructing two plants—in Murmansk

District and Magadan (Russian subarctic north-east)—plus potentially one more in Kamchatka Peninsula, to serve the north Pacific trade routes and to produce hydrogen with electricity from a proposed tidal power plant (Mingazov, 2021),
- *Hydrogen Technology Solutions*, a non-profit research center set up in Moscow in 2021 under the auspices of SK Trade, a large Russian multimodal transportation and engineering group, etc.

The Russian Arctic is well positioned for clean hydrogen production. The region contains vast reserves and potential resources of natural gas and fresh water, with developed infrastructure for both pipeline and seaborne transportation. The NSR makes it possible to export Arctic fuels to either European or Asian markets. In addition, the overwhelming majority of the Russian Arctic Zone, with its isolated settlements and fragile environment, is itself a very promising potential hydrogen market. While poorly covered by the unified (nationwide) or regional power grids, the area possesses of high wind power potential and vast public lands, well suited for constructing windfarms and green hydrogen units, which may represent an environmentally friendly and economically viable alternative to motor fuels imported from the south. As the international pressure on use of heavy fuel oils in the Arctic mounts, hydrogen, as well as LNG, also gets increasingly attractive as a potential bunkering fuel. According to Dr. Krivichev from Moscow State University, investments in a hydrogen-propelled cargo ship to transport containers over the Russian Arctic would pay back in 7 years, generating a net operating income of ₽450 million by 2030 (PORA Expert Center, 2021).

The Murmansk District has made an assumption about the development of carbon-free energy. Currently, hydro energy accounts for 40.2 percent of its electricity production, with another 57.1 percent provided by the Kola nuclear power plant in Polyarnye Zori. The Kola wind farm, to be launched in 2022 in cooperation with Italian Enel, shall become the largest one in Russia beyond the Arctic Circle. In 2021, Rusnano, the state nanotech corporation, and the District Government signed an agreement on creating a carbon-free cluster to reduce greenhouse gas emissions. It is planned to start the production of green hydrogen in 2024 at the Kola nuclear power plant.

The main Russian commercial projects within low-carbon and carbon-free hydrogen and ammonia production in the Arctic Zone are presented in Table 24.2. The figures show that the most promising projects are

Table 24.2 Hydrogen production projects in the AZRF

Region	*Company*	*Energy source*	*Output (tons per year)*	*Year*	*Markets*
Republic of Karelia	En + Group	Green Hydrogen/Ammonia	5200	2024	Russia, Europe
Arkhangelsk District	Regional Development Agency of the Arkhangelsk District	Green Hydrogen	500,000 1 million	2030 2033	Russia, Europe, Asia
Murmansk District	Rusnano	Green Hydrogen	12,000	2024	Russia, Europe
	Rosatom	Low-Carbon Hydrogen	150	2024	Russia, Europe
	H4Energy	Green Hydrogen/Ammonia	17,000 170,000	2024 2030	Russia, Europe, Asia
	H2 Clean Energy	Green Hydrogen	16,000	2025	Russia, Europe
	H2	Green Hydrogen	10,000	2024	Russia, Europe
	Gazprom Energoholding Group	Green Hydrogen/Ammonia	2000 20,000	2024 2030	Russia, Europe
Yamal-Nenets Autonomous Okrug	NOVATEK	Blue Hydrogen/Ammonia	2.2 million (of ammonia)	2027	Russia, Europe, Asia
	Corporation Energy	Blue Ammonia/Hydrogen	2.2 million (of ammonia)	2025	Russia, Europe, Asia
	Corporation Energy	Green Hydrogen	N/a	2025	Russia, Europe, Asia
	Corporation Energy	Blue Ammonia/Hydrogen	2.2 million (of ammonia)	2026	Russia, Europe, Asia
Krasnoyarsk Krai	North Star	Low-Carbon Hydrogen	N/A	2024	Russia, Europe, Asia

Source Compiled by the authors based on Ministry of Industry and Trade of Russia (2021). ***Russian Atlas of Low-Carbon and Carbon-Free Hydrogen and Ammonia Production Projects***. URL: https://minpromtorg.gov.ru/docs/#!atlas_rossiyskih_proektov_po_proizvodstvu_nizkouglerodnogo_i_bezuglerodnogo_vodoroda_i_ammiaka (in Russian)

NOVATEK's Obsky hydrogen and ammonia facility on the Yamal Peninsula (known as Yamal LNG), and a project using electricity from the Mezen tidal power plant, and wind farm in the Arkhangelsk District.

In 2019, Russia also launched a major international, research-related hydrogen project. Capitalizing on its solid R&D base and expertise in the field of Arctic energy, the renowned Moscow Institute of Physics and Technology, together with Russian and international partners, initiated construction of a fully-autonomous international Arctic research station named *Snowflake*, to operate entirely on renewable energy sources and hydrogen energy. The project got a strong approval from all member countries in the Arctic Council in 2020.

In 2024, the *Snowflake* shall start operations in Yamal Peninsula. It will use hydrogen, wind, and sun energy. A second station (*Snowflake-2*) is planned in Murmansk District. The stations shall become testing grounds for advanced hydrogen energy technologies, as well as new platforms for international cooperation between engineers, researchers, scientists, and students working on robust solutions for a sustainable life and work in the Arctic. Their provisional research profiles ensure that they do not overlap, but rather complement each other (Snowflake International Arctic Station, 2021).

The development of the eco-friendly fuels market is a global trend, and Russia is not lagging behind the world leaders. Relying on new industrial capacities, energy-efficient power plants, and large natural gas reserves, it is joining the fight for a niche in the new promising market. The ramp-up of production capacities and the launch of new industrial and research clusters will not only create new jobs and growth poles in the Arctic, but also strengthen Russia's positions in terms of the energy transition. However, the present focus on hydrogen energy development is to a large degree politically motivated, posing a few challenges in industrial safety, technology development, and project management. Unlike traditional energy sources, hydrogen is more sophisticated and potentially dangerous to store and transport. Efficient technologies are yet in the making, and a fair return on investments may be a challenge. Russia, in line with other nations, shall thus make sure that its hydrogen efforts contribute to SDG 7 on ensuring access to affordable, reliable, sustainable, and modern energy for all.

24.4 Conclusions

Nuclear power and hydrogen represent two technologies with a large development potential in the Russian Arctic, especially under the dramatically enhanced domestic climate focus. Given its enormous Arctic resource base and the historically strong R&D potential in both technologies and corresponding industries, Russia has the potential to become a true international trendsetter. This vision is shared by the authorities, the research community, and the business community. Furthermore, while Rosatom has well defined and ambitious long-term plans, it is definitely too early to chart specific scenarios for the hydrogen industry development. It is clear, however, that the future prospects for both industries will depend upon a few key factors.

The overall development path of the global climate debate will be a crucial factor. While virtually all players acknowledge the need to combat global warming, there is no broad consensus on the appropriate way forward. The acceptability of nuclear power and the preferred methods of clean (therein “yellow”) hydrogen production represent some of the hot discussion topics in this regard, and their outcome will largely shape the industry worldwide, including notably in the vulnerable Arctic system. It will be important for Russia to promote its views efficiently, first of all by demonstrating workable, environmentally safe, and economically competitive solutions.

Progress in cutting-edge technologies is therefore a must (Burykin, 2021). Under an optimistic scenario, Russia may also position itself an international supplier of clean energy technologies and services (Henderson & Mitrova, 2020). The Russian Arctic Zone may then turn into a global-scale testing ground for cold climate applications within green energy, the *Snowflake* labs being the first step. In order to succeed, though, the government will have to dramatically enhance R&D efforts (Kopteva et al., 2021) and technology export promotion.

A balanced development of the domestic and export markets will also be of major importance. The future of the Russian nuclear icebreaker fleet is inseparable from the navigation along the Northern Sea Route, the country’s key transportation line in the Arctic. The cargo turnover along the NSR has already grown from 2 million tons in 2010 to 33 million tons in 2020, and according to plans shall reach 130 million tons in 2035. The national cargo base (primarily Arctic oil and LNG projects) is sufficient to support these ambitions goals, but it will be crucial to increase international transit, making the icebreakers an integral element of the global value chains and transportation corridors. In the case of hydrogen, the opposite task seems more relevant—namely to reconcile the far-reaching export goals with stimulating domestic demand. This implies, among other things, a coordinated development of infrastructure for hydrogen production, offloading, and bunkering along the Northern Sea Route to fuel a hydrogen-driven merchant fleet.

The development of the Arctic hydrogen energy shall further fit into Russia's newly revised Arctic strategy. Its 2020 edition, as well as the follow-up government documents and recent pieces of legislation, have a stronger focus on the environment, innovative business development, and international competitiveness. Most of the ongoing and planned large projects, however, remain centered on traditional resource extraction and exports. A rapid development of the hydrogen industry will thus confront governmental and corporate strategists with the need to revise some of their plans. However, this seems inevitable in order to be in line with global market trends.

Last but definitely not least, the above plans, all charted in 2019-2021, require a serious revision in the aftermath of the ongoing crisis in Russia-West relations. In mid-April 2022, President Putin held two consecutive meetings on Russian Arctic and petroleum development with the heads of federal agencies. The key outcomes were that the Arctic projects (including notably oil and gas extraction) should progress as planned, while energy industry should adapt to the EU import sanctions (President of the Russian Federation 2022a, 2022b). Rerouting major volumes of fossil fuels from the Russian Arctic to the domestic and Asian markets is sure to have major impacts on future hydrogen production and exports, as well as on the NSR cargo shipping, eventually amending the patterns of nuclear icebreaker operations.

References

Boyko, A. (2021). Prospects for Nuclear Energy in the 21st Century. *Russia and the Contemporary World*. https://cyberleninka.ru/article/n/perspektivy-yadernoy-energetiki-v-xxi-veke (in Russian). Accessed November 12, 2021.

Burykin, V. (2021). *Lowering Energy Transition Risks for Russia*. Forbes. https://www.forbes.ru/biznes/449879-kak-snizit-riski-energoperehoda-dla-rossii (in Russian). Accessed December 16, 2021.

Go Arctic. (2020). *RHYTHM-200 Suggested for Container Ships and Oil Tankers*. https://goarctic.ru/news/ritm-200-predlozhili-primenyat-na-konteynerovozakh-i-neftyanykh-tankerakh/ (in Russian). Accessed November 12, 2021.

Government of the Russian Federation (2020a). *Energy Strategy of the Russian Federation through 2035. Approved by Directive of June 9, 2020a No. 1523-r.* (in Russian).

Government of the Russian Federation (2020b). *Roadmap on "Development of hydrogen energy in the Russian Federation through 2024". Approved by Directive of October 12, 2020b No. 2634-r.* (in Russian).

Government of the Russian Federation (2021a). *Concept of hydrogen energy development in the Russian Federation. Approved by Directive of August 5, 2021a No. 2162-r.* (in Russian).

Government of the Russian Federation (2021b). *Strategy for socio-economic development of the Russian Federation under low greenhouse gases emissions through 2050. Approved by Directive of October 29, 2021b No. 3052-r.* (in Russian)

Henderson, J., & Mitrova, T. (2020). Implications of the global energy transition on Russia. In M. Hafner, & S. Tagliapietra (Eds.), *The Geopolitics of the Global Energy Transition. Lecture Notes in Energy 73* (pp. 93–114). Springer Nature.

Information and Analytical Centre of State Commission for Arctic Development. (2021). *Russia's Chairmanship Priorities for the Arctic Council 2021–2023*. http://library.arcticportal.org/1993/1/Russias_Chairmanship_Priorities_Arctic_Council_2021-2023.pdf. Accessed December 18, 2021.

International Atomic Energy Agency. (2020). *Advances in Small Modular Reactor Technology Developments*. https://aris.iaea.org/Publications/SMR_Book_2020.pdf. Accessed October 10, 2021.

Konoplyanik, A. (2020). Clean, natural gas-based hydrogen: A new, promising area of Russia-EU cooperation. *Gazprom Journal, 9*, 20–29. (in Russian).

Kopteva, A., Kalimullin, L., Tcvetkov, P., & Soares, A. (2021). Prospects and obstacles for green hydrogen production in Russia. *Energies, 14*, 718.

Ministry of Industry and Trade of Russia. (2021). *Russian Atlas of Low-Carbon and Carbon-Free Hydrogen and Ammonia Production Projects*. https://minpromtorg.gov.ru/docs/#!atlas_rossiyskih_proektov_po_proizvodstvu_nizkouglerodnogo_i_bezuglerodnogo_vodoroda_i_ammiaka (in Russian). Accessed December 15, 2021.

Mingazov, S. (2021). *Polyus CEO Assess Feasibility of His Hydrogen Output Projects*. https://www.forbes.ru/tekhnologii/441853-glava-polusa-ocenil-vozmoznosti-realizacii-svoih-proektov-po-proizvodstvu-vodoroda (in Russian). Accessed December 4, 2021.

Mitrova, T., Melnikov, Y., & Chugunov, D. (2019). *The Hydrogen Economy – a Path Towards Low Carbon Development*. SKOLKOVO Energy Centre, Moscow School of Management.

Novak, A. (2021). Hydrogen: Energy for a 'Clean' future. *Energeticheskaya Politika, 4*, 6–11. (in Russian).

PORA Expert Center. (2021). *Net Income of a Hydrogen Ship on Northern Sea Route May Exceed 450m Rubles in 2030.* https://porarctic.ru/ru/events/chistaya-pribyl-sudna-na-vodorodnom-toplive-na-sevmorputi-mozhet-prevysit-450-mln-rubley-v-2030-godu/ (in Russian). Accessed December 16, 2021.

President of the Russian Federation. (2021). *Second Session of the G20 Summit.* http://www.kremlin.ru/events/president/news/67044 (in Russian). Accessed October 31, 2021.

President of the Russian Federation. (2022a). *Meeting on Arctic Zone Development.* http://www.kremlin.ru/events/president/news/68188 (in Russian). Accessed April 13, 2022.

President of the Russian Federation. (2022b). *Meeting on Current Situation in the Oil and Gas Sector.* http://www.kremlin.ru/events/president/news/68191 (in Russian). Accessed April 13, 2022.

RBC. (2021). *'System Breakup': Key Quotes from RBC's ESG Congress.* https://www.rbc.ru/business/15/10/2021/61696aa09a7947753dbd6f1e (in Russian). Accessed October 15, 2021.

Rosatom. (2021a). *Nuclear Icebreaker Fleet.* https://www.rosatom.ru/en/rosatom-group/the-nuclear-icebreaker-fleet/. Accessed December 16, 2021.

Rosatom. (2021b). *ROSATOM, KAMAZ, Far East and Arctic Development Corporation and Sakhalin Region Development Corporation Join Forces to Develop Environmentally Safe Transport on Sakhalin Island.* https://www.rosatom.ru/en/press-centre/news/rosatom-kamaz-far-east-and-arctic-development-corporation-and-sakhalin-region-development-corporatio/. Accessed December 18, 2021.

Rosatomflot. (2021). http://rosatomflot.ru/flot/atomnye-ledokoly/deystvuyushie/ (in Russian). Accessed December 16, 2021.

Snowflake International Arctic Station. (2021). https://arctic-mipt.com/en. Accessed December 12, 2021.

Voropaeva, E. (2021). *Putin Instructs Evaluating Hydrogen Exports to the EU by Existing Pipelines.* https://www.rbc.ru/economics/30/11/2021/61a605db9a7947f4033b04e5 (in Russian). Accessed December 15, 2021.

Westphal, K., & Zabanova, Y. (2021). Russia in the global hydrogen race. *SWP Comment, 34*, 1–8.

Zhiznin, S. Z., Timokhov, V. M., & Gusev, A. L. (2020). Economic aspects of nuclear and hydrogen energy in the World and Russia. *International Journal of Hydrogen Energy, 45*(56), 31353–31366.

PART V

Regional Framework for the Exploration of Russian Arctic

CHAPTER 25

Norwegian Oil Industry Model: Potential for Norwegian-Russian Cooperation in the Arctic

Tatiana Damaskova and Mariia I. Ermilova

25.1 Introduction

Norway is a small open economy famous for its success story on petroleum resources management. The country is Europe's largest exporter of oil and the world's third-largest exporter of gas, supplying approximately 25 percent of EU gas demand (Grytten & Hunnes, 2021). Norway has been remarkably immune to the so-called oil curse, which has provoked income inequality and poverty in some other countries and

T. Damaskova
Faculty of International Relations, University of Economics, Bratislava, Petržalka, Slovak Republic
e-mail: tatiana.damaskova@euba.sk

M. I. Ermilova (✉)
Department of Financial Management, Plekhanov Russian University of Economics, Moscow, Russia
e-mail: masha080487@mail.ru

V. I. Salygin (ed.), *Energy of the Russian Arctic*,
https://doi.org/10.1007/978-981-19-2817-8_25

even caused corruption and internal conflicts. Since the discovery of oil in 1969, Norway has maintained a healthy democracy and a government free from corruption scandals. All this ranked Norway among the top 10 countries that earned the highest GDP per capita (StatisticsTimes.com, 2021). Overall, Norway has passed the "oil test" successfully. The country is considered a pioneer in safety and environmental protection.

It is obvious that today, when the exploration of the natural resources of the Arctic is given a lot of attention, Russia is considering the possibility of applying various components of the Norwegian model. Both countries are looking for opportunities to exchange experience in research, education, and joint development of environmentally friendly solutions. The paper will examine the best practices of bilateral cooperation in oil and gas exploration in the Arctic region, and where Russia and Norway share challenges and opportunities, especially in the search for environmentally friendly solutions. At the same time, the world faces major challenges in procuring energy for an increasing global population, so hydrocarbon exports will continue to be an important factor during the upcoming decades, which, however, will require the leading environmental technologies and new knowledge. Undoubtedly, the best solutions can be often found in cross-border cooperation.

25.2 Literature Review

The Norwegian model of management of the oil industry, as well as its possible application in other states, is the subject of interest of many researchers. Several scholars analyze the decisive role of the Norwegian state and its political, administrative, and economic functions, the structure of state institutions, and the system of redistribution of natural rent. First, it is worth mentioning the publications of the Norwegian expert and pioneer of the oil industry, Farouk Al-Kasim, a top expert in the oil industry, whom the Financial Times Magazine described as "the engine of Norway's success as an oil power" (Sandbu, 2009). Farouk Al-Kasim's book "Managing Petroleum Resources" is considered a reference work in the field of hydrocarbon resource management. Against the background of the "Norwegian model," Al-Kasim gives a comprehensive explanation of how Norway avoided the "oil curse," which was so destructive for many oil-rich countries. In his paper "Petroleum exploration in a resource management perspective" he argues that "an appropriate concession policy is the main tool for developing the entrepreneurial potential of the host country, both in the oil sector and beyond." Al-Kasim emphasizes that licensing as a key component in exploration, is implemented as part of an overall resource management strategy (Al-Kasim, 2006).

Another paper written by scientists from Norway and Netherlands, Ola Honningdal Grytten and John Arngrim Hunnes, contributes to the knowledge on how environmental ethics values and historical circumstances influence public policy. They analyze the success of petroleum resource management in the small open economy of Norway. "The study is conducted by mapping policy decisions and the arguments behind them regarding environmental and ethical issues" (Grytten & Hunnes, 2021). In addition, it describes a model of Ricardian resource rent and investigates the set of values that were in place before petroleum production started in the 1970s. Furthermore, it argues that a principal goal was to build a "qualitatively better society for the benefit of the people" with a historical retrospective in the background. The main findings of the research say: "The success lies in understanding the ethics behind the environmental resource rent harvesting of this non-renewable natural resource." It was further emphasized that "trust between the state and its citizens played a key role in shaping policy" (Grytten & Hunnes, 2021) An interesting study was accomplished by a research team including Farouk Al-Kasim, Tina Søreide, and Aled Williams, entitled "Corruption and reduced oil production: An additional resource curse factor?" This paper provides a broad literature view on the "resource curse" phenomenon, as well as empirical results from in-depth interviews with oil sector experts and employees in order to assess the relationship between corruption and low oil production. Therefore, many results show that weak governance and corruption are important causes that contribute to the large differences in wealth observed in oil-producing countries. In addition, it examines how weak governance and corruption affect the management of revenues and expenditures, as well as the potential increase in oil wealth (Al-Kasim et al., 2013).

Research conducted by Mark Thurber, David Hults, and Patrick Heller analyzes the extent to which other countries can benefit from the "Norwegian model" of separated government functions. Their research is based on a comparative analysis of ten countries that have used different hydrocarbon management approaches. The main finding is "that separation of functions is not a prerequisite to successful oil sector development." Therefore, those countries "with vibrant political competition but limited institutional capacity pose the most significant challenge for oil sector reform: Unitary control over the sector is impossible but a separation of functions is often difficult to implement" (Thurber et al., 2011).

In her paper "Model of state management of petroleum sector case of Norway," Katarzyna Dośpiał-Borysiak, an academic researcher from Poland, deals with questions of whether competence and the success of the petroleum industry arise from socio-political conditions and the precisely built management model; then she scrutinizes the "continuity and predictability of the petroleum sector results which leads to social and political trust." The main conclusion by Dośpiał-Borysiak is that the Norwegian model of separation of duties is difficult to implement without other prerequisites features that cannot be easily copied by other countries. These require a long tradition and high level of democratic cooperation, socio-democratic traditions of strong public involvement, mature institutional capacity (mainly a competent bureaucracy), and a highly developed model of cooperation between government, business, and research institutions (Dośpiał-Borysiak, 2017).

The applicability of the "Norwegian model" for the Russian oil and gas industry is questioned in the works by Kryukov and Krivorotov (2020), Krivorotov (2020), Belyakova (2020), and Polovtseva (2020).

In the article "Norwegian Oil and Gas Complex Management Model," Russian experts Valery A. Kryukov and Andrey K. Krivorotov draw some general conclusions about the applicability of the Norwegian oil model in Russia by distinguishing characteristic components. According to these scientists, the active political role of the state as the owner of natural resources is a key prerequisite. They further emphasized that trust between the government and foreign investors, as well as political stability, are key success factors that work along with high taxes and the strictest environmental requirements.

> Giant oil corporations, accustomed to operating by their own rules around the world, in Norway quickly became convinced that ignoring the demands of the local government (or simply buying them off) would not work. At the same time, the state pays due attention to investors. The traditional Scandinavian culture of compromises has successfully worked here—pursuing their own interests, understanding and respecting the legitimate interests of the other side, on this basis looking for mutually acceptable options. (Kryukov & Krivorotov, 2020)

Therefore, the researchers concluded that the experience of Norway is of great interest for Russian oil and gas sector. They highlight the similarities, such as communality and the desire for justice, a powerful economic and social function played by the state, the ability to perform an economic and technological "miracle" at the right time, etc.

25.3 Results

25.3.1 History of the Petroleum Industry in Norway

The establishment of the oil industry in Norway has a relatively short but distinctive history, which significantly affected the prosperity of this small European state. Norway's economic policy on the management of natural hydrocarbon resources is deservedly considered a "best practice" model, especially for emerging economies based on the oil industry. In this respect, Norway had the advantage of long-standing democracy, a well-established economic model, and an extended and successful tradition of governance. Norway indeed had a "good starting point in terms of its own capabilities and management system" (Benghida, 2017)—arguably, extremely so. Finally, yet importantly, another significant factor was the "lucky coincidence" that Farouk Al-Kasima had moved from Iran to Norway shortly before the discovery of the first oil field.

The famous story begins in 1959 with the discovery of large natural gas fields on the continental shelf in the Netherlands in Groningen, which made the region of the North Sea the center of interest in oil and gas exploration in Europe. In 1962, Phillips Petroleum, the American company, approached the Norwegian government with a proposal to cover the exploration and development of the entire continental shelf of the country. The offer from Phillips Petroleum was seen as an attempt to gain exclusive rights, which the Norwegian government reasonably could not agree with (Norwegian Petroleum Directorate, 2021).

Before the start of the exploration of the shelf, Norway needed to solve three main issues: defining the border between Norway and the United Kingdom in the North Sea; developing a legal basis for the oil exploration and production; and ensuring Norway's economic sovereignty. The first task was solved in March 1965, when Norway and Great Britain signed a maritime border agreement in December, and a similar agreement was concluded with Denmark. The second issue was solved already in 1963 when the Storting (Norwegian Parliament) passed the law on research and development of submarine natural deposits. Following the introduced legislation, the full right to natural underwater resources belonged to the state. (Norwegian Ministry of Petroleum and Energy, 2011). The third problem was solved by bringing together a licensing system for companies that wanted to operate in the Norwegian petroleum sector. Norway opted for an administrative licensing method that ensured a level playing field for all companies and allowed Norway to control the activities of foreign oil companies.

In 1965, the first round of concessions took place. According to its results, all leading positions went to foreign companies that had the necessary technical base, trained staff, and sufficient capital (Norwegian Petroleum Directorate, 2021). The first well was drilled in the summer of 1966, but it was dry. Norway waited until 1969 for the first finding in the Ekofisk Oil Field. Mining began on June 15, 1971, and several major discoveries took place in the following years. In the beginning, exploration was only allowed on the fields south of latitude 62°N. Each of the marine areas is divided into quadrants. The quadrants in the North Sea are numbered from 1 to 36, while the quadrants in the Norwegian Sea and the Barents Sea are numbered according to longitude and latitude. Each quadrant is further divided into 12 blocks. The rules ensured that only limited numbers of blocks were open for biding in the licensing round. Production was concentrated on a large area on the Norwegian continental shelf, such as Ekofisk, Statfjord, Oseberg, Gullfaks, and Troll. These fields are still very important for the development of the Norwegian oil industry. The historical map in Fig. 25.1 shows the licensing position from the first round in 1965 (Norwegian Petroleum Directorate, 2021).

At the early stage, the specialized expert foreign corporations preferred to explore and develop the Norway's first productive fields. The state oil company Statoil was founded in 1972 and owned at least 50 percent of each mining license. The 50 percent stake rule was later changed so that the Storting could assess whether the state's participation rate should be lower or higher depending on different conditions.

25.3.2 *The Role of the Norwegian State in the Management of the Oil Industry*

"The Ten Oil Commandments" endorsed by Storting in 1971 became the general rulebook for the oil industry management for the subsequent 50 years (Fig. 25.2).

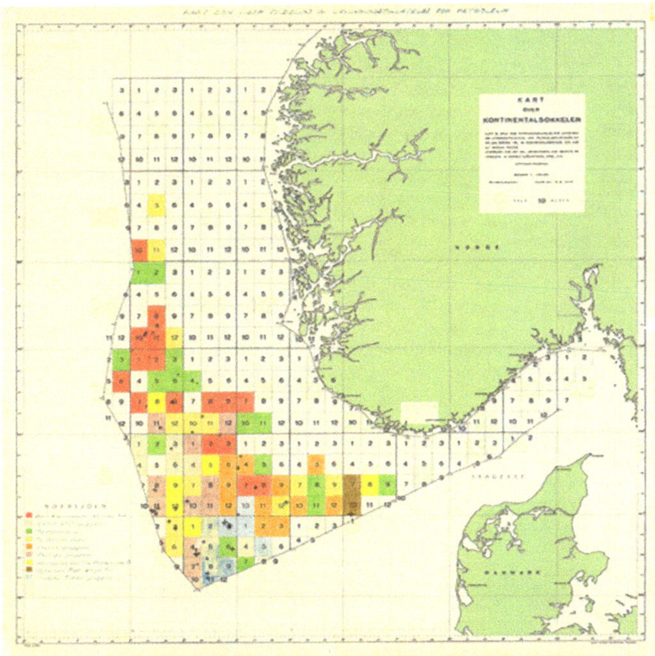

Fig. 25.1 The historical map. The licensing positions (*Source* Adapted by the authors from Norwegian Petroleum Directorate [2021]. Norway's petroleum history. URL: http://www.norskpetroleum.no/en/framework/norways-petroleum-history/)

These "oil commandments" point out that the petroleum policy must be comprehensive and that national management and control are important to ensure that management of the resources benefits the entire Norwegian society and building a "qualitatively better society." The special tax system was the regulatory tool that allow the state to collect the "natural rent." As oil prices rose in 1973, oil companies' profits also increased, but the income tax was on the level of 50.8 percent, which seemed to be too low of a percentage. Thus, the state, with all the ownership rights to Norwegian oil, had the full right to reform the distribution of economic rents from oil production. Three main decisions have been taken in this area. Firstly, 70 percent of profits in the oil industry have been transferred to the state as the owner of natural resources. Secondly,

Box 1.1 The 10 Oil Commandments

The 10 Oil Commandments are Chapters in a declaration of principles underpinning Norwegian oil policy, submitted by the Standing Committee on Industry in a Storting White Paper dated 14 June 1971. These principles have subsequently been dubbed the 10 Oil Commandments, and represented a clarification of what was needed to make sure that the oil activities would "benefit the entire nation":

1. That national supervision and control of all activity on the Norwegian Continental Shelf must be ensured.
2. That the petroleum discoveries must be exploited in a manner designed to ensure maximum independence for Norway in terms of reliance on others for supply of crude oil.
3. That new business activity must be developed, based on petroleum.
4. That the development of an oil industry must take place with necessary consideration for existing commercial activity, as well as protection of nature and the environment.
5. That flaring of exploitable gas on the Norwegian Continental Shelf must only be allowed in limited test periods.
6. That petroleum from the Norwegian Continental Shelf must, as a main rule, be landed in Norway, with the exception of special cases in which socio-political considerations warrant a different solution.
7. That the State involves itself at all reasonable levels, contributes to coordinating Norwegian interests within the Norwegian petroleum industry, and to developing an integrated Norwegian oil community with both national and international objectives.
8. That a state-owned oil company be established to safeguard the State's commercial interests, and to pursue expedient cooperation with domestic and foreign oil stakeholders.
9. That an activity plan must be adopted for the area north of the 62nd parallel which satisfies the unique socio-political factors associated with that part of the country.
10. That Norwegian petroleum discoveries could present new tasks to Norway's foreign policy.

Fig. 25.2 The 10 Oil Commandments (*Source* Adapted by the authors from Norwegian Ministry of Finance [2021]. *The Government Pension Fund 2020. Meld. St. 32 [2019–2020] Report to the Storting*. URL: https://www.regjeringen.no/contentassets/4184d59ae0ea48d6894838ace1d6998f/en-gb/sved/executive-summary.pdf)

a transition took place from a two-tier taxation system (license payments plus income tax) to a three-tier system (special oil income tax payment added). Thirdly, a new term—"standard price" of oil—was introduced in order to prevent tax manipulation in the event of artificial price reductions. In 1975, the "Oil and Gas Industry Taxation" law was passed. The corporate income tax for all sectors was 25 percent, while oil and gas activities were subject to an additional special tax of 53 percent (Volkov, 2019).

In 1975, a decision was taken on an additional geological survey of the North Sea, north of the 62nd parallel. In 1976, the parliament decided on three rounds of licensing, the main condition of which was to provide Statoil with at least a 50 percent share in each concession. In the same year, British Petroleum discovered *Statfury's* largest field. The Statfury project was operated by American company Mobil together with oil giants such as Royal Dutch Shell, Esso, and Conoco (Norwegian Petroleum Directorate, 2021). The first oil pipelines were being built at the same time. Construction of a pipeline to the United Kingdom called *Norpipe* began in 1975. In the same year, the Norwegian Stat Pipe pipeline project was launched. In 1977, a gas pipeline was built to Germany. The period from 1980 to 1986 is considered to be the boom of the Norwegian oil industry, as these years became a period of maximizing the economic rent of the oil sector against a background of rising international prices and industrial production. The development of the share in GDP of revenues from oil and gas activities, state budget revenues, and export supplies from 1971 to 2022 are presented in Fig. 25.3.

In June 1980, in response to rising global oil prices following the oil crisis and rising "super-profits" of multinational companies, Norway tightened its tax regime. The increase in the tax rate was also explained by the government's efforts to eliminate foreign debt, which was created in the initial phase of the development of the oil and gas industry. The

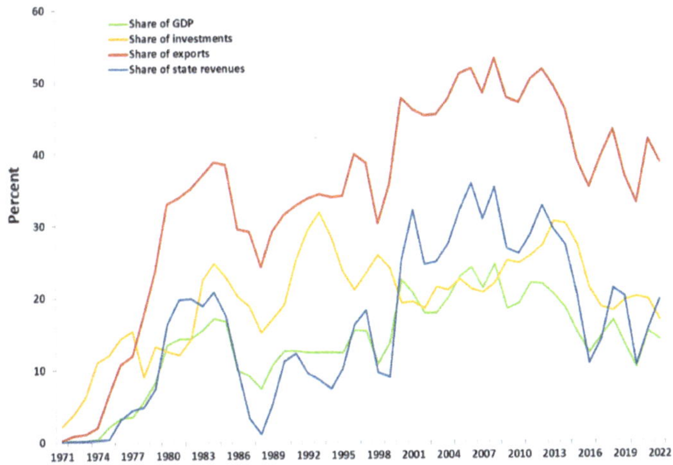

Fig. 25.3 Macroeconomic indicators for the petroleum sector in 1971–2022 (*Source* Adapted by the authors from Norwegian Petroleum Directorate [2021]. *Norway's Petroleum History*. Retrieved from: http://www.norskpetroleum.no/en/framework/norways-petroleum-history/)

changes affected the special tax rate, which increased up to 85 percent (Volkov, 2019).

The question of the rational use of oil revenues has been increasingly arising. The economy was also in danger of overheating. Initially, oil revenues went to social needs and the repayment of foreign debt. Once the debts were paid government came with a proposal to create a special oil fund modeled on the fund established in 1970 in Kuwait. The fund's resources were to be used to invest in government securities of other countries, as well as in shares of the most stable foreign companies. The first attempt to create such a fund—in 1982—failed. Later, in 1990, a "fund of future generations" was established, which still successfully operates today.

Under Norwegian law, the ownership of oil and gas resources belongs to the state. In practice, this right is executed by the state bodies, such as the Ministry of Petroleum and Energy and its subordinate structures, among which the Directorate for Petroleum has a special place.

From the very beginning, the Norwegian state intended to participate directly in the oil sector to protect important public interests and ensure the most efficient use of natural resources. Therefore, the organizational structure and areas of responsibility should be clearly defined. The schedule below visualizes the structure of the Norwegian authorities (Fig. 25.4).

The Storting sets the major directions for oil activities in Norway; it has the power to introduce and change legislation, as well as the power to take approve all key decisions, including key development projects. The Storting also supervises the government and public administration. Although the structure of the Storting has changed over time, the overall framework of Norwegian oil policy has remained stable. For many years, there was a political consensus on the main directions of oil policy, which allowed the industry a predictable and stable framework for its long-term operations (Norwegian Petroleum Directorate, 2021).

The Government has executive powers and carries out the realization of oil policy. The Government reports to the Storting. Ministries and their subordinates have operational responsibilities. The separation of responsibilities was crucial in the implementation of the oil policy. The Ministry of Petroleum and Energy is responsible for the petroleum sector. Further, the Ministry is acting on behalf of the state as the main shareholder in Equinor, Gassco AS, and Petoro AS. The Norwegian Directorate of Petroleum is a subordinate agency of the Ministry of Petroleum and

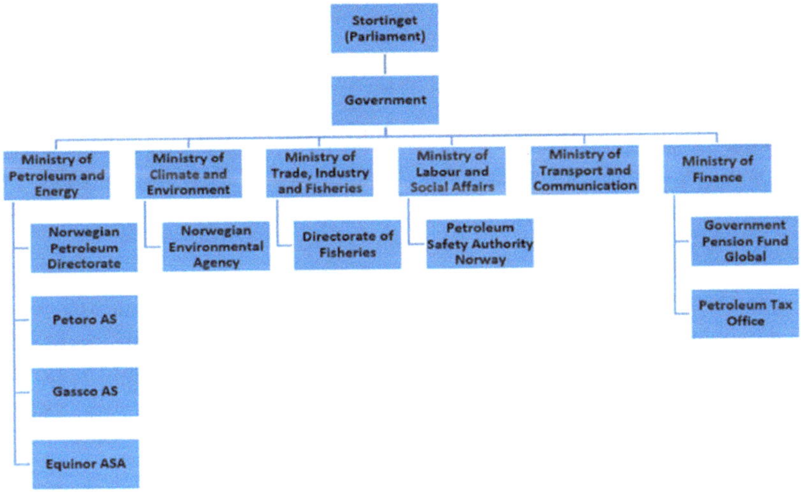

Fig. 25.4 Organogram of Norwegian institution (*Source* Adapted by the authors from Norwegian Petroleum Directorate [2021]. Norway's petroleum history. Retrieved from: http://www.norskpetroleum.no/en/framework/norways-petroleum-history/)

Energy. It plays a key role in oil management and is an important advisory body to the ministry. The directorate exercises administrative authority over oil exploration and production on the Norwegian continental shelf and has the power to make regulations and decisions in accordance with oil legislation.

The Ministry of Labour and Social Affairs has overall responsibility for the working environment and safety and emergency preparedness in the petroleum sector (Norwegian Petroleum Directorate, 2021).

The Norwegian Oil Safety Authority is a subordinate agency of the Ministry of Labour and Social Affairs; it holds responsibility for all safety issues, as well as "emergency preparedness to deal with both accidents and willful acts such as sabotage, and the working environment throughout the petroleum industry" (Norwegian Petroleum Directorate, 2021).

The Ministry of Finance is responsible for the tax system in the oil sector.

The next regulatory body is the Norwegian Customs Directorate, which has responsibility for the correct settlement of customs duties.

The Ministry of Finance is also responsible for managing the Global Government Pension Fund.

Operational Norges Bank has control of the Government Pension Fund.

The Ministry of Transport and Communications is responsible for preparedness and response to acute pollution in Norwegian waters. Its subordinate agency, the Norwegian Coastal Administration, is responsible for government preparedness and response to oil spills.

The Ministry of Trade, Industry, and Fisheries is involved in licensing procedures to assess the possible risks on the side of fishing industries.

Finally, yet importantly, the Ministry of Climate and Environment and its subordinate, the Norwegian Environment Agency, carry out the policy on climate and environmental protection in Norway, and is responsible for inspecting and enforcing state policy under the Pollution Control Act. (Norwegian Petroleum Directorate, 2021).

25.3.2.1 State Oil Companies

Petoro AS is a wholly state-owned company that manages the commercial aspects of state direct financial participation, through which the Norwegian state directly participates in oil activities on the Norwegian continental shelf. Petoro holds a state share license for the production and ownership of fields and pipelines on the Norwegian shelf and related offshore installations. Petoro's main goal is to maximize government revenue from the portfolio. The company was founded in 2001 and has almost 70 employees. The Ministry of Petroleum and Energy is responsible for the administration of Petoro's state property.

Today *Equinor ASA* (formerly Statoil) is a company with extensive international activities in more than 30 countries. It employs about 21,600 people. Equinor mainly deals with oil and gas extraction. This Norwegian giant operates about 70 percent of all oil and gas production on the Norwegian shelf. The company is listed on the Oslo and New York stock exchanges. The Board of Directors of Equinor is responsible for the strategic development of the company. The state is a major stakeholder of Equinor ASA with 67 percent of the shares. The Ministry of Petroleum and Energy is in charge of managing the state's shares. "Equinor's state-owned goal is to maintain a knowledge-based and high-tech company headquartered in Norway. Equinor should operate exclusively on a commercial basis" (Norwegian Petroleum Directorate, 2021).

Another state company, Gassco AS, was founded in 2001. The company is the operator of the integrated system for transporting gas from the Norwegian continental shelf to other European countries. It does not generate a profit or loss from its activities (Gassco, 2021).

25.3.2.2 State Pension Fund

The Government Pension Fund Global (GPFG) has declared an ambitious mission: to protect and build financial wealth for future generations (Norges Bank, 2021). The annual publication of the annual management report is an anticipated event in Norway. Fund management and investment risk management can provide good examples for pension funds in other countries. From its inception until the end of 2020, the fund has produced revenue of NOK 6427 billion, which corresponds to a return of 6.3 percent per year. After deducting operational costs and inflation, the yearly revenue was on the level of 4.4 percent. During the last ten years, the fund's accumulated return has been NOK 5681 billion, which is a return of 7.9 percent per year. After taking into account administration costs and inflation, annual revenue has been 6.2 percent over the last ten years. According to the 2020 annual report, the GPFG had a market value of NOK 10,914 at the end of 2020. Despite the extraordinary conditions associated with the coronavirus pandemic and the sudden movement in the financial markets, investments in the GPFG returned to 10.9 percent in 2020. According to the CEO's statement, "The oil fund is a generation fund, but also a reserve fund. In 2020, we saw how important the fund is when Norway is hit by the crisis. During the year, we transferred NOK 298 billion from the fund to cover the state budget deficit. During the same period, the fund received a dividend, coupon, and rental income of NOK 218 billion from its investments" (Norges Bank, 2021).

25.3.3 History of Relations Between Russia and Norway

During the last century, several historical milestones had a decisive impact on the relations and cooperation between Russia and Norway. With the discovery of oil and gas fields in the 1960s, the continental shelf in the Barents Sea became particularly important as a promising area for oil production. In 1967, negotiations between Norway and Russia began at a diplomatic level, to decide on a dividing line that defines their boundaries on the continental shelf. At the outset, both parties agreed to observe a transitory moratorium on geological exploration in the disputed area

(Wither, 2021). However, due to the controversial interpretations of international law and practice in the matter, as well as for political reasons, the negotiations were delayed by 40 years.

In 2010, an important turn of events brought about a change in the relationship between Russia and Norway. On September 15, 2010, the Russian Federation and the Kingdom of Norway concluded an agreement specifying the delimitation of sea spaces and cooperation expected of the nations in the Barents Sea and the Arctic Ocean (Moe et al., 2011). This was called the Murmansk Treaty.

After signing the agreement, both sides expressed their readiness to expand cooperation in the Arctic on many subjects, including the oil and gas industry. Many reflections on the Murmansk Treaty underlined the prospects for bilateral cooperation opened up by the agreement, as well as its international significance, as a precedent for the peaceful settlement of territorial disputes between the nations based on generally accepted norms of international law (Henriksen & Ulfsten, 2011). It is not coincidental that this agreement was signed when there was significant international attention given to the Arctic. The increasing sea-ice loss, the opening of the region to maritime transport, and easier access to natural resources (particularly potential petroleum deposits) were probably some of the main factors that pushed both parties to sign this agreement at that time. The treaty helped the coastal states of both nations to sort out their differences in meeting these new challenges and in ensuring the consolidation of their sovereign rights over natural resources and jurisdiction over international shipping. It also removed overlapping claims and clarified the rights and obligations of third states with respect to the disputed area (Henriksen & Ulfsten, 2011).

The development of relations between Russia and Norway in the context of global development reached another milestone in 2014, when the annexation of Crimea led to an escalation of relations between Russia and the West, followed by the imposition of sanctions against Russia. The sanctions were especially harmful for the technologies and equipment supply for the Russian oil and gas industry, which further led to the significant slowdown of cooperation in the oil and gas sector, particularly between Russia and Norway.

25.3.4 Cooperation Between Russia and Norway in the High North: Assumptions Versus Reality

In 2014–2015, a working group of scientists and experts from Norway and Russia held several workshops to discuss and define possible scenarios of cooperation between the two countries concerning oil and gas exploration in the Arctic. They applied the well-established methodology used by Royal Dutch Shell, which helped them to identify a number of assumptions and major uncertainties that, according to experts, "determine the future of the development of oil and gas fields in the Barents Sea. After examining all the factors, the three scenarios were developed in respect to Russian-Norwegian cooperation in the future" (Bourmistrov et al., 2015). Assumptions were made about the following factors in order to arrive at the below figuration.

- level of technology advancement achieved in the Arctic Zone;
- future price of oil and gas;
- development of the Lofoten area;
- the amount of oil found;
- growth recorded by Asia in the field;
- the extent of Russia's orientation toward Asia;
- the amount of gas found;
- unconventional methods used for gas extraction outside the US;
- Norwegian-Russian relations as a consequence of Russian-Western relations;
- restrictions laid down in the global climate policy.

Schematic visualizatio'ts is shown in Fig. 25.5.

Six years later, it is now possible to analyze the extent to which the above assumptions have been met and how this affects the cooperation between Russia and Norway. General assessment can be made of the developments that have taken place since the completion of the study in 2015. Climate change has become a top priority on the global political agenda. Most countries have pledged to achieve net-zero emissions by 2050, Russia and China by 2060, and India by 2070 (Carbon Brief, 2021; Trenin, 2021).

- There is no improvement in the relations between Russia and the West, nor has there been any easing of sanctions;

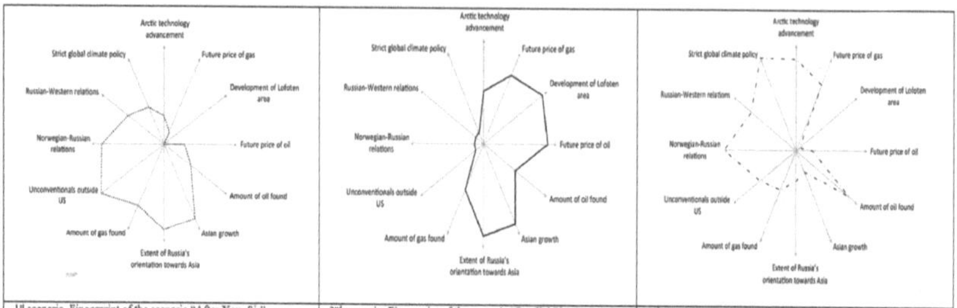

1st scenario. Fingerprint of the scenario "After You, Sir"	2nd scenario. Fingerprint of the scenario "Parallel Play"	3rd scenario. Fingerprint of the scenario "Let's Dance"

Fig. 25.5 Three scenarios of Russian-Norwegian cooperation (*Source* Adapted by the authors from Bourmistrov et al. [2015]. International Arctic Petroleum Cooperation: Barents Sea scenarios. London and New York: Routledge Taylor&Francis Group)

- The export of oil and gas from Russia to China is growing. In December 2019, the first gas pipeline from Russia to China, called the "Power of Siberia," was brought into operation (International Energy Agency, 2021);
- Asia shows steady economic growth after the pandemic recession;
- As a result of economic and political uncertainty, there has not been registered a significant development of technologies in the Arctic Zone;
- Oil and gas development in the Lofoten area has been discontinued due to environmental reasons;
- No new significant sites have been discovered for oil or gas extraction in the Arctic in Norway or Russia. Moreover, as officially stated by Kristin Westvik (area director for Exploration and Production North Equinor), "future value creation will largely come from increased recovery from existing fields, and connection of new discoveries close to existing infrastructure. It is therefore uplifting once again to be able to demonstrate resources that can strengthen the service life of producing fields in the Norwegian Sea. Such near-field discoveries are profitable, robust against fluctuations in oil/gas prices, they have a short payback period and low emissions" (Equinor, 2021b).

As can be seen in the chart below, after a decline mainly triggered by the 2020 pandemic recession, oil prices have already recovered to pre-pandemic levels, and gas prices recently hit record highs (Fig. 25.6).

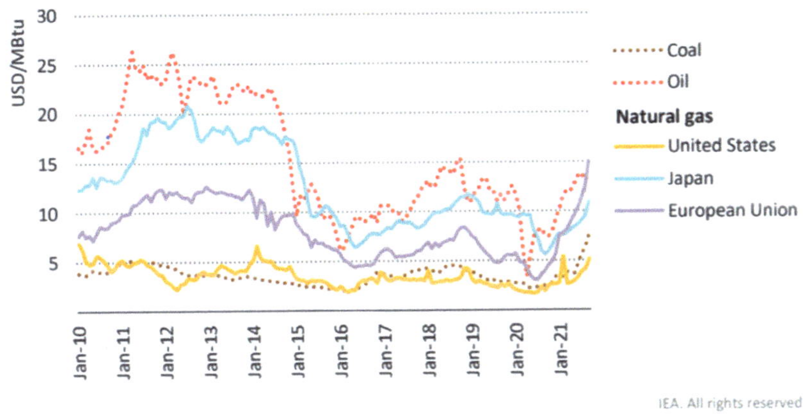

Fig. 25.6 Hydrocarbon price development (*Source* Adapted by the authors from International Energy Agency [2021]. World Energy Outlook 2021. URL: https://iea.blob.core.windows.net/assets/888004cf-1a38-4716-9e0c-3b0e3f dbf609/WorldEnergyOutlook2021.pdf)

The updated graph below provides a general picture of the current state of the various factors that can affect possible bilateral cooperation between Norway and Russia (Fig. 25.7).

Considering the above-mentioned development, Norway's future oil exploration in its Arctic coastal waters and its related cooperation with Russia in this regard has become less likely. "Race for the Arctic" and "New Cold War" have become the common headlines published by newspapers when they cover Arctic affairs. However, the reality confirms the opposite. The region possesses a strong track record of peace and cooperation (Bourmistrov et al., 2015). Nevertheless, we are of the opinion that despite the current political crisis, continuing the tradition and maintaining good cooperation in the Arctic Zone can help both countries in preventing extreme scenarios and finding a compromise middle ground. Therefore, since 2014, tensions between Russia and the Western countries have not affected the signing of significant international agreements covering major issues such as shipping and scientific research. Russia's chairmanship of the Arctic Council over the next two years and the new U.S. administration's commitment to multilateralism and security dialogue may offer a chance for improved cooperation between Russia and Norway in the High North (Wither, 2021). Arctic cooperation with

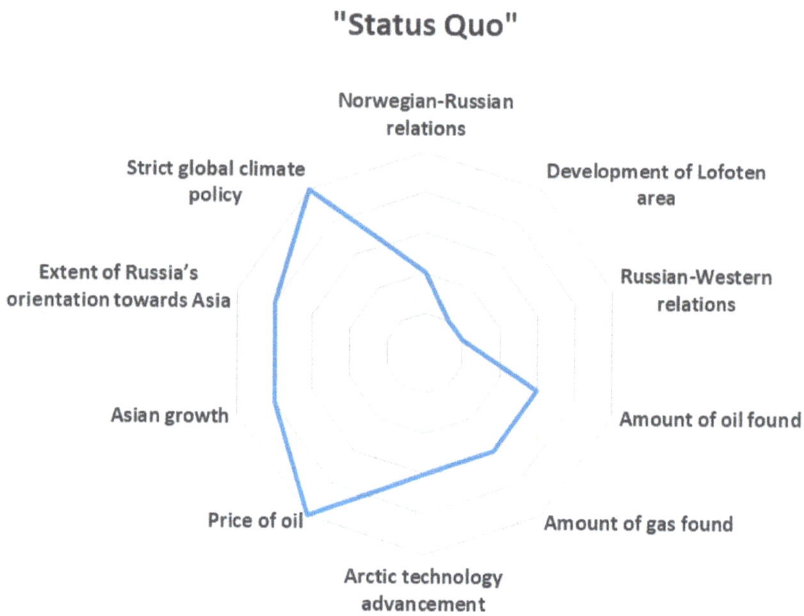

Fig. 25.7 Current state of factors (*Source* Compiled by the authors)

the support of international organizations engages various bodies such as local and regional level administrations and authorities, ethnic groups, and cultural communities. The Arctic Council and other regional organizations have become the ambassadors of local interests when it comes to the preservation of the environment, human wellbeing, and the pursuit of business activities. The global tensions prevailing between Russia and the West during recent years have brought cooperation in the Arctic to a new level and strengthened regional and cultural ties with Norway. Below, we have provided an overview of the international organizations operating in the Arctic, and the upcoming activities of the member countries, in which Russia and Norway play an important role (Arctic Council, 2021).

The Arctic Council is a leading intergovernmental forum that promotes cooperation, coordination, and interaction among the Arctic states, Arctic indigenous peoples, and other Arctic inhabitants on common Arctic issues, particularly issues related to sustainable development and environmental protection in the Arctic. It was formally established in 1996 (Arctic Council, 2021). The members of the Arctic Council are the eight Arctic States: Canada, the Kingdom of Denmark, Finland, Iceland, Norway, the Russian Federation, Sweden, the United States, and the six Indigenous Permanent Participant (PP) organizations. The PP organizations represent Indigenous Peoples in the Arctic Council, either by representing (a) a single Indigenous People resident in more than one Arctic State, or (b) more than one Arctic Indigenous People resident in a single Arctic State. The six PP organizations are the Aleut International Association, Arctic Athabaskan Council, Gwich'in Council International, Inuit Circumpolar Council, Russian Association of Indigenous Peoples of the North, and Saami Council. The PPs take an active part in all the deliberations and activities of the Arctic Council and are served by the Indigenous Peoples' Secretariat.

There are six working groups within the Arctic Council, each of which focuses on a particular set of issues, as is evident from their names: Arctic Contaminants Action Program; Arctic Monitoring and Assessment Programme; Conservation of Arctic Flora and Fauna; Emergency Prevention, Preparedness, and Response; Protection of the Arctic Marine Environment;—and Sustainable Development Working Group. In addition, the Arctic Council has more than 35 observer states and organizations. All Arctic Council decisions and statements require agreement from all the eight Arctic States (Arctic Council, 2021).

On May 20, 2021, Russia took over as the chair of the Arctic Council. The chairmanship program includes more than a hundred events in ten main areas such as economics, ecology, human capital development, and support for the indigenous small-numbered peoples of the Arctic. The central event of the chairmanship will be the International Arctic Forum, which will take place in April 2022. The International Arctic Forum has been held since 2010. Since then, the number of attendees of this forum has increased from 300 in 2010 to over 3600 in 2019. The forum has become a communication platform for international bodies to discuss topical issues of developing the economic potential of the Arctic territories in compliance with environmental standards, as applicable to the unique nature of the Arctic. It is believed that this forum is an exceptional platform for engaging in dialogue to reach a mutual understanding (International Arctic Forum, 2021).

The Barents Euro-Arctic Council and Barents Regional Council are organizations that were formed to ensure cooperation between countries in the Barents Euro-Arctic Region. They were launched in

1993 at two levels: intergovernmental Barents Euro-Arctic Council and interregional Barents Regional Council. The overall objective of these organizations is the sustainable development of the Barents Sea area. Membership in these organizations is based on adherence to the regional principles of ethnic origin applicable in the northernmost parts of Finland, Norway, Sweden, and North-West Russia. The Barents region includes Finnish Lapland and Oulu Region, Kainuu and North Karelia, Norwegian Nordland and Troms og Finnmark, Russian Arkhangelsk Region, Murmansk Region, Karelia, Komi and Nenets, and Swedish Norrbotten and Västerbotten. Indigenous Peoples from the Barents Region having representation in these organizations are Sami (in Norway, Sweden, Finland, and Russia), Nenets (in Russia), and Veps (in Russia). The region now comprises 13 member counties, covering an area of approximately 1.75 million sq. km, and having nearly 5.3 million inhabitants (Barents Euro-Arctic Council, 2021).

The Northern Dimension is a joint policy formed by four equal partners: the European Union, the Russian Federation, Norway, and Iceland. The Northern Dimension policy provides a platform for practical cooperation through thematic partnerships in the fields of environment, transport and logistics, culture, and public health and social wellbeing. The Northern Dimension Institute, the Northern Dimension Business Council and the Northern Dimension Parliamentary Forum complement the cooperation (Barents Cooperation, 2021).

25.3.5 Research Cooperation Between Russia and Norway

Cooperation between Russian and Norwegian researchers has already built solid ground. Therefore, the year 2005 was the starting point for a tradition of good cooperation between researchers from both countries (Research Council of Norway, 2012). Therefore, they can get together and lay the foundations for scientific collaboration on different topics. The one that attracts the foremost attention is of course the Arctic and its resources. In this respect, during 2004–2013, the PETROMAKS Programme of the Research Council of Norway was the contact point and driver, while the Petroleum Institute and RAS played the same role in Russia. This program was one of the seven large programs of the Research Council of Norway and embraces a large part of the oil and gas research of the Research Council. PETROMAKS included both long-term pure research, competence training, applied science, and the development of new technologies. Through reinforced development of knowledge, development of the industry, and international competitiveness, the program set out to contribute to creating an added value for society when using the oil and gas resources. The program started in 2004 and continued until the end of 2013. During that period got funds of NOK 2 billion (Research Council of Norway, 2012). The beginning of cooperation was mainly focused on oil and gas, a sector in the High North where both countries have common problems and opportunities. The focus has been on the environment, geosciences, and technology for field development in the Arctic.

On the Optimal Management of Petroleum Resources, a number of fruitful projects were accomplished. A few examples of good practice cooperation include the following. "Biota Guard Arctic" is a project on monitoring the environment in the Barents Sea with mollusks; by equipping mollusks and other organisms with instruments, leakages from the oil and gas industry can be effectively controlled. "Petro Arctic" is an example of another successful project on offshore and coastal technology for petroleum production and transport from Arctic waters. The outcome from project "GlaciBar" gives better knowledge about the geological development of the Barents Sea over the last five million years, with special emphasis on the ice ages. All in all, from 2006 to 2013, PETROMAX funded 24 Norwegian research projects in cooperation with Russia (Research Council of Norway, 2012). Unfortunately, the cooperation was affected by political tensions between the West and Russia.

The new Research Council of Norway Program (NORRUS program) was established for the period from 2017 to 2023 and identified several major and complex challenges, including the further strengthening of Norwegian-Russian research groups and institutions, as well as joint efforts to ensure the value of Norwegian-Russian research and research in the High North. and motivating politicians and creating more effective inputs for politicians.

As stated in the NORRUS Plus white paper (Research Council of Norway, 2017), the international research landscape is changing rapidly. To solve the worldwide problems currently faced, global cooperation is needed. While Norway continues to prioritize European cooperation as one of its main arenas for international cooperation, the Research Council is committed to expanding bilateral partnerships with strong and emerging research countries outside of Europe. Russia has been identified as one of the eight priority countries of the Research Council outside the EU/EEA.

A plan for bilateral research cooperation with Russia was drawn up to highlight strategic priority areas for Norwegian-Russian research cooperation (Belogurov et al., 2020; Inozemtsev, 2019; Kasatkin & Romanenko, 2019). The NORRUSS program aims to expand, strengthen, and update the research framework that has been undertaken by research groups so far on a wide range of issues related to Russia and the Arctic (Kasatkin et al., 2021).

This NORRUSS program aims to deepen knowledge about the Arctic and Russia and related topics, as well as to initiate discussion in the wider public sphere, government bodies, and at the policy-making level of the Russian and Norwegian governments. The following activities will be carried out to achieve the goals and objectives (Research Council of Norway, 2017). The primary objectives are the following: to expand knowledge on geopolitical and foreign policy issues of relevance to the High North and the Arctic; to develop long-term strategic expertise in Norway on Russia concerning both its domestic situation and its foreign relations; and to generate research findings of high scientific quality and strengthen the Norwegian research community. The secondary objectives are: to promote international research cooperation and networking with established research institutions; to promote a multi-disciplinary approach when appropriate and to encourage the use of innovative methodologies; to support projects within different disciplines such as political science, sociology, economics, law, history, media studies, social policy, and climate and environmental studies; to actively disseminate research findings to political authorities, public administration, trade, and industry, media as well as other groups involved in High North and Arctic issues; and to instigate Norwegian-Russian dialogue and cooperation.

The comprehensive list of programs that can be coordinated with the NORRUS program can be found below (Research Council of Norway, 2017).

- The Polar Research Programme (POLARPROG) generates knowledge to enhance our understanding of the climate system and ecosystems in the Arctic and for use in the context of policy implementation, responsible resource management, and industrial activity;

- The Large-Scale Programme on Climate Research (KLIMAFORSK) is a broad-based, long-term program aimed at generating new, future-oriented knowledge of national and international significance. There is long-standing cooperation between Norwegian research groups and Russian researchers;
- The Large-Scale Programme for Petroleum Research (PETRO-MAKS 2) generates new knowledge and technology to facilitate the optimal utilization of Norwegian petroleum resources and enhance the competitiveness of the Norwegian continental shelf;
- The Norwegian Programme for Research Cooperation with China (CHINOR) facilitates Sino-Norwegian cooperation through collaboration with thematic research programs at the Research Council;
- The Norwegian Programme for Research Cooperation with India (INDNOR) promotes cooperation between India and Norway in the following thematic priority areas—international political issues, climate, the environment, clean energy, and social development;
- The Programme for User-driven Research-based Innovation (BIA) funds industry-oriented research and has no thematic restrictions. This broad-based program supports high-quality R&D projects with good industrial and socio-economic potential;
- The Large-scale Programme on Health, Care and Welfare Services (HELSEVEL) encompass health and care services, social and welfare services, and child welfare. Services research and service innovation are the key components of the program;
- The Programme for Research and Innovation in the Educational Sector (FINNUT) funds research across a wide scope from early childhood education and care to higher education and adult learning.

25.3.6 *Climate Change and the Transition to a Green Economy*

In recent times, climate change is driving a faster transition to a green low-carbon economy than ever. The implementation of renewable energy sources is gaining momentum. Conference of the Parties (COP) is the climate change conference held by the United Nations. The latest COP-26 conference, held this year in Glasgow, attracted worldwide attention. At this conference, most countries reaffirmed their commitment to the goals set in Paris in 2015, and their intention to accelerate the introduction of environment-friendly initiatives and reach net-zero emissions by mid-century.

Norway's Prime Minister Jonas Strøre "came to Glasgow with three firm commitments" (Støre, 2021). Firstly, Norway will double its climate finance contribution to more than 1.6 billion dollars by 2026 and will collaborate with development partners to finance their transition to renewable energy. Secondly, the Norwegian Government will prioritize working with business, industry, and social partners on the transition to a renewable, circular, and sustainable future. This transition will mean a change in Norway. However, the ambition goes even further: the aim is to develop and export new technology that can be useful beyond Norway's borders. Norway is in a position to take the lead in developing ocean-based solutions, such as offshore wind energy, green transport, carbon capture, recovery, and storage, hydrogen, and electric mobility. Thirdly, the financial sector has a role to play. The Norwegian public pension fund Global is the largest sovereign wealth fund in the world, making investments globally. The Norwegian government aims to make it the leading fund in responsible investment and climate risk management. The Norwegian pension fund has a newly adopted strategic intent, according to which if the Global Fund will invest in a company in the future, that company must take biodiversity and sustainable use of ecosystems into account in their business activities. Biodiversity will become a priority area of Global in the future.

Taking into consideration Norway's position as a world leader in the transition to a green economy, we see it as a major potential for cooperation in the Arctic. What is more, there are several projects already in progress.

Therefore, on September 28, 2021, Equinor and Rosneft agreed on cooperation in carbon management. "The agreement builds on the two companies' strategic partnership in Russia. Drawing on this long-standing cooperation and aiming to support the goals of the Paris Agreement, Equinor and Rosneft will join forces to develop low carbon solutions and reduce the carbon footprint from joint projects" (Equinor, 2021a, 2021b). In accordance with the agreement, two companies will demonstrate their common effort and search for opportunities in such areas as a "reduction in flaring and methane emissions, energy efficiency, and reporting of greenhouse gas emissions." Equinor and Rosneft will also assess possible collaboration on renewable energy, carbon capture, utilization and storage, and low-carbon hydrogen solutions. The agreement set the goal of identifying low-carbon technologies in joint exploration and production projects in Russia, as well as engaging in joint sustainable development activities based on the United Nations Sustainable Development Goals and recognized environmental, social, and governance (ESG) frameworks (Equinor, 2021a, 2021b).

25.4 Conclusions

In conclusion, Norway has over 50 years of successful history in creating and developing the oil industry. As an advanced and stable democracy, with a long and successful tradition of governance that ensured effective cooperation between the state and commercial oil companies, it had a solid starting point. Thanks to a transparent system of taxation and the redistribution of earnings, Norway has created considerable reserves for future generations and, equally importantly, for meeting the new requirements for environmental protection. The crucial prerequisite of Norwegian successes was the solid ethical basis and strong political consensus in respect to the main mission of petroleum industry. The fundamental goal to ensure that resource management benefits the entire Norwegian society and building a qualitatively better society was not just a formal slogan. Historical relationship between politicians and citizens, the government, commercial companies, and foreign investors is based on long-standing trust. In addition, the Norwegian oil industry sets the highest standard in terms of environmental protection and became the best practice example of how oil extraction does not leave a pollution footprint in seawaters. The Norwegian state gratefully shares its oil expertise with the rest of the world and invests significant funds to the developing countries. That is why cooperation between Russia and Norway has always had great prospects. Moreover, there is already a successful tradition of appreciative cooperation based on mutual interest.

References

Al-Kasim, F. (2006). *Managing Petroleum Resources: The 'Norwegian Model' in a Broad Perspective*. Oxford: Institute for Energy Studies. OSL Number: 891547589.

Al-Kasima, F., Søreide, T., Williams, A. (2013). Corruption and reduced oil production: An additional resource curse factor? *Energy Policy*, Vol. 54, pp. 137–147.

Arctic Council. (2021). [cited 2021.11.15]. https://arctic-council.org/about/

Barents Euro-Arctic Council. (2021). [cited 2021.11.15]. https://www.barentscooperation.org/en/Barents-Euro-Arctic-Council

Belogurov, A. Y., Voevoda, E. V., Inozemtsev, M. I., Romanova, E. A., & Khokhlova, N. I. (2020). Development of education districts in the strategy of strengthening the axiological foundations of the Russian Education Space. *Vysshee Obrazovanie v Rossii, 29*(1), 25–36. [cited 2021.11.15]. https://doi.org/10.31992/0869-3617-2020-29-1-25-36

Belyakova, M. Y. (2020). The Role of the State in the Development of the Oil and Gas Complex of Norway. *Neftegaz.ru.*, 11, November 2020. [cited 2021.11.15]. https://magazine.neftegaz.ru/articles/gosregulirovanie/638996-rol-gosudarstva-v-razvitii-neftegazovogo-kompleksa-norvegii/

Benghida, S. (2017). Norwegian oil management structure: Farouk Al-Kasim innovative solution. *International Journal of Civil Engineering and Technology, 8*, 535–542.

Bourmistrov, A., Mellemvik, F., Bambulyak, A., Gudmestad, O., Overland, I., Zolotukhin, A. (2015). *International Arctic Petroleum Cooperation: Barents Sea Scenarios*. Routledge Taylor&Francis Group.

Carbon Brief. (2021). *COP26: Key outcomes for food, forests, land use and nature in Glasgow*. [cited 2021.11.15]. https://www.carbonbrief.org/cop26-key-outcomes-for-food-forests-land-use-and-nature-in-glasgow

Dośpiał-Borysiak, K. (2017). Model of state management of petroleum sector – Case of Norway. *International Studies, 20*(1), 97–112. [cited 2021.11.15]. https://doi.org/10.1515/ipcj-2017-0019

Equinor. (2021a). *Equinor and Rosneft Agree to Cooperate on Carbon Management*. [cited 2021.11.15]. https://www.equinor.com/en/news/2021a0928-agreement-carbon-management.html

Equinor. (2021b). *New oil discovery in the Norwegian Sea*. [cited 2021.11.15]. https://www.equinor.com/en/news/2021b1028-new-oil-discovery-norwegian-sea.html

Gassco. (2021). Official website. [cited 2021.11.15]. https://www.gassco.no/en/about-gassco/operator-duties/

Grytten, O., & Hunnes, J. (2021). Ethics, resource rent, environment and petroleum policy: The case of a small open economy. *Environmental Economics, 12*(1), 76–89. [cited 2021.11.15]. https://www.businessperspectives.org/index.php/journals/environmental-economics/issue-376/ethics-resource-rent-environment-and-petroleum-policy-the-case-of-a-small-open-economy

Henriksen, T., & Ulfsten, G. (2011). Maritime delimitation in the Arctic: The Barents Sea Treaty. *Ocean Development & International Law, 42*, 1–21. [cited 2021.11.15]. https://doi.org/10.1080/00908320.2011.542389

Inozemtsev, M. I. (2019). Individual granting of academic degrees model: Case of MGIMO-University. *Vysshee Obrazovanie v Rossii, 28*(12), 151–158. [cited 2021.11.15]. https://doi.org/10.31992/0869-3617-2019-28-12-151-158

International Arctic Forum. (2021). Official website. [cited 2021.11.15]. https://forumarctica.ru/en/

International Energy Agency. (2021). *World Energy Outlook 2021.* [cited 2021.11.15]. URL: https://iea.blob.core.windows.net/assets/888004cf-1a38-4716-9e0c-3b0e3fdbf609/WorldEnergyOutlook2021.pdf

Kasatkin, P. I., & Romanenko, A. V. (2019). The Rhetoric of the Political Leaders of Russia and the U.S.A. Comparative Analysis. *Polis, 5*, 167–180. [cited 2021.11.15]. https://doi.org/10.17976/jpps/2019.05.12

Kasatkin, P. I., Salnikova, L. S., & Fatykhova, V. M. (2021). Scientific, technical and educational cooperation in the EAEU. In N. A. Piskulova (Eds.), *The Economic Dimension of Eurasian Integration*. Palgrave Macmillan. https://doi.org/10.1007/978-3-030-59886-0_7

Krivorotov, A. K. (2020). Norwegian model of oil and gas complex management. *Ministry of Energy of the Russian Federation. Energy Policy*. [cited 2021.11.15]. https://energypolicy.ru/a-k-krivorotov-norvezhskaya-model-upra/regiony/2020/14/11/

Kryukov, V. A., & Krivorotov, A. K. (2020). Introducing Norwegian Oil and Gas Model in Russia: Institutional Challenges. *Journal of Siberian Federal University. Humanities & Social Sciences, 13*(4), 529–537.

Moe A., Fjærtoft D., Øverland I. (2011). Space and timing: Why was the Barents Sea delimitation dispute resolved in 2010? *Polar Geography, 4*(3), 145–162. [cited 2021.11.15]. https://doi.org/10.1080/1088937X.2011.597887

Norges Bank. (2021). *Government Pension Fund Global. Annual Report 2020.* [cited 2021.11.15]. https://www.nbim.no/en/publications/reports/2020/annual-report-2020/

Norwegian Ministry of Finance. (2021). *The Government Pension Fund 2020. Meld. St. 32 (2019–2020) Report to the Storting*. [cited 2021.11.15]. https://www.regjeringen.no/contentassets/4184d59ae0ea48d6894838ace1d6998f/en-gb/sved/executive-summary.pdf

Norwegian Petroleum Directorate. (2021). *Norway's Petroleum History*. [cited 2021.11.15]. http://www.norskpetroleum.no/en/framework/norways-petroleum-history/

Polovtseva, M. (2020). Never too late: Russian Hydrocarbon Development in the Arctic. *The Arctic Institute*. [cited 2021.11.15]. https://www.thearcticinstitute.org/russian-hydrocarbon-development-arctic/

Research Council of Norway. (2012). *Norway-Russia. Research and Cooperation in the High North.* [cited 2021.11.15]. https://www.forskningsradet.no/siteassets/publikasjoner/1253979429937.pdf

Research Council of Norway. (2017). *Work Programme 2017–2023. 2017. The High North and Russia – NORRUSS Pluss.* [cited 2021.11.15]. https://www.forskningsradet.no/siteassets/publikasjoner/1254032508371.pdf

Sandbu, M. (2009). *The Iraqi who Saved Norway from Oil. Financial Time Magazine.* [cited 2021.11.15]. https://www.ft.com/content/99680a04-92a0-11de-b63b-00144feabdc0

StatisticsTimes.com. (2021). *GDP (Nominal) Per Capita Ranking.* [cited 2021.11.15]. https://statisticstimes.com/economy/projected-world-gdp-capita-ranking.php

Støre, J. G. (2021). *Statement at the UN Climate Change Conference in Glasgow.* https://www.regjeringen.no/no/aktuelt/statement-at-the-un-climate-change-conference-in-glasgow/id2882242/

Thurber, M. et al. (2011). Exporting the "Norwegian Model": The effect of administrative design on oil sector performance. *Energy Policy, 39*, 5366–5378. https://www.researchgate.net/publication/227415788_Exporting_the_Norwegian_Model_The_effect_of_administrative_design_on_oil_sector_performance/stats

Trenin, D. (2021). After COP26: Russia's Path to the Global Green Future. *Carnegie Moscow Center.* [cited 2021.11.15]. https://carnegiemoscow.org/commentary/85789

Volkov, A. M. (2019). *Economy of Norway: Study Guide.* Moscow: Moscow State Institute of International Relations (University) of the Ministry of Foreign Affairs of the Russian Federation, Department of World Economy.

Wither, J. K. (2021). Can the high north security dilemma be mitigated? *George C. Marshall European Center for Security Studies.* [cited 2021.11.15]. https://www.marshallcenter.org/en/event/can-high-north-security-dilemma-be-mitigated

Index

V. I. Salygin (ed.), *Energy of the Russian Arctic*, https://doi.org/10.1007/978-981-19-2817-8

FSC
www.fsc.org
MIX
Papier | Fördert gute Waldnutzung
FSC® C083411

Kolibri 360 GmbH
Ferdinand-Jühlke-Straße 7
99095 Erfurt, Deutschland
produktsicherheit@kolibri360.de